Computer Networks and the Internet

Gerry Howser

Computer Networks and the Internet

A Hands–On Approach

 Springer

Gerry Howser
Kalamazoo College
Kalamazoo, MI, USA

ISBN 978-3-030-34498-6   ISBN 978-3-030-34496-2   (eBook)
https://doi.org/10.1007/978-3-030-34496-2

This Springer imprint is published by the registered company Springer Nature Switzerland AG
The registered company address is: Gewerbestrasse 11, 6330 Cham, Switzerland

This book is dedicated to my muse, first reader, and loving spouse Patricia Berens.

It is also dedicated to all my students who acted as guinea pigs through this idea in its many half-baked forms. It has been great fun.

Preface

"Any sufficiently advanced technology is indistinguishable from magic."

Arthur C. Clark [21]

The Internet

Everyone uses the Internet[1], so of course everyone knows how it works: from the user's point of view. However, I have found over the years that few people really understand what happens behind the scenes. Oddly enough it is not all "smoke and mirrors" or some arcane knowledge that can only be understood by a chosen few. Anyone with the ability to plug in a few cables and edit a text file (see Section 8.9) can build a self–contained Internet or Intranet.

The goal of this book is to provide enough background into the inner–workings of the Internet to allow a novice to understand how the various protocols on the Internet work together to accomplish simple tasks such as a search. The hope is that in building an Internet with all the various services a person uses everyday, one will gain an appreciation not only of the work that goes on unseen but also of the choices made by the designers to make life easier for the user. This has not always been the case in the computer industry.

Hopefully you will find this book useful in many different ways. It can be used as a step–by–step guide to build your own Intranet. It can also be used as a text for a course in Internet protocols and services. Or it can be used as a reference guide for how things work on the global Internet[2].

---

[1] Throughout this book, Internet will be used to refer to the global network we all know and love and internet (lowercase) will be used to refer to any generic Internet or intranet that does not require access to the public Internet to fully function.

[2] This book draws heavily on my experience teaching CompTIA Network+ [23] classes using Tamara Dean's excellent book [27].

To the instructor

This book is designed for dual purposes. Each chapter consists of background information on a specific topic or Internet service and where appropriate a final section on how to configure a Raspberry Pi to provide that service. If these configuration sections are skipped, This book can be used for a course on the Internet and routing.

When used with the suggested equipment, the main part of this book can be used for background material for a hands–on lab course in building a fully–functional Internet using inexpensive Raspberry Pi's. If you have access to a number of "white box" computers running Linux (Debian [28] is a good choice), this book can be used with minor adjustments to build an Internet of Linux boxes.

One possible approach to using this book would be to assign the chapters to be read before class. Class time would be used to answer questions from the reading and go over the chapters that relate specifically to the configuration of the Raspberry Pi. The bulk of the class time should be reserved for actually configuring the network in a lab setting. This has proved successful in the past[3].

Additional resources can be found at:

https://www.springer.com/book/9783030344955/

https://www.gerryhowser.com/book/9783030344955/.

To the student or hobbyist

I hope that this textbook provides you an enjoyable introduction to the inner workings of the Internet. If you already have some familiarity with a topic, you will find the chapters organized so that you can skim introductory sections and proceed quickly to the more advanced material. My intent is to provide you with a clear text that you will find useful in building your own networks and as a first reference for understanding the many Internet protocols.

This book is designed as a project for groups of four students each with their own Raspberry Pi; however, smaller groups can easily run all of the required protocols on as few as one Raspberry Pi. In fact, you will be encouraged to install and configure all of the services so that the group can still function when a member is unavailable. While it may be possible to use just the configuration sections to build an Intranet, it is still best to read the background material first.

At the end of each chapter are exercises relevant to that topic. As usual the easier exercises are first with progressively more challenging problems as the numbers grow larger. You will find solutions to some of the exercises at the end of this book.

---

[3] This book was inspired by courses taught at Loy Norrix High School and Kalamazoo College. Both are in Kalamazoo, Michigan.

What are the prerequisites for this book?

- You should have some familiarity with computers beyond simply using applications, but you can get by without it.
- Programming experience is helpful but not necessary. The same is true of experience installing and configuring software.
- You should be comfortable with the Internet as a user.
- You must be willing to think before you start making changes. Raspian is a Linux distribution and as such it is sometimes difficult to reverse changes made in haste. If you backup each configuration file before you change it you can always back-out any changes.
- You *must* be curious and fearless. Remember: the worst that can happen is you may need to reinstall the operating system. If there is a chance of harming your hardware, you will be warned in advance.
- Simple solutions are usually the fastest, least difficult to understand, and least prone to fail.
- In networking the goal is usually to move data as fast as possible (high through-put) and correctly as long as that does not slow things down. This seems counter intuitive at first, but the end–points of the conversation are tasked with handling errors, not the network.

Additional resources can be found at:
https://www.springer.com/us/book/9783030344955/
https://www.gerryhowser.com/book/9783030344955/.

To the professional

You should find this book useful as an overview to how the Internet works and how many of the protocols work. However, this is not an exhaustive reference to the Internet as the Internet is growing and changing at a staggering rate. Indeed, the only true references for the Internet, the final authority as it were, are the current RFCs which can only be found on the web. The most reliable place to look is on the IETF website https://tools.ietf.org/rfc/index.

If this book is used as a guide to set up an Intranet, please pay close attention to the sections marked "**Security**". These actions should be taken along with any other security actions required by your organization[4].

---

[4] In my opinion, perfect security is not possible if your network is connected to anything.

Acknowledgments for the first edition

I would like to thank the anonymous first readers of this book. Their suggestions made this a better work. Thank you.

This work would not be possible without the help of my many students over the years. This course was first taught as a second year program under the Kalamazoo Regional Education Service Area (KRESA) as part of Education for Employment (EFE). These poor students were subjected to working with antiquated equipment, Linux (which they were *not* usually familiar with), very limited outside resources, and many difficult challenges[5]. They loved it.

A more structured version of this course was taught in 2016 at Kalamazoo College in Kalamazoo, Michigan as *Building the Internet in a Room* using Raspberry Pi computers as described in this book. Apparently all went well as some students wanted to take the course again.

To all these students I would like to say: you put a lot of sweat into the classes upon which this book is built. I can't thank you enough.

Kalamazoo, Michigan   *Gerry Howser*
                                                                       Fall, 2019

---

[5] Things never worked out as planned. That was part of the attraction and challenge.

Contents

List of Acronyms

A
A: Administrative Authority record (IPv4)
A-PDU: Application Layer PDU
AAAA: Administrative Authority record (IPv6 or NSAP)
ABR: Area Border Router
ACK: Acknowledge transmission
AFI: Authority and Format Identifier (NSAP)
AFXR: Asynchronous Full Transfer
ANSI: American National Standards Institute
API: Application Program Interface
APIPA: Automatic Private IP Addressing
ARP: Address Resolution Protocol
ARPA: Advanced Research Projects Agency
ARPANET: Advanced Research Projects Agency Network
AS: Autonomous System
ASCII: American Standard Code for Information Interchange
ASIC: Application Specific Integrated Circuit
ASN: Autonomous System Number
ATM: Asynchronous Transfer Mode

B
BDR: Backup Designated Router
BGP: Border Gateway Protocol
BIND: Berkeley Internet Name Domain service
BIOS: Basic Input/Output System
BITNET: Because It's Time Network
BNA: Burroughs Network Architecture
BOOTP: Bootstrap Protocol
Bps: Bytes per second
bps: Bits per second
BTOS: Burroughs Task Operating System

C
CAT: Category (Structured Wiring)
CIDR: Classless Inter–Domain Routing
CNAME: Canonical Name
CPU: Central Processing Unit
CRC: Cyclical Redundancy Check
CSMA/CA: Carrier Sense Media Access/Collision Avoidance
CSMA/CD: Carrier Sense Media Access/Collision Detection

D
D-PDU: Data Link Layer PDU
DARPA: Defense Advanced Research Projects Agency
DDNS: Dynamic Domain Name System
DDOS: Distributed Denial of Service attack
DFI: DSP Format Identifier
DHCP: Dynamic Host Configuration Protocol
DIG: Domain Information Groper
DIS: Designated Intermediate System
DMZ: Demilitarized Zone
DNS: Domain Name Service
DNSSEC: Secure Domain Name Service
DOS: Disk Operating System
DOS attack: Denial of Service attack
DR: Designated Router
DS0:   Data Stream Zero
DS1:   Data Stream 1
DS3:   Data Stream 3
DSP: Domain Specific Part

E
EBGP: External BGP session
EIA: Electronic Industries Alliance
EIGRP: Enhanced Internal Gateway Routing Protocol
ESMTP: Enhanced Simple Mail Transfer Protocol

F
FAT: File Allocation Table (16 bit version)
FAT32:   File Allocation Table (32 bit Version)
FCS: Frame Check Sequence
FDDI: Fiber Data Distribution Interface
FIFO: First In, First Out
FQDN: Fully Qualified Domain Name
FRR: Free Range Routing
FTP: File Transfer Protocol

G

GUI: Graphical User Interface

H
HDMI: High Definition Multimedia Interface
HTML: HyperText Markup Language
HTTP: Hyper–Text Transfer Protocol
HTTPS: Secure Hyper–Text Transfer Protocol

I
IANA: Internet Authority for Names and Addresses
IBGP: Internal BGP session
IBM: International Business Machines
ICANN: Internet Corporation for Assigning Names and Numbers
ICMP: Internet Control Message Protocol
ID: System Identifier (NSAP)
IDI: Initial Domain Identifier (NSAP)
IDP: Initial Domain Part (NSAP)
IEEE: Institute of Electrical and Electronics Engineers
IETF: Internet Engineering Task Force
IFXR: Incremental Zone Transfer
IGRP: Internal Gateway Routing Protocol
IHU: I Hear U message
IMAP: Internet Message Access Protocol
Internet: Interconnected Networks
Intranet: Private Internet
IOS: Internet Operating System
IP: Internet Protocol
IPng: Internet Protocol, Next Generation
IPv4:   Internet Protocol, Version 4
IPv6:   Internet Protocol, Version 6
IPX: Internetwork Packet Exchange
IS: Intermediate System
IS–IS: ISIS Inter–Area Routing
ISIS: Intermediate System to Intermediate System
ISO: International Standards Organization
ISP: Internet Service Provider

L
L2TP: Layer 2 Tunneling Protocol
LAMP: LAMP Web Server
LAN: Local Area Network
Layer 1:   Physical Layer
Layer 2:   Data Link Layer
Layer 3:   Network Layer
Layer 4:   Transport Layer
Layer 5:   Session Layer

Layer 6:   Presentation Layer
Layer 7:   Application Layer
LED: Light Emitting Diode
LIFO: Last in – First out
LSA: Link State Announcement
LSD: Link State Database
LSP: Link State Packet Pseudonode

M
MAC: Media Access Control
MIME: Multipurpose Internet Mail Extensions
MobileIP: Cellular IP
MODEM: Modulator/Demodulator
MPLS: Multi-Protocol Label Switching
MST: Minimum Spanning Tree
MTA: Mail Transfer Agent
MX: Mail Exchange Resource Record (DNS)

N
N-PDU: Network Layer PDU
NAK: Negative Acknowledgment
NAT: Network Address Translation
NetBEUI: NetBIOS Extended User Interface
NetBIOS: Network BIOS
NGO: Non-Governmental Organization
NIC: Network Interface Card
NIST: National Institute of Standards and Technology
NNTP: Network News Transfer Protocol
NS: Name service
NSAP: Network Service Access Point
nslookup: Name Service Lookup
NTP: Network Time Protocol

O
OC1:   Optical Carrier 1
OC12:   Optical Carrier 12
OC24:   Optical Carrier 24
OC3:   Optical Carrier 3
OS: Operating System
OSI: Open Systems Interchange
OSPF: Open Shortest Path First (IPv4)
OSPFv3:   Open Shortest Path First (IPv6)

P
P-PDU: Presentation Layer PDU
PC: Personal Computer

PDU: Protocol Datagram Unit
PHP: PHP: Hypertext Preprocessor
ping: Echo Request and Echo Response
POP3:   Post Office Protocol
PPP: Point–to–Point Protocol
PPTP: Point–to–Point Tunneling Protocol
putty: Public TTY Client for Windows
PXE: Preboot eXecution Environment

Q
QoS: Quality of Service

R
RARP: Reverse Address Resolution Protocol
RD: Routing Domain Identifier
RFC: Request For Comments
RIP: Route Interchange Protocol
RIPng: Route Interchange Protocol for IPv6
RIPv1:   Route Interchange Protocol, Version 1
RIPv2:   Route Interchange Protocol, Version 2
RJ45:   Registered Jack 45
RR: Resource Record
RSVP: Resource Reservation Protocol

S
S-PDU: Session Layer PDU
SDA: SD Association
SDH: Synchronous Digital Hierarchy
SEL: NSAP Selector
SLIP: Serial Line Internet Protocol
SMTP: Simple Mail Transfer Protocol
SOA: Start Of Authority
SOHO: Small Office/Home Office
SONET: Synchronous Optical Network
SPF: Shortest Path First
SPX: Sequenced Packet Exchange
SQL: Standard Query Language
ssh: Secure Shell (ssh)
sudo: sudo

T
T-PDU: Transport Layer PDU
T1:   T–Carrier 1
T2:   T–Carrier 2
T3:   T–Carrier 3
TCP: Transaction Control Protocol

TCP/IP: Transaction Control Protocol over IP
TDM: Time Division Multiplexing
TFTP: Trivial File Transfer Protocol
TIA: Telecommunications Industry Association
TLD: Top Level Domain
TOR: The Onion Router
TTL: Time To Live

U
UDP: User Datagram Protocol
URL: Universal Resource Locator
USB: Universal Serial Bus

V
VERP: Variable Envelope Return Paths
vi: vi text editor
VLAN: Virtual Local Area Network
VLSM: Variable Length Subnet Mask
VOIP: Voice Over Internet Protocol
VPN: Virtual Private Network
vtysh: Virtual Terminal Shell

W
WAMP: Windows web server
WAN: Wide Area Network
WAP: Wireless Access Point
WiFi: Wireless Network
WLAN: Wireless Local Area Network
www: World Wide Web

X
XAMP: Cross–platform web server

List of Algorithms

List of Figures

List of Tables

Chapter 1
Introduction

Introduction

This book is designed to discuss networking from the simplest network of two devices through the behind the scenes elements of the modern Internet[1]. While it is completely possible to learn all of this from the web, it is hoped that the combination of text and hands–on will give the reader a better appreciation for how all the magic happens than could be obtained piecemeal from web–surfing.

The first part of the book, **The IP Network**, will give the reader some background into how devices communicate and the importance of the Physical Layer, the Data Link Layer, and the Network Layer of communications. Great care has been taken to approach these layers from a non–vendor specific viewpoint. In addition, the reader should get a feel for why corrupted packets can be dropped without dealing with the impact on the transfer of information. One of the upper layers, the Transport Layer, must deal with missing packets, not the network.

The second part of the book, **The Router**, explains the two main functions of a router and how they are accomplished. As in Part 1, the viewpoint is non–vendor specific and treats all routable protocols as equal. The emphasis is on how the router moves packets from known network to known network, not as much on how the router knows these networks.

The third part of the book, **Dynamic Networks**, covers the true heart of the Internet and how networks deal with the fact that everything changes on the Internet. Even rather small networks such as the Laboratory Network of Raspberry Pi microcomputers can be too complex for network administrators to easily handle. Only by using a routing protocol to dynamically learn the network can one hope to have a working, resilient network of any size[2].

---

[1] Throughout this book the term **Internet** will be used to denote the public Internet we all know and use. Private internets will be denoted by the terms **internet** (with a lowercase "i") and **intranet**.

[2] I speak from first–hand experience having watched a network administrator *manually* configure a large network with many subnetworks. It was not pretty.

© Springer Nature Switzerland AG 2020
G. Howser, *Computer Networks and the Internet*,
https://doi.org/10.1007/978-3-030-34496-2_1

The last part of this book, **Internet Services**, is an exploration of some of the most common services provided by the Internet and a few that are not as common but interesting to explore such as Telnet.

Finally, those interested in hacking the Internet either as a "black hat" or "red hat" will be disappointed. There is little here to help you, but security is discussed in various places where it makes sense for a private internet. Security on the true Internet would be another book.

The plan is to post errata, an FAQ, and other things of interest as time goes on.

I hope this little trip from connecting two computers together through building LANs to building a fully functional Intranet is informative, interesting, and as much fun as I have had in working through the different protocols and lab networks.

Enjoy!

Part I
The IP Network

Overview

"Begin at the beginning," the King said, gravely, "and go on till you come to an end; then stop."

Lewis Carroll, *Alice's Adventures in Wonderland* [9]

Few people today would even think about using a computer that could not connect to a network, but how does this connection happen? To start with, there must be a network and the computer must have the specific hardware to connect to the network. This hardware could be an RJ45 jack attached to the computer's motherboard, a separate card installed in an expansion slot, an infrared send/receive unit, or a dongle with a built in RJ45 jack. All of these are referred to as NICs and follow standards from ANSI[3], IEEE[4], IETF[5], NIST[6], and TIA[7]/EIA[8]. Bear in mind that WiFi and Bluetooth both are network connections; therefore, the WiFi and Bluetooth adapters are both **NIC**[9]**s**. In fact, most computers have more than one NIC and can attach to multiple networks at the same time.

The first task in building an **Internet** is to build a network of computers and other devices.

---

[3] American National Standards Institute

[4] Institute of Electrical and Electronics Engineers

[5] Internet Engineering Task Force

[6] National Institute of Standards and Technology

[7] Telecommunications Industry Association

[8] Electronic Industries Alliance

[9] Network Interface Card

Chapter 2
The OSI Seven Layer Model

Overview

The seven layer OSI[1] Model is an excellent theoretical framework for discussing messaging between processes running on different physical systems[2]. This exchange of messages is key for both peer–to–peer and client/server networks. Without a clear model, such as the OSI model, interoperability would be much more difficult. The current status of the lower three layers allows the networking professional to pick and choose cabling and device hardware from multiple vendors with the firm expectation any combination of media and device hardware will work as desired. In the early days of networking, this was not always guaranteed.

2.1 Analog Signals

There are two main methods to use electromagnetic signals to transmit data. The earliest electronic signals were modulated waves which smoothly varied the amplitude of the wave much like the waves created when a stone is thrown in a pond, see Figure 2.1. These analog signals have some distinct advantages due to their ability to exactly reproduce the input signal. However, as analog signals propagate, two bad things happen. The amplitude, or strength, of the wave is attenuated as it passes through the medium. As the amplitude of the signal decreases, it approaches the amplitude of any background noise. Sooner or later, the signal gets lost in the noise.

---

[1] Open Systems Interchange

[2] There are exceptions such as Asynchronous Transfer Mode (ATM) but even then the OSI Model is useful.

© Springer Nature Switzerland AG 2020
G. Howser, *Computer Networks and the Internet*,
https://doi.org/10.1007/978-3-030-34496-2_2

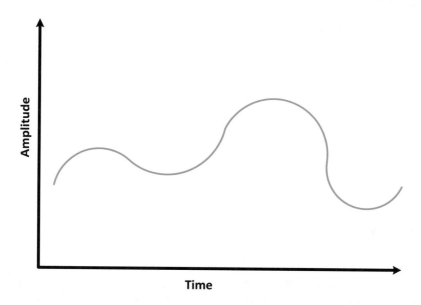

Fig. 2.1: An Analog Signal

The second issue is also caused by the media. Along with attenuation, the media will tend to spread out the signal and cause the waveform to deteriorate. Metallic media such as copper will also pick up any stray electromagnet signals in the area and add them to the signal, even if the media is well shielded. This is why speaker wires tend to pick up a 60 cycle hum on older stereos and TVs.

A solution to the problem of attenuation is install a device to boost the incoming signal amplitude back to an acceptable level. Unfortunately, this also amplifies the background noise. Expensive electronics can suppress some of the noise, but not completely. As the speed of the data transmission increases, these problems tend to get much worse.

2.2 Digital Signals

In order to amplify the signal without amplifying the noise, a different method of transmitting the data must be found. Frequency modulation, or FM, helps for lower data rates such as audio but does not work as well for high–speed data. Converting the signal from analog to digital, see Figure 2.2, not only helps with the problems of amplifying a signal but is semantically closer to the goal of transmitting binary data such as used by digital computers. The signal is sampled by small time slices and the mean amplitude for each time slice is recorded as an integer value. The smaller the time slices the closer the digital signal represents the analog signal, but it *cannot* exactly duplicate the incoming analog signal. This is why some musicians prefer older amplifiers and vinyl records as they are both analog and can to an extent support the claim to truer reproduction.

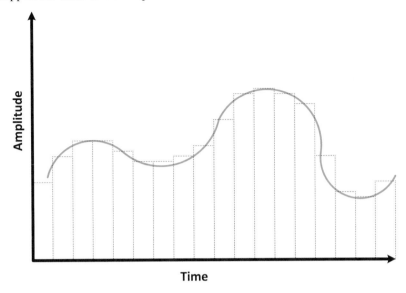

Fig. 2.2: A Digital Signal

Computer data is already stored as binary digits so sending it as a series of binary amplitudes of "one" or "zero" does not lead to any loss of data quality. When the signal is amplified, it is known that the value was originally an integer so the output signal is sent as a set of integers. The noise is typically much smaller and gets truncated, or rounded, out of existence when the amplified value is set to the closest integer. Any spreading of the signal disappears in much the same manner as long as the amplifier can determine the start/stop of each digit. The transmission error rate is extremely low for digital signals sent over media specifically designed for the top speed of the data such as structured wiring in a building or long–range Telco facilities.

2.3 Asynchronous and Synchronous Communications

In this text the terms "message" and "conversation" have distinct meanings. A message is a single communication from a sender to a receiver while a conversation is a series of messages sent back and forth between the endpoints of the conversation. Messages are most often made of digital signals, electrical or optical, but do not have to be unless we are talking about computer networks. Another issue is the related terms asynchronous and synchronous.

2.3.1 Synchronous

synchronous translates directly as "same time" or "at the same time" and refers to the fact that synchronous communications require either a single clock or two clocks that have the same time in terms of time from the start of the communication. This is typically done by sending a known preamble to the message to allow the receiving clock to synchronize to the timing of the bits in the message. For example, an Ethernet frame has an eight octet preamble of "10101010...10101011"[3] to allow the receiver to synchronize to the start/stop of the bits before any useful data is sent.

Another use of the term synchronous is to describe any data communication that is governed by a common time. If a sender must send every so many milliseconds, then when that time expires it must send a message of all nulls, or binary "00", to signify the communications are still happening but there is no data to send. This is common in communications such as SONET[4] or other Telco offerings.

The third use of the term is to signify that the communication relies upon previous knowledge or agreement between the sender and receiver as to what a specific message means in order to facilitate either faster or more secure communications. This sense of the term will not be used in this text unless explicitly noted. The reason this will not be used in this text is because this sense of synchronous is found mostly in operating systems and encryption[5].

2.3.2 Asynchronous

asynchronous translates to "not the same time" or "not synchronous" and will be used in that sense in this text. asynchronous communications typically have fewer constraints and require more effort and machine "smarts" to manage which can lead to more expense. However, asynchronous messaging is more efficient due to the fact that an endpoint need only transmit when there is data to send and does not need to wait for the clock to reach a specific point before beginning to transmit. In many cases asynchronous communications is more desirable than synchronous communications.

In some cases asynchronous is used to signify there is no need of prior agreement or knowledge for the communications to make sense, much like usage three for synchronous. This usage will be avoided in this text as well.

---

[3] This series of 64 binary digits is known as a semaphore.

[4] Synchronous Optical Network

[5] I would prefer a different term but the usage is too ingrained to be changed now. Besides it would require an additional and superfluous term and there are enough of those in networking already.

2.4 The Seven Layer OSI Model

The most useful theoretical model is the seven layer OSI model introduced by the ISO[6] in 1984. This model is very useful when talking about issues with the various services provided by devices on the Internet[7] but with one or two exceptions the full OSI model has rarely been implemented [312]. While the model was being developed, UNIX Operating System installations were developing a set of protocols for file transfer (FTP[8]), remote access by a "dumb" terminal (Telnet), and email (SMTP[9]). This suite of services was intended to be useful across many interconnected networks using TCP/IP[10]. As is often the case in the computer industry, custom overtook the proposal and TCP/IP became the *de facto* standard on the Internet. This section will address the differences, and similarities, between the OSI model and TCP/IP.

Table 2.1: The Names and Functions of the OSI Layers

Layer	Name	PDU	Function
1	Physical	Bits	Places bits on the wire
2	Data Link	Frames	Sends messages across the local network
3	Network	Packets	Sends packets across the networks
4	Transport	T-PDUs	Provides guaranteed delivery
5	Session	S-PDUs	Manages sessions between endpoints
6	Presentation	P-PDUs	Encoding, encryption, and compression
7	Application	A-PDUs	Manages service advertisements and connections to API Microcomputers

Traditionally communications between devices such as computers or routers has always been discussed in terms of the seven layer OSI[11] model which is interesting since very few vendors have ever introduced a NIC that uses the full OSI Model to communicate. However, there are some important advantages to using the OSI model when discussing electronic communications.

- Each layer establishes a virtual connection with the other endpoint of the communication by exchanging PDUs[12], see Table 2.1.
- The hardware responsible for connecting to a network is called a NIC. A device must have a separate NIC for each network connection.

---

[6] International Standards Organization

[7] Interconnected Networks

[8] File Transfer Protocol

[9] Simple Mail Transfer Protocol

[10] Transaction Control Protocol over IP

[11] Mnemonic for Network Engineers: Please Do Not Tell Sales People Anything, or Please Do Not Tell Silly People Anything.

[12] Protocol Datagram Units

- Networks are built upon Layers 1, 2, and 3 which are the same for virtually all electronic communications. These are sometimes called the Communications Layers.
- The OSI model uses strict structured programming techniques which limits the scope of most problems to one or two of the layers at a time.
- The interfaces between layers are very strictly defined to enhance interoperability.
- The OSI provides a very clear explanation of why a device can use any standard NIC to communicate with any other standard NIC regardless of the vendor(s).
- Many engineers, especially those working for telecommunications companies, will discuss services and problems by the layer at which those services operate.
- The OSI model provides a common "language" across different vendors, services, and communications standards.
- The structured nature of the layers allows us to concentrate on a single layer without needing to know *anything* about the other layers. This is why we can run the same Internet Protocols over wireless, copper wires, or fiber optics without making changes to the IP[13] configuration.
- When there is a failure in a network the most effective way to trouble–shoot the problem is to work the OSI model from layer 1[14] on up to layer 7.
- Even protocols that do *not* follow the OSI model, such as TCP/IP can be easily discussed in terms of the OSI model.

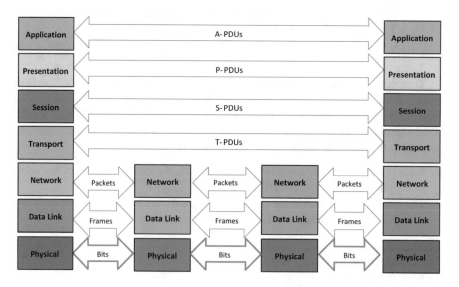

Fig. 2.3: Network Architecture Based Upon the OSI Model

---

[13] Internet Protocol

[14] When a network goes wrong, it's always the Physical Layer.

2.5 Communications Layers

The communications layers predate both the OSI Model and TCP/IP even though these layers were formalized when the ISO introduced the OSI Model. Layer 1, the Physical Layer, dates to the earliest electronic communications or earlier. This layer is tasked with transmitting the atomic information[15], usually bits, between two devices. It is at this level the actual network connections are made. Only the Physical Layer can transmit information with all higher layers transmitting information by establishing virtual connections as if some "media" existed to connect the layers.

The Data Link Layer, one of the Communications Layers, was developed when devices were required to share physical media. In order to dedicate media to a pair of devices, each device is required to have two connections: one for transmission to the other device and one to receive from the other device. Indeed, each bidirectional connection is actually a pair of unidirectional connections configured in opposite directions to avoid the disastrous possibility of both devices attempting to transmit at the same time. While such a dedicated connection may be required in some cases, the amount of media and the number of physical connections, called NICs, for any device quickly becomes prohibitive for a network of any size. To build anything but the smallest network requires shared media and Data Link addressing, or MAC[16] addressing.

With MAC addressing it is possible to send a message over shared media to one or all of the devices in the network. A message directed to a single device, or NIC, is called a unicast while one to all devices is sent to a special MAC address, the Layer 2 Broadcast address, and is called a Broadcast. It is even possible to send a single copy of a message over the shared media and have only selected devices process it. This is called a multicast. Having an address on each message also allows a NIC to ignore any message not addressed to it or the Broadcast address.

A group of layer 2 devices that can exchange unicasts and Broadcasts is called a LAN[17]. Many devices have been developed to extend the range of a LAN by interconnecting separate LANs to overcome restrictions of the underlying media. Unfortunately all Broadcasts must be received by all devices or the Layer 2 network is broken which limits the size and scope of the network to a single LAN or "broadcast domain." Even worse, as the number of devices on a LAN increases, the probability that two devices will attempt to send at the same time, a collision, goes up quickly. In practice it is impractical for more than 25 to 50 devices to share a single piece of physical media. LANs larger than this can only be created using Layer 2 devices to limit the size of collision domains.

To overcome the problems with huge "broadcast" domains and the problems with processing so many Broadcasts, Layer 3 or the Network Layer was developed. Like Layer 2 networks, Layer 3 networks are formed by exchanging virtual messages

---

[15] Atomic information is information that cannot be broken down into smaller units. For our purposes this means "bits" or binary digits.

[16] Media Access Control

[17] Local Area Network

which are called packets at Layer 3. Devices that connect two Layer 3 networks together must somehow transfer these packets from one network to the next while *not* transferring any Broadcasts. Small Layer 3 networks can be connected as in Figure 2.3 to form public Internets or private Intranets[18]. In fact, using relatively unsophisticated Layer 3 devices networks of any desired size can be created. However, sophistication is needed because the Internet is extremely volatile and humans are not competent to exchange the information required to make the frequent (more than daily) reconfigurations required to allow devices on any given Layer 3 network to communicate with all other devices on the Internet. Methods exist to exchange the required information, but this is not practical nor desirable. We will see that there are better solutions to the problems of sending packets from device to device than configuring every intermediate device to know the status of all the Layer 3 networks that are part of the Internet. This is the goal of this book, to explore how to build a practical Internet.

2.6 Layer 1: The Physical Layer

Layer 1, the Physical Layer, is responsible for putting bits onto the media and retrieving them from the media. In a lab setting, the media will be electrical wires and the bits will be generated as square wave voltages whereas in a house the media might be WiFi. Fortunately, we will not need to specify what media is being used in almost all cases. This will greatly simplify the discussions of protocols.

Table 2.2: Some Common Physical Media

Name	Media	Usage
Signal Fire	Light	Predetermined messages
Smoke Detector	Sound	Predetermined message
Infrared	Non–visible light	Laptop–to–laptop close range
CAT 1, 2, 3, 4	Low Grade Copper	Voice and less than 10 megabit
CAT 5, 5e, and 6	Structured wiring	10, 100 megabit, and gigabyte
Multimode Fiber	Light via glass	short range high speed data
Single Mode Fiber	Light via glass	Long range high speed data
WiFi	Electromagnetic waves	Mobile devices

Currently, wired Ethernet (IEEE standard 802.3) and WiFi (various IEEE 802.11x standards where x is: a, b, g, n, ...) are the most common network connections. Actually these are the Layer 2 protocols that use RJ45 jacks and antennas respectively, but these are the vernacular for wires and wireless and the distinction is not normally important unless the actual standards are being discussed.

---

[18] Private Internets

An interesting question for any Layer 1 protocol is: How do we represent a "zero" bit versus a "one" bit. The naive choice would be to use zero volts or no signal to represent a "zero", but this presents a major problem. We need to be able to determine a "zero" versus *loss of signal* or transmission error. The most common way to do this is to send a "zero" at some standard voltage[19] and a "one" at some other standard voltage.

As long as the media is dedicated to only one sender and receiver, this will work very well for small networks. A message sent from one NIC to only one other NIC is called a unicast. Unfortunately, this presents another problem as this technique requires a send NIC and a receive NIC for each connection. Obviously this does not scale well even for small networks, so we need some way to share media and address messages to the proper NIC so that other NICs on the same network can ignore them. Some method must be found to send a message from one NIC to all the other NICs on the network (a Broadcast).

One rather interesting possibility would be to use different voltages for each destination NIC, but this rapidly becomes too complicated for practical uses. There is no known practical, fast method to address a stream of bits at Layer 1, nor would one be wanted. At Layer 1 speed is of the utmost importance. To solve this problem, we need to build a logical network at Layer 2 that will work independently of Layer 1 so that it can run over any physical media.

Table 2.3: Some Common Layer 1 Devices

Name	Usage
Repeater	Detects bits and re–sends the bits at the proper levels
Modem	Modulates/Demodulates (converts) digital signals to analog signals
Hub	Detects the bits on a NIC and re–sends them out all connections at the proper levels and timings
Patch Panel	A method of wiring to facilitate reconfiguring the network
Wiring Closet	A secure location for patch panels

2.7 Shared Media

Directly connecting each device with each other device[20] presents a problem that quickly becomes unmanageable. The number of connected pairs in a network of n devices is given by the formula:

---

[19] This is called the "carrier" voltage. The sending end will "raise carrier" to signify the start of a message.

[20] This is a full mesh network.

$$\#\text{of pairs} = \frac{n(n-1)}{2}. \tag{2.1}$$

While this does not seem too bad for small networks, using this method to fully connect a network of 20 devices requires each device to have 19 connections and a total of 190 connections in the network. Clearly this is not sustainable for even small networks. The only workable solution is to somehow share media between connections. A simple, elegant solution to this problem, TDM[21] was developed by the telephone industry in the 1870's and put into heavy use during the middle of the 1900's.

2.7.1 Time Division Multiplexing

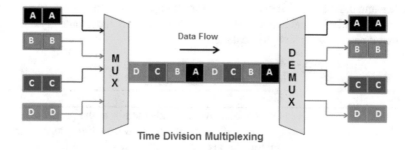

Fig. 2.4: Four Conversations Over a Single Wire Using TDM

TDM, also known as multiplexing, is a method to share a high–speed connection between a number of low–speed connections in such a way that each of the low–speed connections acts as if it has full bandwidth between the endpoints. The high–speed media is shared by giving each connection a time slice of the bandwidth in a round robin fashion.

The endpoints of the low–speed conversations are connected to a device called a multiplexer. The multiplexer assigns the high–speed connection to each of the slow–speed connections for an equal slice of time in a round robin as in Figure 2.4. This is done fast enough so that each conversation gets the bandwidth it would get if it were directly connected.

Some of the most common TDM services provided by ISPs[22] or Telcos in the United States is given in Table 2.4. Telcos typically build their networks based upon very few building blocks for economical reasons. A voice conversation is carried

---

[21] Time Division Multiplexing

[22] Internet Service Providers

over a DS0[23] of 64kbits/second. Twenty–four DS0s voice or data channels are combined via TDM into a single T1[24] and multiple T1s are typically combined to form a T3[25] at 45 Mbits/second. For most Telcos the T3 is the basic building block of the network with smaller bandwidth allocated at the customer's premises.

Fiber connections provided by Telcos are also TDM high–speed connections build of multiple OC1[26] or OC3[27] data streams. Local ISPs might run SONET connections at OC12[28] while larger ISPs might have connections at much higher speeds such as OC12 or above.

Table 2.4: Some Common TDM Telco Services

Name	Speed	Usage
T1	1.544 Mbit/s	Typical voice service and low speed data
T2	6.312 Mbit/s	Multiple T1s services
T3	44.736 Mbit/s	Multiple T1 services[a]
OC1	51.84 Mbit/s	The basic block of data for SONET
OC3c	155.52 Mbit/s	Three OC1 frames concatenated to form the payload of a single OC3 frame.
OC12	622.08 Mbit/s	OC12 lines are commonly used by ISPs as wide area network (WAN) connections, but not as backbone connections.
OC24	1244.16 Mbit/s	Used for large ISP backbones.
OC48	2488.32 Mbit/s	Not in common use except possibly by large ISPs
OC1920	99.5328 Gbit/s	Obviously not in common use[b].

[a]Telcos typically run a T3 to a business and then break it down to single voice connections.
[b]Not supported on the Pi as it might have problems with speeds in excess of 1 Gbit/sec.

It would seem that TDM is capable of handling all the shared media needs of modern networking, but there is a major issue. What happens to the available bandwidth in Figure 2.4 if the pair denoted as "A" does not need to communicate at this time? The bandwidth is *dedicated* to that conversation and can only be used by that conversation. This means that a large amount of bandwidth can be idle when using TDM and therefore wasted[29].

---

[23] Data Stream Zero

[24] T–Carrier 1

[25] T–Carrier 3

[26] Optical Carrier 1

[27] Optical Carrier 3

[28] Optical Carrier 12

[29] What is even worse is that each conversation *must* transmit when it is its turn so the wasted bandwidth is carried as a string of nulls or binary zeroes.

Another disadvantage of TDM for some networks is that to achieve better speeds the equipment is built to combine a specific number of low–speed conversations at a specific speed into one specific high–speed connection. For example, a voice connection over a T1 gets exactly 64kbits/second regardless of what is carried over the other 23 channels. This could mean a lot of binary zeros to fill the T1. Most data conversations are not good candidates for TDM at the local network level unless the data stream is constantly transmitting. While there are sophisticated multiplexers that can help redistribute this wasted bandwidth, there is a better way.

2.7.2 Layer 2 as an Alternative to TDM

For local networks, OSI Layer 2 provides a method to share media without expensive multiplexers and with much more flexibility. The solution is to provide each endpoint NIC with a unique address and uses frames to encapsulate messages as the data payload much like an envelope encloses a letter. Addressing is outside of the encapsulation and the payload does not need to be examined until the message arrives at the proper destination.

2.8 Layer 2: The Data Layer

The Data Layer is responsible for sending frames across a local network or LAN. The exact format and size of a frame depends upon the Layer 2 protocol in use on the LAN. For the purposes of this book, the only Layer 2 protocol of interest is Ethernet and other protocols are similar at Layer 2.

Because it is not possible to mark a message with a specific destination at Layer 1, the Data Layer contains protocols for Layer 2 addresses with the most common being the MAC address[30]. The MAC address is unique[31] to each NIC and is physically connected with the hardware. For this reason it is often called the hardware address or layer 2 address.

At Layer 2 the networks are called LANs and are logical networks built over the physical network of media. For our purposes, the association between two wireless devices will be considered as physical media and these networks are LANs. Often Layer 2 LANs exist only as part of the configuration of Layer 2 hardware and do not

---

[30] The MAC address is six bytes (or octets) long. This is much longer than was needed at the time Layer 2 networks were developed by Xerox and it is not clear why such a long address was chosen. If a reasonable size had been chosen, the networking world would have been in deep trouble.

[31] The first three bytes of the MAC denote the manufacturer and the second three bytes are used to form a unique MAC. Unfortunately, some manufacturers reuse the same last three bytes from time to time. Two identical MACs on the same LAN cause problems that are extremely difficult to detect and correct. If this ever happens to you complain to the manufacturer and salesman. It may not help, but duplicate MACs should not be tolerated.

always correspond to the actual media. These networks are virtual[32] and are called VLANs[33][34].

Errors on modern networks are not very common, but errors still occur. When a frame is too short, a "runt", or too long, a "giant", or the FCS[35], points to a transmission error, the NIC must somehow handle the situation[36]. It cannot simply pass the frame as if it were correct and the communication layer protocols do not have any retry methods built into them. The correct action in the case of error frames is to drop the frame. No effort is made to notify the sender of the error; the NIC simply processes the next frame[37].

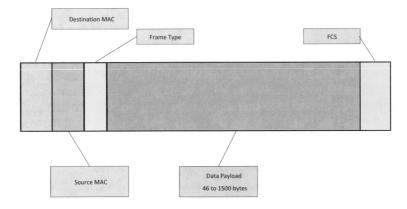

Fig. 2.5: IEEE 802.3 Ethernet Frame

---

[32] Any network that cannot be physically seen and traced out (other than WiFi) is a virtual network. Rule of thumb: If it goes away when power is turned off, it is virtual.

[33] Virtual Local Area Networks

[34] For our purposes, LANs, VLANs, and WLANs are essentially the same.

[35] Frame Check Sequence

[36] Some vendors take advantage of the requirement that a NIC discard giant frames and use giant frames for management messages between Layer 2 devices.

[37] Devices may attempt to use the FCS to correct single bit errors or the device may keep track of the error for its own purposes. Many L2 switches will monitor error rates and change their switching mode if the rate changes. This is beyond the scope of this book, but this is related to "cut-through" switching and "store–and–forward" switching.

Table 2.5: Fields in an Ethernet Frame

Name	Length	Usage
Preamble	2 Bytes	Frame synchronization
Destination MAC	6 Bytes	The "to" address
Source MAC	6 Bytes	The "from" address
Frame Type	2 Bytes	
Data Payload	46 to 1500 Bytes	The data, typically a Layer 3 packet
FCS	4 Bytes	Checks for errors in the rest of the frame

LANs are formed by exchanging well formatted streams of bits called frames. While the exact format of the frame depends upon the particular Layer 2 protocol used, all frames have a similar structure. The most common Layer 2 protocol is Ethernet (IEEE 802.3) and has the format shown in Figure 2.5. Ethernet frames are variable in length with a minimum length of 64 bytes and a maximum length of 1516 bytes[38]. If the data to be transmitted is less than 46 bytes the data payload is padded to make the length 46 bytes. If the message is too long to fit in the longest allowed packet, it is split into multiple messages by one of the upper layers.

Table 2.6: Some Common Layer 2 Devices

Name	Usage
L2 Hub	Send input frames out all other hub NICs
L2 Switch	Sends input frames out the NIC where destination MAC is detected

2.8.1 Configure the Pi for Layer 1 and Layer 2

Plug a CAT[39] 5e cable into the RJ45[40] jack of the Raspberry Pi. For additional network connections required for routing, plug a USB/NIC dongle into a vacant USB[41] port on the Raspberry Pi and then plug a CAT 5e cable into the RJ45 jack of the dongle. That is all that is needed, the NICs will do the rest.

---

[38] Actually there is a preamble for frame synchronization and some stop bits at the end. The NIC is designed to handle these issues.

[39] Category (Structured Wiring)

[40] Registered Jack 45

[41] Universal Serial Bus

2.9 Layer 3: The Network Layer

If two Layer 2 LANs are connected using any Layer 2 or Layer 1 device, the result is simply a larger LAN which means more broadcasts. However, if instead the devices embed the messages of a purely logical network in the Layer 2 frames, devices can continue to exchange messages *without* the requirement to share all Layer 2 broadcasts. At Layer 3 it is possible to interconnect as many networks as are needed in whatever manner makes the most sense. The task of Layer 3 is to correctly move messages as packets from a source Layer 3 network to a Layer 3 destination network as efficiently as possible.

2.9.1 Layer 3 Addresses

During the early years of networking, a number of layer 3 addresses were developed such as AppleTalk addresses by Apple, IPX[42] by Novell, NetBIOS[43][44] addresses by Microsoft, and IP addresses by the Internet community. Each address type has its advantages, but each requires a separate Layer 3 protocol to be running on each NIC. For this reason, Apple and Novell both stopped development on their proprietary networking and shifted development to IP addressing[48].

This allows us to concentrate on the IPv4 and IPv6 addresses as the only layer 3 addresses. Bear in mind that any discussion about interconnecting networks at layer 3 apply equally to IP, IPX, and AppleTalk even if these address schemes are not explicitly mentioned.

Table 2.7: Fields in a Typical IP Packet

Name	Length	Usage
IP Header	12 Bytes	Header Information
Source IP	4 Bytes	The "to" address
Destination IP	4 Bytes	The "from" address
Frame Type	2 Bytes	
Data Payload	46 to 1500 Bytes	The data, typically a Layer 3 packet
FCS	4 Bytes	Checks for errors in the rest of the frame

---

[42] Internetwork Packet Exchange

[43] Network BIOS

[44] NetBIOS is a network aware version of the BIOS[45] which is an integral part of the Microsoft DOS[46] OS[47]

[48] Microsoft invested too much in NetBIOS as the means for file and device sharing in its Windows operating system to be able to drop NetBIOS, but had incorporated IP into its server products from the beginning.

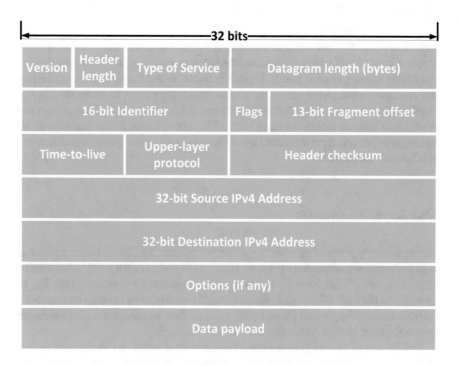

Fig. 2.6: A Simple IPv4 Packet

The only layer 3 address that will be running on the Pi NICs will be an IP address. This greatly simplifies matters and does not limit the services on the network.

Table 2.8: Some Common Layer 3 Devices

Name	Usage
L3 Forwarder	Routes packets between Forwarder's interfaces.
	Does not "learn" the network.
L3 Switch	Uses the current best route to send packets to next hop.
	Uses a routing protocol to automatically learn the network.
L3 Router	Uses the current best route to send packets to next hop.
	Uses a routing protocol to automatically learn the network.

2.10 Upper Layers

While all NICs operate at layers 1–3, the layers 4–7 are active only on the end–points of the conversation. These layers handle all the functions required to transfer information that are not directly involved in moving a message over the network.

The upper layers are not directly involved in physically, or logically, building a network.

2.11 Layer 4: The Transport Layer

The Transport Layer is responsible for providing either guaranteed delivery via a connection oriented conversation or "best try" delivery by a connectionless conversation. Connectionless delivery is fast and simple. Packets are sent our over Layer 3 and *no* efforts are made to correct errors or missing packets. Some uses of connectionless delivery are given in Table 2.9.

Table 2.9: Some Common Uses of Connectionless Transport

Usage	Comments
Video	Error correction is needless because video is "on the fly."
Audio	Error correction is needless because audio is "on the fly."
Telephones	Errors cause such small issues correction is pointless.
Push Services	Error can be corrected with the next push.

Connection oriented transport is more complex than connectionless transport. Layer 4 is responsible for verifying that all messages sent by one end of the conversation are correctly received by the other end. Because the Transport Layer checks each message for correctness, the communications layers are relieved of the need to correct each bad packet or frame. In fact, the communications layers simply "drop", or ignore, any packet or frame that has errors. Likewise, a layer 3 device does not track whether or not a packet ever arrives at the final destination. At layers 1–3 the only concern is speed. Data is moved as quickly as possible and the end–points of the conversation are responsible for correct delivery.

Table 2.10: Some Common Uses of Connection Oriented Transport

Usage	Comments
Software downloads	Uncorrected errors would force a new download.
Secure communications	Errors could be enemy action.
Banking	Money is involved.
Device negotiations	Must insure device agreement.

2.12 Layer 5: The Session Layer

The Session Layer handles requesting a session, approving a session, negotiating parameters for the conversation, detecting a failed session, restarting a failed session, and gracefully terminating a session. This layer is also responsible for "keep alive"s and "heartbeat"s to insure a session does not time out unexpectedly.

Layer 5 also handles the allocation of resources on each end of the conversation and the proper release of those resources when the session terminates. This de–allocation, or release, of resources is critical in preventing "memory leaks"

2.13 Layer 6: The Presentation Layer

My experience with teaching networking is that the Presentation Layer is the most unfortunately named of the seven layers as it has *nothing* to do with how the data is presented to the user; rather it is tasked with how the messages are presented to the end–points of the conversation. This layer handles encoding and decoding the messages, e.g into ASCII[49] or some other agreed upon coding. The Presentation Layer also handles any message compression or encryption.

2.14 Layer 7: The Application Layer

Contrary to what some people think, applications such as a web server do not function at the Application Layer. Instead, this layer is responsible for guiding messages to the correct application and for service announcements. This will be covered in more detail in chapter 6.5.

---

[49] American Standard Code for Information Interchange

2.15 TCP and UDP Upper Layers

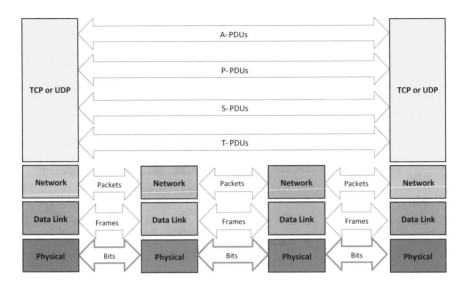

Fig. 2.7: Internetworking with the TCP/IP Model

TCP[50] and UDP[51] were developed about the same time as the OSI model but did not follow the OSI model for the upper layers. In fact, the upper layers in TCP and UDP are not clearly differentiated at all. Fortunately, for the purposes of this book, there is no real need to try to unravel the threads of the various upper layer processes in TCP for connection–oriented transports or UDP for connectionless, best effort transport, see Figure 2.7.

2.16 Mapping OSI and TCP Stacks to Client/Server Processes

One critical missing piece is how messages get from an OSI or TCP/UDP stack to the correct application, or process, on a device. Fortunately, all stacks use the same method to perform this mapping so we no longer need to distinguish between them and can simply talk about stacks. This mapping is done by attaching a two byte port or socket address to each layer 3 packet. A process sends a message with a specific socket number to a device which has a process "listening" to that same socket number. Sockets can be pre–defined, such as 80 for **httpd**, or sockets can be negotiated as part of establishing a session. A typical mapping of processes to sockets for a Raspberry Pi (192.168.1.31) is given below.

---

[50] Transaction Control Protocol

[51] User Datagram Protocol

```
pi@customPi:~ # sudo netstat -aptun4
Active Internet connections (servers and established)
Proto Recv-Q Send-Q Local Address              Foreign Address
             State        PID/Program name
tcp      0       0 0.0.0.0:110                  0.0.0.0:*
             LISTEN       584/dovecot
tcp      0       0 0.0.0.0:143                  0.0.0.0:*
             LISTEN       584/dovecot
tcp      0       0 192.168.1.31:53              0.0.0.0:*
             LISTEN       530/named
tcp      0       0 127.0.0.1:53                 0.0.0.0:*
             LISTEN       530/named
tcp      0       0 0.0.0.0:22                   0.0.0.0:*
             LISTEN       582/sshd
tcp      0       0 127.0.0.1:953                0.0.0.0:*
             LISTEN       530/named
tcp      0       0 127.0.0.1:3306               0.0.0.0:*
             LISTEN       731/mysqld
tcp      0     400 192.168.1.31:22
   192.168.1.109:39881       ESTABLISHED 924/sshd: pi [priv]
udp      0       0 0.0.0.0:50597                   0.0.0.0:*
                       343/avahi-daemon: r
udp      0       0 192.168.1.31:53                 0.0.0.0:*
                       530/named
udp      0       0 127.0.0.1:53                    0.0.0.0:*
                       530/named
udp      0       0 0.0.0.0:68                      0.0.0.0:*
                       529/dhcpcd
udp      0       0 0.0.0.0:5353                    0.0.0.0:*
                       343/avahi-daemon: r
pi@custompi:~ #
```

The terms socket and port are used to refer to a two octet field, or 65,636 possible ports, that is the link between an IP packet and a process. Packets sent to another device by IP address are labeled with a port number to allow the other endpoint to guide the packets to the proper process. Interestingly enough, neither NIC is at all involved in this. The sending process appends the port number which is either a well known port assigned by the IANA[52] or a port number agreed upon by the two end-points during session negotiations. The receiving NIC passes incoming messages to a FIFO[53] Queue associated with that port number. When ready, a process dequeues messages from the queue for processing in the same order as received. This is what we mean when we say a process "listens" to the port.

The use of sockets between the application and the stack allows for different types of mappings depending upon the requirements of the application. The simplest case is when the conversation is between two devices that are each running one process as in Figure 2.8. Usually the situation is more complex such as One–to–Many[54] as in Figure 2.11, Many–to–One as in Figure 2.10, Many–to–Many as in Figure 2.12.

---

[52] Internet Authority for Names and Addresses

[53] First In, First Out

[54] Obviously, a one–to–many relationship in networking looks like a many–to–one from the other endpoint. As with many things, it all depends upon how you look at it.

2.16.1 One–to–One Conversations

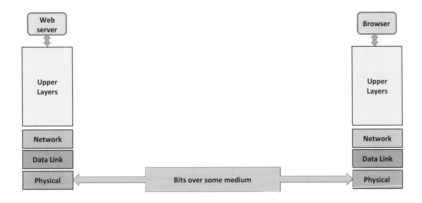

Fig. 2.8: A Typical One–to–One mapping

The simplest type of data conversation is when one process is exchanging messages with a remote device running exactly one network process. In this type of conversation, each NIC sends messages to the other endpoint of the conversation by address without needing to take into account the fact that the other endpoint most likely is running multiple networked processes. TCP and UDP will still use sockets because that is a fundamental feature of both protocols, but we do not need to concern ourselves with that.

2.16.2 Many–to–One Conversations

Fig. 2.9: A Typical Many–to–One Mapping for Processes on the Same Device

The first example, see Figure 2.9, is when a single request from a user generates conversations between a process such as a web browser and multiple devices on the network. Suppose a web page is requested from `http://www.yahoo.com`. The browser might first need to find the web server by asking for help from a name service and then sending the web page request to a web server process on the same physical device. In this situation, the web browser process is carrying on conversations with multiple servers. From one viewpoint, this looks like a many–to–one conversation. The same type of conversation goes on in the background if the requested web page includes information from multiple web servers. If the two processes are located at the same network address, the port or socket appended to the address allows the receiving NIC to guide the conversation to the proper service or process.

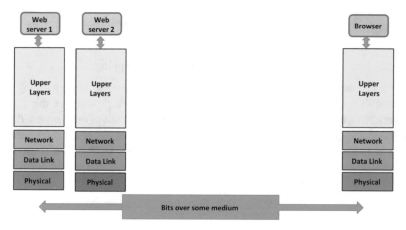

Fig. 2.10: A Typical Many–to–One Mapping

A more common situation, see Figure 2.10, is when multiple physical devices must be contacted to fill requests from a single process such as when a user is browsing the web. Web page requests go out from the same address and socket (80 in this case) to multiple services at multiple addresses. These services will all be expecting requests with a socket of 80, but the device addresses will be different.

2.16.3 One–to–Many Conversations

Again, looking from the server point of view, it is very uncommon to have a service answering requests from only one client process. It is much more common, and desirable, for a service to fill requests from many, many client processes. Figure 2.11 represents the situation where a web service is responding to requests from many different devices. All of the requests are sent with a socket of 80 but from different addresses. The service responds to the requests one at a time with the requester's address and a socket of 80. The address guides the response to the correct device.

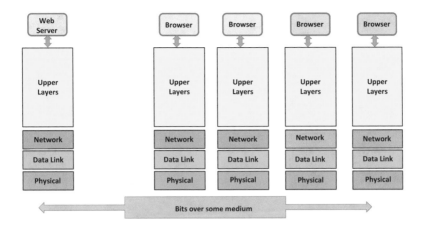

Fig. 2.11: Another One–to–Many Mapping

Another common example of many–to–one mapping occurs with streaming media and broadcasts. One process streams the media by a multicast at Layer 2 or Layer 3 and user processes have the option of joining the multicast group. This allows the server to transmit one copy of the data and have it automatically delivered to multiple users. This has the added benefit of potentially reducing traffic on the network.

2.16.4 Many–to–Many Conversations

The most common situation on a network is many–to–many conversations between devices. Clients request resources with the services' addresses and the appropriate socket number and services then responds with the requester's address and the appropriate socket. This has worked well from many different addressing schemes to the point where, on the Internet there are about 2000–4000 "well known" ports or sockets. A process and service can also have a conversation on a negotiated port number higher than 4000 and less than 65636. If you decide a client and service need more than 65,636 ports, you have a much more severe problem than port numbers and need to re–think the whole thing!

The possible existence of multiple users contacting multiple processes is an important issue to bear in mind when designing a network or a network process. Targeting a one–to–one mapping creates the potential for a major overhaul later in the life of the application which is always more expensive in terms of time and money than an open–ended design from the beginning. Furthermore, designing for many–to–many mappings from the beginning leads to a more robust application with fewer hidden bugs and restrictions.

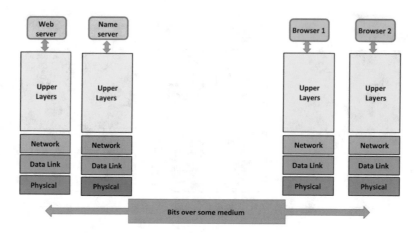

Fig. 2.12: A Typical Many–to–Many Mapping

Projects

1. Apply the OSI Model to explain the following communications:

   a. Calling your aunt Bea on the telephone.
   b. Gondor using signal fires to call for help from Rohan. Is this synchronous or asynchronous in terms of the third meaning in Section 2.3.1?
   c. Obtaining a webpage with a browser.

Exercises

2.1 What are some of the effects of the OSI Model requirement that each layer have one way in and only one way out?

2.2 Why should you care about the OSI Model if most of the devices on the Internet use TCP/IP which does not follow the model?

2.3 Give a non-computer network example of each of the following.

   a. A connection–oriented conversation.
   b. A connectionless conversation.
   c. Error correction. (Hint: look in your wallet.)
   d. A "keep alive" for a session.
   e. Any of the three Presentation Layer functions.
   f. Guiding a conversation from one sender to many receivers.

2.4 If a packet is damaged beyond repair, the Layer 3 device simply ignores it and goes on.

   a. If the conversation uses guaranteed delivery, how does the receiver determine a packet has been missed?
   b. How does this affect guaranteed delivery? What must the receiver do and what layer does it?
   c. What must the receiver do if the conversation does not implement guaranteed delivery?

2.5 Give at least two reasons why a session must be terminated gracefully.

Further Reading

The RFCs below provide further information about the OSI model and it pertains to the Internet. This is *not* an exhaustive list and most RFCs are typically dense and hard to read. Normally RFC are most useful when writing a process to implement a specific protocol.

RFCs Directly Related to This Chapter

OSI	Title
RFC 1070	Use of the Internet as a subnetwork for experimentation with the OSI network layer [80]
RFC 1888	OSI NSAPs and IPv6 [139]

Other RFCs Related to OSI

For a list of other RFCs related to the OSI Model but not closely referenced in this chapter, please see Appendix B.

Chapter 3
The Physical Layer

Overview

As stated before, see 2.1, the physical layer has one simple, and *extremely* important function: to move bits from one NIC to another as quickly as possible. How this is done depends solely upon the media used to connect the two NICs. For copper wiring this is done by varying the voltage between one voltage for a binary "0" and another for a binary "1". For wireless or fiber similar techniques are used. The speed at which a message can be transmitted is based solely upon how fast one NIC can produce a varying signal that can be correctly interpreted by the other. This speed is measured in terms of raw bits and can vary from a few hundred bps[1] to many millions of Bps[2].

3.1 The Network Interface Card

In order for a device to connect to any network, the proper NIC is required. Some connections, such as high speed fiber, can carry multiple networks and a NIC connected to such media may function as a number of logical NICs. [3]. This can be an Ethernet connection over copper or fiber, an infrared connection, or a wireless connection. A NIC is primarily a layer 1 device, but it must also handle layers 2 and 3; therefore, the choice of NIC determines the type of physical network to which it can connect and vice versa.

Any NIC can handle a single sending connection and a single receiving connection which allows for bidirectional communications. Notice that in Figure 3.1 the "send" of one NIC is connected to the "receive" of the other NIC. This pairing of

---

[1] Bits per second

[2] Bytes per second

[3] Unless explicitly stated otherwise, logical NICs will be considered as completely separate standard NICs

© Springer Nature Switzerland AG 2020
G. Howser, *Computer Networks and the Internet*,
https://doi.org/10.1007/978-3-030-34496-2_3

unidirectional messaging provides bidirectional communications. While it is possible for some media to truly provide bidirectional communications, usually this done by such a pairing of unidirectional connections[4]. This greatly simplifies things and automatically provides for concurrent messages in both directions. An examination of a fiber jumper or an Ethernet cable will show how "one" cable is really multiple media.

3.2 Communications Between Two NICs at Layer 1

An interesting question for any Layer 1 protocol is: How do we represent a "zero" bit versus a "one" bit. The naive choice would be to use zero volts or no signal to represent a "zero", but this presents a major problem. We need to be able to determine a "zero" versus *loss of signal* or transmission error. The most common way to do this is to send a "zero" at some standard voltage[5] and a "one" at some other standard voltage, see Figure 3.2.

Messages sent from a NIC can be broken down into three types. A message sent from one NIC to only one other NIC is called a unicast, a message sent to all NICs on a network is called a Broadcast, and a message sent to only some of the NICs is called a multicast.

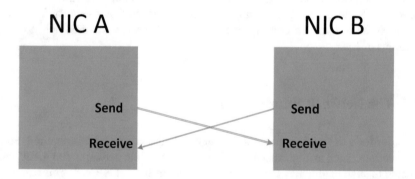

Fig. 3.1: Bidirectional Communications Between Two NICs Using Paired Unicasts

---

[4] A prime example is a satellite–based internet connection. The connection from the satellite is usually much faster than the connection into the internet.

[5] This is called the "carrier" voltage. The sending end will "raise carrier" to signify the start of a message.

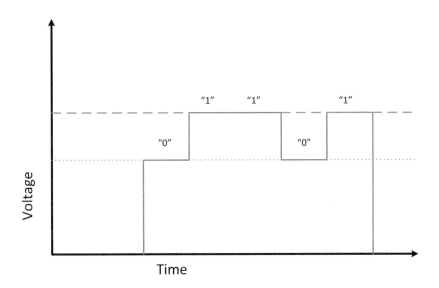

Fig. 3.2: The Message 01101 Over Copper Wire

As long as the media is dedicated to only one sender and receiver, this will work very well for small networks. Unfortunately, this presents another problem as this technique requires a send NIC and a receive NIC for each connection. Obviously this does not scale well even for small networks, so we need some way to share media and address messages to the proper NIC so that other NICs on the same network can ignore them. Some method must be found to send a message from one NIC to all the other NICs on the network (a Broadcast) or only some of the NICs (a multicast).

One rather interesting possibility would be to use different voltages for each destination NIC, but this rapidly becomes too complicated for practical uses. There is no known practical, fast method to address a stream of bits at Layer 1, nor would one be wanted. At Layer 1 speed is of the utmost importance. To solve this problem, we need to build a logical network at Layer 2 that will work independently of Layer 1 so that it can run over any physical media.

3.3 Cables and Signaling

The Physical Layer is concerned with sending information over the media *within* the rules for the media and the appropriate protocol. The large number of different types of media combined with the various protocols in use on those media leads to an overwhelming number of possible ways to send information. Market forces limit the actual number of choices available. No matter how well a physical media works, it is usually the case that the most popular ones are the most cost effective[6].

All media have a common issue in attenuation. As a signal travels over any media, the characteristics of the media cause the signal to degrade. The strength, or amplitude, of the signal decreases and background noise tends to increase. For electrical signals, this tends to be caused by the electromagnetic properties of the metallic wire. For fiber optics the problem is related to the media absorbing the lights to some extent. For whatever reason, all signals are attenuated and can suffer other issues as they are transmitted and thereby limiting the effective range of the media.

Within reason, this book will cover those media which are readily available on the Raspberry Pi with a short discussion on fiber optic communications cables.

3.3.1 Copper Wire

The terms "copper wire" and "wire" will be used here to refer to any metallic conductor surrounded by an insulating jacket. Signals over copper wire are usually done as a square wave of varying voltage such a in Figure 3.2. The electrical characteristics of the wire and how it is terminated determine the raw speed of the connection as modern electronics are capable of producing and detecting square wave signals much faster than copper wires can carry the signal.

The method of running and terminating the wires is so critical to the final speed that there are a number of companies in any area that specialize in structured wiring. For speeds above 10 megabits per second, it is imperative that wiring is done to the Category 5 (or better) standards and each cable run is tested. One often overlooked problem with sending high speed signals over a conductor is that any straight conductor with a varying voltage is an antenna and will radiate a signal. Using the proper insulating jacket and terminations will minimize the unwanted radiating of electromagnetic waves and of receiving the same kind of waves as interference.

Signals over wires are distance limited by three things; the speed of light, attenuation, and distortion. The speed of light in copper is a physical characteristic beyond our control. Voltage signals are actually light waves at a low frequency and cannot move information faster than light. The internal characteristics of the wire will cause some of the signal to be dissipated as heat and side currents. No matter how well the wire is designed for a given signal the amplitude of the wave will decrease with distance. This is attenuation.

As the wave moves down the wire, the internal impedance, capacitance, and reluctance of the wire will distort the square shape of the wave. When the distortion is great enough, the receiving NIC is unable to properly determine the bits in the message and will ignore the message.

---

[6] A prime example of this is the prevalence of Ethernet wiring instead of IBM Token Ring. Ethernet is an open standard while Token Ring was proprietary for many years. Popular equipment becomes cheaper if anyone can attempt to manufacture it.

3.3.2 Glass Fiber or Fiber Optics

A glass fiber is made up of three parts. The glass fiber itself is inexpensive silica glass, but of extremely high purity. The fiber is surrounded by a cladding which is then surrounded by a protective jacket. In order to know which fiber is which in a bundle of large numbers of individual fibers, the jacket is color–coded using the same TIA colors. Multi–mode fiber jumpers are often have orange outer jackets while single mode jumpers usually have yellow outer jackets.

Glass fibers can carry bits as changes to a beam of light. The fiber and its cladding act as something called a "wave guide" to keep the signal withing the fiber cable until it reached the end. This is why fiber optics can be routed in curves and bends as long as the bends are not extreme enough to cause the glass to fracture or crack[7].

Glass fiber can be classified as multi–mode or single mode. Multi–mode fiber uses an infrared LED[8] and sensor to send messages down the fiber. The interface between the glass and cladding act like a mirror and bounce the light back and forth until it reaches the end. Because of this, multi–mode fiber is limited to short distances of up to 600 meters and speeds up to about 1 gigabit per second. Multi–mode fiber is usually used in campus–sized networks and device interconnections. While termination of any fiber is expensive relative to copper, there are a plethora of connectors available for fiber; however, one must be careful to use multi–mode connectors on multi–mode fiber and single mode connectors on single mode fibers.

Single mode fiber typically uses an LED generated infrared laser to carry the stream of bits. Because of the characteristics of laser light, single mode fiber can carry signals long distances at high speeds up to 30–40 terabits per second. Terabit speeds are possible in theory and the 30–40 terabit upper limit is questionable. This is the limit for a single laser frequency at a given polarity. It may be possible to achieve much higher speed by multiplexing laser frequencies and polarities.

3.3.3 Wireless

It may seem odd to discuss wireless networks as having a physical layer. Indeed, wireless signals cannot be seen or felt but the electromagnetic waves are subject to physical restrictions just like copper or glass. Diffraction of the signal, interference with other devices, and range issues are best understood as physical layer issues and wireless is prone to the same attenuation as all media.

---

[7] All fiber cables have a minimum safe bend radius. If bent sharper than the bend radius at any time, the fiber will crack and develop "micro–fractures" which cause back scattering and loss of signal."

[8] Light Emitting Diode

3.4 Repeaters and Hubs

As we have seen, all media are subject to attenuation and distortion of the signal with distance. In fact, all media have standard maximum and minimum lengths due to electrical, optical, or other characteristics. However, many networks have viable cable runs longer than allowed by the standard. This is accomplished by the use of repeaters. Repeaters are powered devices with two NICs. Signals received at one NIC are simply sent out the other NIC at the proper levels and waveform. This allows a single repeater to double the allowed distance for a given media. Some repeaters, called transceivers, work in pairs by sending a signal from one physical media, say copper, over a different longer range media such as fiber and then back to copper at the other transceiver, as in Figure. A dial–up modem does exactly this.

Sometimes it is desirable to have all messages sent to all the devices on the network which can be done by building a device with a large number of specialized repeaters. A signal detected on any NIC relays the message (at the proper voltages and waveform) out all the other NICs. Such a device is very useful in networks that use media that is shared at Layer 2.

There are many practical advantages of building the physical network using hubs located in wiring closets. Longer wiring runs are prone to all kinds of problems and not all are simply distance related. Shorter runs are easier to test and repair and often are initially cheaper to install. Using hubs also facilitates changes to the network and *all* networks change.[9]

3.5 Shared Physical Media

How can a message be sent to a destination over physical media that is shared by multiple devices? We have *no* simple way to mark a bit with a destination address. As we have seen before, we could use a different set of voltages for each destination but the complexities of such a schema would at best lead to a very expensive NIC. There is no workable way to mark bits with a destination address at Layer 1. In any working network, all bits can be detected at the physical layer, but the meaning of the stream of bits may not be easily discernible[10].

---

[9] In the telecommunications world these are known as "moves and changes."

[10] Contrary to some opinions, even fiber optical connections can be monitored for bit traffic. This means that even if traffic cannot be understood on a connection, it can be detected which does give an adversary some information.

3.6 The Raspberry Pi and Layer 1

The Raspberry Pi come equipped with one RJ45 Ethernet NIC or jack. Newer Pi Microcomputers also come with an on-board wireless NIC and a number of USB ports which can be used to connect Ethernet dongles for more NICs. As with most physical layer connections, all you need to do to connect the Pi is plug in the proper cable. Fortunately, it is not possible to plug the cable into the wrong place.

> Consult the Pi documentation for your specific model/dongle before connecting to a power–over–Ethernet cable. The Pi could suffer serious damage. There is no advantage to connecting the Pi to power–over–Ethernet as the Pi cannot be powered that way.

The most common wired connection is a CAT 5 or higher cable plugged into the RJ45 jack on the Raspberry Pi. There are two standards for connecting the RJ45 plugs as shown in Figure 3.3.

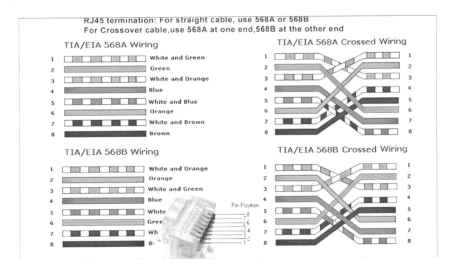

Fig. 3.3: Wiring a Common Network Cable

Projects

3.1 Research how a MODEM[11] works and *why* they are required.

3.2 (Optional) Find out how the EIA/TIA color code standard allows one to quickly separate pair 73 from a cable with 800 pairs.

3.3 If you purchased a Raspberry Pi as part of a kit, it should have come with an Ethernet cable. Determine if the cable follows the EIA/TIA 568A standard or the 568B standard.Telecommunications Industry Association

3.4 If you have access to the required equipment and materials, make an Ethernet crossover cable or a standard Ethernet cable. Be prepared to explain what you did to make the cable.

Exercises

3.1 Most of the communications at Layer 1 are digital but some are analog.

    a. Does Figure 3.2 show digital communications or analog communications?
    b. Explain what is meant by digital communications.
    c. Explain what is meant by analog communications.

3.2 Give an example of an analog signal.

3.3 Give an example of a digital signal.

3.4 When might two computers communicate with digital signals?

3.5 When might two computers communicate with analog signals? (Hint: The movies *Wargames* [46], *Hackers* [49], and *Sneakers* [307] might help.)

3.6 Some engineers consider a wiring closet or patch panel as a Layer 1 device and some do not. Support your opinion.

---

[11] Modulator/Demodulator

Chapter 4
The Data Link Layer

Overview

At the Physical Layer, messages consist of bits placed on the media but this technique does not scale to even small–to–moderate size networks. To build a network of any size, the media must be shared between multiple devices rather than dedicated to a single pair of devices. In order to do this, each device must have an address and there must be protocols to insure that each device can eventually send a message with a reasonable chance of that message reaching the correct destination.

Since every device on a shared media sees every message, it is possible to use this fact to allow a device to send a single copy of a message to all the devices on the network at the same time. With a the help of applications running on these devices it is possible to send a single message to a special address that a predetermined group of devices would process and devices not in the group would ignore.

These things are possible on Layer 2 networks or LANs.

4.1 Broadcasts, Unicasts, and Multicasts

Messages sent from a NIC can be broken down into three types. A message sent from one NIC to only one other NIC is called a unicast, a message sent to all NICs on a network is called a Broadcast, and a message sent to only some of the NICs is called a multicast. The actual type of a message is denoted by the destination Layer 2 address, or MAC address, as in Figure 4.1. Each and every NIC has an address that is physically associated with only that NIC and is usually a chip on the NIC or somehow burned onto the chip when it is produced[1]. A message with a valid destination MAC is to be processed only by that NIC and is a unicast. A message with the special Broadcast address is to be processed by every NIC on the

---

[1] Some devices can be configured to send and receive at a different MAC address but this is beyond the scope of this text and should be avoided whenever possible.

network and is a Broadcast. Likewise, if the destination address is one of the special
MAC addresses that is to be processed by a group, but not all, of the devices on the
network it is a multicast message.

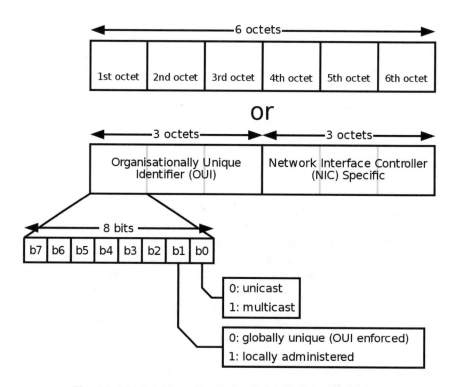

Fig. 4.1: MAC Address Format and Message Type Indicators

4.2 Frames

The bits transmitted at Layer 1 can be logically grouped into Layer 2 PDUs called
frames, see Figure 2.5. These frames consist of a destination MAC address, a source
MAC address, a length field, a variable length data payload (usually a packet or
Layer 3 PDU), and a protection field known as a CRC[2].[3] Because the data payload
has a variable length, the length of an Ethernet frame must be no less than 64 bytes
and no greater than 1516 bytes. Table 4.1 [27] give the most common IEEE[4] Layer 2

---

[2] Cyclical Redundancy Check

[3] A good analogy is a parcel post package.

[4] Pronounced "Eye Triple E". The IEEE maintains many hardware and protocol standards for
networking, computers, and media.

standards. We are most interested in IEEE 802.3 (wired Ethernet) and IEEE 802.11x (Wireless) where x is a, b, g, or n.

Table 4.1: Selected IEEE 802 Standards

Standard	Name	Topic
802.1	Internetworking	Routing, bridging, and network–to–network communications
802.1	Logical Link Control	Error and flow control over data frames
802.3	Ethernet LAN	All forms of Ethernet media and interfaces
802.4	Token Bus LAN	
802.5	Token Ring LAN	All forms of Token Ring media and interfaces
802.6	Metropolitan Area Network	MAN technologies, addressing, and services
802.7	Broadband Technical Advisory Group	Broadband networking media, interfaces, and other equipment Institute
802.8	Fiber Optic Technical Advisory Group	Fiber optic media used in token–passing networks like FDDI[a]
802.9	Integrated Voice/	Integration of voice and data traffic over a single network medium
802.10	Network Security	Network access controls, encryption, certification and other security topics
802.11	Wireless	Standards for wireless networking for many broadcast frequencies and usage techniques
802.14	Cable broadband LANs and MANs	Standards for designing networks over coaxial cable based broadband connections.

of Electrical and Electronics Engineers

[a] Fiber Data Distribution Interface

4.2.1 Runts, Giants, and Super–Frames

All Layer 2 frames, have a well–defined range of sizes and devices can ignore any frames outside of this range as an error frame. No action need be taken and usually these frames are dropped. Frames that are too small are called "runts" and those that are too large are called "giants". Some devices take advantage of this by intentionally sending management messages using frames that are larger than the protocol allows. Normal devices drop these "super–frames" while the manufacturer's devices properly decode the frame as a management message. One side effect of this is that most Ethernet protocol decoders (sniffers) will not decode super–frames but simply list them as errors.

4.3 Local Area Networks or LANs

The purpose of sharing media and messages is to form small networks such as a computer lab. The next few sections will explain the conditions which must be met to create what is known as a LAN.

LANs can be built upon a number of different physical network topologies and each Layer 2 protocol is designed to operate over a specific set of logical topologies regardless of the actual physical topology. For example, most Ethernet LANs are physically wired as star networks while Ethernet requires a logical bus network. FDDI and Token Ring on the other hand require a logical ring topology but can be wired as star networks with the proper equipment as the hub of the physical network. Unless there is some significant detail that requires examination, we will not concern ourselves with the physical or logical topology of the network.

The two LANs that are common today are defined by the standards wired Ethernet 802.3 and wireless 802.11.

4.3.1 Broadcast Domains

All of the devices in a network that can detect a Broadcast make up a Broadcast domain or LAN. If two devices cannot share Broadcasts, *they are not in the same local area network*. This is a two–edged sword. All of the devices on a LAN must detect, and deal with, every Broadcast or the LAN is broken. As LANs grow larger, the number of Broadcasts tends to increase as well. To make matters worse, some programmers are lazy and Broadcast messages that could just as easily be sent as unicasts or multicasts. Since each NIC must process each Broadcast, often only to have it be ignored by the device after working its way up the TCP/IP stack, this wastes processing time and will eventually stop needed messages from being delivered.

4.3.2 Collision Domains

Shared media creates two major problems: collisions and starvation. Collisions occur when two or more NICs "raise carrier" and transmit bits at the same time. Starvation occurs when a device is not able to send a message due to too many other messages being sent over the shared media. Ethernet prevents starvation of a conversation by limiting the length of a frame and with collisions by a technique known as CSMA/CD[5].

---

[5] Carrier Sense Media Access/Collision Detection

4.3.2.1 CSMA/CD

Collisions on Ethernet are dealt with by using a protocol known as CSMA/CD which follows Algorithm 1. When no device is transmitting, a NIC can send a message. If a device detects a carrier voltage, then the device waits until no carrier is present before sending. A collision can occur when two NICs do not detect any carrier voltage and attempt to start sending at the exactly the same time. Each NIC will sense that the bit pattern on the line does not match what it is sending, immediately stop sending, and wait a *random* time before attempting to send again. The reason why the wait time is random is left for an exercise, see Exercise 1.

Algorithm 1 Carrier Sense Media Access with Collision Detection

---

```
 1:  procedure CSMA/CD
 2:      while There is a frame to send do
 3:          if No carrier voltage is sensed on the media then
 4:              Send bits while monitoring message for other bits
 5:          else
 6:              Wait a random time for no carrier
 7:          end if
 8:          if Collision sensed then
 9:              Reset to the beginning of the current frame
10:              Wait a random time and attempt to transmit the frame
11:          end if
12:      end while
13:  end procedure
```

4.3.2.2 CSMA/CA

Another method to deal with collisions on shared media with many devices is CSMA/CA[6], which is not used with Ethernet. This method is very similar to CSMA/CD, but instead of dealing with collisions, CSMA/CA relies upon avoiding collisions altogether as in Algorithm 2.

4.4 Bridges and L2 Switches

As Ethernet networks add devices, the possibility of a collision increases to the point where a single shared media with about 50 devices is not practical. Also, a single run of copper media is limited to about 300 feet. Both of these problems can be overcome by the use of a Layer 2 bridge or a Layer 2 switch. Similar problems occur with wireless networks, but are not dealt with in this book.

---

[6] Carrier Sense Media Access/Collision Avoidance

Algorithm 2 Carrier Sense Media Access with Collision Avoidance

---
```
 1: procedure CSMA/CA
 2:     while There is a frame to send do
 3:         if No carrier voltage is sensed on the media and there are no outstanding send requests
    then
 4:             Send "send request" short message
 5:             Send bits
 6:         else
 7:             Wait a random time for no carrier or the current frame to end
 8:         end if
 9:     end while
10: end procedure
```

4.4.1 Extending Broadcast Domains

In order to be a part of a LAN, a device must be able to exchange Broadcasts and unicasts with all other devices on the network; however, on larger networks this leads to all NICs dealing with a significant number of unicasts that are meant for other devices. The frames being exchanged by other devices force a NIC to remain idle until the media is available. The congestion on the media becomes worse and worse and the percentage of messages that are Broadcasts becomes higher and higher. The solution is to extend the Broadcast domain while minimizing the other background traffic on the LAN or VLAN. There are two types of devices to do this: bridges or Layer 2 switches.

Table 4.2: Example Bridge MAC Address Table

Known MAC	Port
MAC1	NIC A
MAC2	NIC A
MAC3	NIC B
MAC4	NIC B

A bridge, see Figure 4.2, is a device with two NICs which connect to two separate segments of an Ethernet network and keeps a table of all known MAC addresses on the LAN, see Table 4.2. The bridge monitors each segment and transfers frames from one segment to the other *only* when the destination MAC address has been discovered on the *other segment*. The following examples and Figure 4.2 will make this functioning clear.

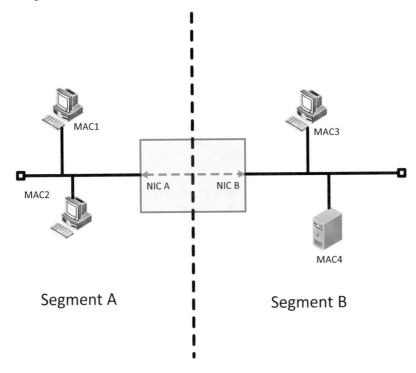

Fig. 4.2: A Typical Layer 2 Bridge

Table 4.3: Example Bridge MAC Address Table

Known MAC	Port
MAC1	NIC A
MAC2	NIC A
MAC4	NIC B

1. Bridge is powered on and sets up an empty bridge table with no known MAC addresses other than **NIC A** and **NIC B**.
2. The PC with MAC address **MAC1** sends a message to the PC with MAC address **MAC2**
3. The bridge does not find **MAC2** in the bridge table, so the frame is relayed out **NIC B** and **MAC1** is entered into the table as on segment A
4. The server, MAC address **MAC4**, sends a message to **MAC1**.
5. **MAC1** is in the bridge table as being on segment A, so the bridge relays the frame out **NIC A** and enters **MAC4** in the table as on segment B
6. **MAC4** sends a Broadcast. All Broadcasts are automatically relayed to the other segment.

7. **MAC2** sends a message to **MAC1**. Since the bridge table already has **MAC1** on the same segment as the message is detected, the bridge adds **MAC2** to the table as being on segment A and does nothing else.
8. At this point, the bridge table looks like Table 4.3.
9. As frames are sent, the bridge continues to learn the network.[7]

4.4.2 Limiting Collision Domains

Because unicasts are not forwarded to the other segment unnecessarily, the traffic on the network is restricted somewhat to a single segment. This has the desired side–effect of reducing the probability of a collision. Broadcasts are always relayed so that the LAN functions correctly. Because a bridge is also a Layer 1 hub, the overall length of the network can also be increased[8]. Because a collision cannot occur between NICs on two different bridged segments, these segments represent two different collision domains.

If a bridge is constructed with more than two ports or NICs, frames can be "switched" between segments with little or no delay. Such a "multi–tailed" bridge is called a switch[9]. Like a two–port bridge, each port limits collisions to the NICs on that port. Large numbers of devices can be easily moved from one port to another simply by unplugging a cable from one port and plugging it into another. The ability of a bridge or switch to dynamically maintain the bridge table is critical to managing changes to the network and avoiding problems. When carried to the extreme of one device per switch port, collisions do not happen as the collision domains now consist of a switch port and one device[10].

4.5 Connecting Layer 2 Networks

What happens if messages need to be traded between two different LANs? If the LANs are connected with a bridge or switch, frames can travel back and forth but Broadcasts will also be relayed between the two LANs. If the LANs can exchange

---

[7] In modern bridges, the table "ages" over time so that MAC addresses that send or receive any messages are eventually removed from the table. Should a MAC address show up on a different port than what is in the table, the table is updated to reflect the changed network. Between the time a NIC is moved from one bridge segment to another, some frames may be lost, but this problem is handled by the Transport Layer. In this way, the bridge table reflects changes to the network.

[8] A bridge usually stores the frame completely before forwarding it to another segment. This has the added benefit of relaxing the requirement that both segments be the same media type and speed.

[9] The term "switch" will be understood to mean a Layer 2 switch. A Layer 3 switch will always be referred to as a Layer 3 switch or a L3 switch.

[10] Technically, collisions can still happen but the rate is negligible

Broadcasts, the two LANs are the *same* LAN. There is no way to exchange frames between two LANs without making them one LAN.

4.5.1 Broadcasts in Networked Switches

As a network grows, more and more switches are required. It is desirable to have multiple connections between switches so if one connection goes down, messages can take a different path through the network. Consider a Broadcast in the network shown in Figure 4.3.

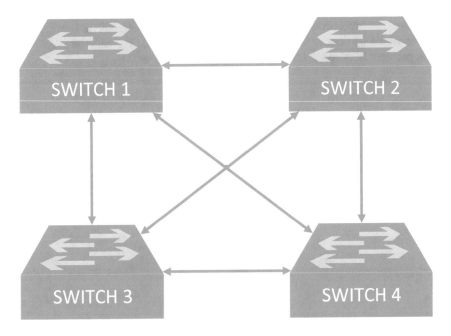

Fig. 4.3: A Typical Switch Network

Suppose a server connected to Switch 1 Broadcasts the availability of a printer. Switch 1 must relay the Broadcast out all the ports other than where the server is connected. This sends the Broadcast to Switch 2, Switch 3, and Switch 4. Each of those relays the Broadcasts out all other ports. For example, Switch 2 relays the Broadcast to Switch 1, Switch 3, and Switch 4. In a few milliseconds, thousands or millions of Broadcasts have been sent and each Broadcast causes more Broadcasts. The switches and network media become saturated and no traffic can move through the network. This is called a "broadcast storm" and brings the network completely down almost instantly.[11]

---

[11] If a number of switches are available, this is an interesting demonstration, especially if there are Windows devices on the network as Windows is *chatty*.

To eliminate Broadcast storms and still allow for redundant links between switches, the Spanning Tree protocol [25] was developed. When a new link is detected, or an existing link goes down, the switch initiates the protocol. A switch is elected as the "root" switch of the spanning tree and the switches detect and turn off all redundant links. The result is a tree network that cannot have a Broadcast storm yet still has redundancy that is held in reserve.

4.5.2 Layer 2 Networks Cannot be Connected

There is no practical way to connect two Layer 2 networks at Layer 2 because of the Broadcast problem. Indeed, were the Internet to have been built at Layer 2, each device would be flooded with Broadcasts and no information could be exchanged. This problem is solved at Layer 3 (see Chapter 5).

4.6 The Raspberry Pi and Layer 2

Configuring the Pi for Layer 2 is extremely easy. Each NIC has a unique MAC address. All that needs to be done is to plug the proper Ethernet CAT5 cable into the Pi. For additional connections to other LANs, a dongle can be plugged into an available USB port and then a CAT5 cable plugged into the dongle.[12]

---

[12] Thanks to the engineers at XEROX during the early days of Ethernet, we do *not* need to configure anything at Layer 2.

Projects

1. Connect a Raspberry Pi to some other device by the built in RJ45 connector. Observe and document the LEDs.
2. Investigate how a switch network with redundant connections (for fault tolerance) is susceptible to a broadcast storm.
3. If you have a sniffer available, monitor and decode frames on your network.

> **Warning:** In many states it is illegal to use a sniffer on a network that you do not own without permission.

Exercises

4.1 Why is the wait time a random number in line 10 of Algorithm 1?

4.2 Is it possible for a Layer 2 device to act at Layer 1 at the same time?

4.3 Can a Layer 2 device be constructed that does not act at Layer 1?

4.4 Why is there an upper limit to the length of an Ethernet frame?

4.5 Why is there a lower limit to the length of an Ethernet frame?

4.6 A general rule of thumb for trouble–shooting is to work up the OSI Model from Layer 1 to Layer 7[13]. Explain how this relates to Layer 2 devices.

4.7 Draw five conversations (A,B,C,D,E) over TDM with no data to transmit for conversation D.

---

[13] When there is a failure, it is always the Physical Layer.

Further Reading

The RFC below provide further information about the Data Link Layer. This is a
fairly exhaustive list and most RFC are typically dense and hard to read. Normally
RFC are most useful when writing a process to implement a specific protocol.

RFCs Directly Related to This Chapter

Layer2	Title
RFC 0826	An Ethernet Address Resolution Protocol: Or Converting Network Protocol Addresses to 48.bit Ethernet Address for Transmission on Ethernet Hardware [67]
RFC 0894	A Standard for the Transmission of IP Datagrams over Ethernet Networks [71]
RFC 0895	Standard for the transmission of IP datagrams over experimental Ethernet networks [72]
RFC 1132	Standard for the transmission of 802.2 packets over IPX networks [83]
RFC 1577	Classical IP and ARP over ATM [115]
RFC 1972	A Method for the Transmission of IPv6 Packets over Ethernet Networks [148]
RFC 2358	Definitions of Managed Objects for the Ethernet-like Interface Types [170]
RFC 2464	Transmission of IPv6 Packets over Ethernet Networks [173]
RFC 3378	EtherIP: Tunneling Ethernet Frames in IP Datagrams [202]
RFC 3619	Extreme Networks' Ethernet Automatic Protection Switching (EAPS) Version 1 [216]
RFC 3621	Power Ethernet MIB [217]
RFC 3635	Definitions of Managed Objects for the Ethernet-like Interface Types [218]
RFC 3637	Definitions of Managed Objects for the Ethernet WAN Interface Sublayer [219]
RFC 3817	Layer 2 Tunneling Protocol (L2TP) Active Discovery Relay for PPP over Ethernet (PPPoE) [223]
RFC 4448	Encapsulation Methods for Transport of Ethernet over MPLS Networks [238]
RFC 4638	Accommodating a Maximum Transit Unit/Maximum Receive Unit (MTU/MRU) Greater Than 1492 in the Point-to-Point Protocol over Ethernet (PPPoE) [243]
RFC 4719	Transport of Ethernet Frames over Layer 2 Tunneling Protocol Version 3 (L2TPv3) [247]

Layer2	Title
RFC 4778	Operational Security Current Practices in Internet Service Provider Environments [248]
RFC 5692	Transmission of IP over Ethernet over IEEE 802.16 Networks [261]
RFC 5828	Generalized Multiprotocol Label Switching (GMPLS) Ethernet Label Switching Architecture and Framework [262]
RFC 5994	Application of Ethernet Pseudowires to MPLS Transport Networks [264]
RFC 6004	Generalized MPLS (GMPLS) Support for Metro Ethernet Forum and G.8011 Ethernet Service Switching [265]
RFC 6005	Generalized MPLS (GMPLS) Support for Metro Ethernet Forum and G.8011 User Network Interface (UNI) [266]
RFC 6060	Generalized Multiprotocol Label Switching (GMPLS) Control of Ethernet Provider Backbone Traffic Engineering (PBB-TE) [267]
RFC 6085	Address Mapping of IPv6 Multicast Packets on Ethernet [268]
RFC 8388	Usage and Applicability of BGP MPLS-Based Ethernet VPN [295]

Other RFCs Related to Layer 2

For a list of other RFCs related to the Data Link Layer but not closely referenced in this chapter, please see Appendix B and Appendix B.

Chapter 5
The Network Layer

Overview

Computer–to–computer connections are made by connecting physical media at Layer 1 or the Physical Layer. The problems encountered are well understood and improvements in speed and reliability have removed virtually all problems, other than connector or "wire" problems, from consideration. Once the building wiring is in and tested, it is simply a matter of plugging in jumper cables to build the physical network. To construct a network of any size or complexity, a logical network of shared media and addressable devices must be built on top of the Layer 1 wiring.

With MAC addressing in place, unicasts and broadcasts can be sent between devices to carry messages back and forth; however, it is not possible to connect two Layer 2 LANs together without making them one big network. The issue now before us is how to trade all the required messages between multiple LANs without being overwhelmed by broadcasts. To do this, we must look at Layer 3, the Network Layer.

5.1 Layer 3 Logical Networks

What is needed is a device that has a NIC in each of the LANs to be connected so the device can forward messages from one LAN to the other only when needed and not forward broadcasts. When a device is installed on a LAN, it already has a unique address in the MAC address. Bridges and switches can learn the network and which MAC addresses are on each of its ports. This works well for small LANs but there is a limit to how many devices can effectively function on a LAN no matter how much thought went into designing the LAN. What is needed is a simple way to group addresses into networks that the designer can control.

Many schemes have been introduced to assign network addresses that can be grouped somehow to form logical networks at Layer 3 such as Novell's IPX, Ap-

© Springer Nature Switzerland AG 2020
G. Howser, *Computer Networks and the Internet*,
https://doi.org/10.1007/978-3-030-34496-2_5

ple's AppleTalk, and IP (IPv4 and IPv6). Other Layer 3 addressing schemes have been used in proprietary networks which add a network prefix to the device address or use the device name as the address. Some work better than others for building interconnected networks[1].

5.2 Flat Addressing

Many early network addressing schemes were developed before large intercon-nected networks were first proposed. These were often "flat" addressing, such as NetBIOS, where a single part address was assigned to each device. Another exam-ple was BTOS[2] developed by Convergent Technologies and Burroughs, which sim-ply added a network prefix to every file and resource. To send a message to another device, a NIC had to be aware of the name of the destination. This works well for distributed file systems on a LAN or sharing resources between Layer 2 networked computers. However, there is no way to determine what Layer 3 network a device is on from the address. Unfortunately, this restricts knowledge of the device to the LAN and does not facilitate sharing messages or resources across interconnected networks. This is why a Windows shared printer cannot easily be accessed from the Internet.

Flat addressing does not add any real functionality to the interconnected networks we are interested in building, so it will not be mentioned again.

5.3 Network Addressing and Host Addressing

What is needed is a two part addressing scheme where one part is the unique device and the other is the address of the network where the device is located. No longer must a device be able to share unicasts and broadcasts to be on the same network. At Layer 3 two devices are on the same network if, and only if, they share the same network address part.[3]

A NIC is usually capable of running multiple OSI Layer 3 versions by running additional low-level software, but this is a large burden on the electronics and soft-ware. As we will see, this is no longer an issue but it was a serious consideration in the earlier days of networking.

---

[1] While the OSI calls a collection of networks in an organization an "Intermediate System", we will call these either "areas" or "Autonomous Systems".

[2] Burroughs Task Operating System

[3] For sanity, it is customary to give each interconnected LAN a separate Layer 3 network address part. There are devices that can get around this custom, but things get very confusing. It is best to stick with this custom: Each LAN gets its own Layer 3 network address.

5.4 IPX and AppleTalk

Two of the most successful Layer 3 addressing schemes were Novell IPX and AppleTalk. Both were two–part addressing schemes and could easily be used to group
any number of devices on a LAN into a single Layer 3 network. Large LANs could
be divided up into smaller LANs and interconnected by devices that could quickly
decide if a packet needed to be forwarded to another LAN or not. One of the most
important benefits of breaking up large LANs was that broadcasts could be kept
local and did not have to be forwarded between LANs. The ability of these new
devices, called routers, to limit the scope of broadcasts means that extremely large
networks are possible by interconnecting a number of small networks.

By the 1990's it was becoming obvious that a single, global network was going to
be developed in the near future and the apparent winner would be the Internet. The
competition was effectively removed when then Senator Al Gore pushed for funding
of the present Internet as a research network to connect the existing supercomputer
sites to allow for more effective utilization and to save on travel for researchers.
Mr. Gore never claimed to be the "Father of the Internet" but other people have
correctly noted that without him it might have been a long time before a single
network and protocol set would win the battle to connect everyone. [4] For all intents
and purposes, only IP is allowed and functional over the Internet. So to have access
to a local, non-IP network and the Internet, a device needed to run the local Layer
3 protocols and IP on the same NIC. This lead to more difficulties in configuration
and even some slowing down of both protocol suites. During the era of the Apple
MAC and DOS, configuring a PC to run DOS and IP was an interesting endeavor
which often required running proprietary IP stacks and OS shim (OS shim)s.

Novell and Apple both moved their full support behind running their network
operating systems over the IP suite of Layer 3 protocols. This meant that a NIC only
needed to run a single set of Layer 3 protocols which was best for everyone. At the
time of this writing, both IPX and AppleTalk [107] [274] are legacy protocols and
rarely used even though Novell's Network OS is still very prevalent in government
and educational settings.

Table 5.1: AppleTalk and IPX Addressing

Name	Company	Network address	Bytes	Host address	Bytes
AppleTalk	Apple	Zone	2	Host ID & Socket	2
IPX	Novell	Network Number	4	MAC	6

---

[4] Indeed, Mr. Gore *is* the "Father of the Internet" in my opinion.

5.5 IPv4 Addressing

Table 5.2: IANA Assigned IP Versions

Version Name	Use	RFC
0	Reserved	rfc4928 section 3
1	Reserved	rfc4928 section 3
2	Unassigned	
3	Unassigned	
4 IP	Internet Protocol	rfc791
5 ST	ST Datagram Mode	rfc1190
6 IPv6	Internet Protocol version 6	rfc1752
7 TP/IX	TP/IX: The Next Internet	rfc1475
8 PIP	The P Internet Protocol	rfc1621
9 TUBA	TUBA	rfc1347
10	Unassigned	
11	Unassigned	
12	Unassigned	
13	Unassigned	
14	Unassigned	
15	Reserved	

Since the early 1990's, the most prevalent network addressing schema is IP and specifically IPv4[5], see Table 5.2. Device addresses are four octets, or 32 bits, with the leading bits denoting the network address and the trailing bits denoting the individual device or host address. The Internet carries both IPv4 and IPv6[6] but most traffic is currently IPv4. While all modern devices can run either, or both, versions of IP, it is obvious that the main Internet is transitioning to IPv6.

When two LANs are connected at Layer 2 the result is *one* large LAN. However, if we build Layer 3 networks over both LANs they can be connected by a router that has an interface in both LANs as in Figure 5.1.

---

[5] Internet Protocol, Version 4

[6] Internet Protocol, Version 6

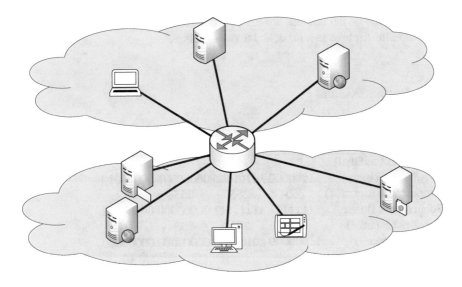

Fig. 5.1: Two Layer 3 Networks Connected by a Router

5.6 Classful IPv4

Table 5.3: Classful IPv4 Addressing

Class	First Octet	Natural Subnet Mask	Number of Hosts	Usage
A	1 to 126	255.0.0.0(/8)	16 million	Large networks
B	128 to 191	255.255.0.0(/16)	65534	Medium networks
C	192 to 223	255.255.255.0(/24)	254	Small networks
D	224 to 239	255.255.255.255(/32)	none	IP Multicasts
E	240 to 254	varies	n/a	Experimental

Table 5.4: Classful IPv4 Prefix

Class	First Octet	Prefix
A	1 to 126	0
B	128 to 191	10
C	192 to 223	110
D	224 to 239	1110
E	240 to 254	11110

IP addresses naturally fall into five classes denoted by A, B, C, D, and E as in
Table 5.3 but most networks are only concerned with one of the classes A, B, or C.
Addresses consist of four bytes, usually entered in "dotted decimal", which give the

network address and the host address and what is called a subnet mask to identify how many bits of the address are the network address. All devices on a network must have the same subnet mask and network address.

5.6.1 Dotted Decimal

	Octet 1	Octet 2	Octet 2	Octet 4
IP Address (decimal)	192	168	1	10
IP Address (binary)	11000000	10101000	00000001	00001010
Subnet mask (decimal)	255	255	255	0
Subnet mask (binary)	11111111	11111111	00000000	00000000
Network (decimal)	192	168	255	0
Network (binary)	11000000	10101000	00000000	00000000
Host (decimal)				10
Host (binary)				00001010
IP Broadcast (decimal)	192	168	1	255
IP Broadcast (binary)	11000000	10101000	00000001	11111111

Fig. 5.2: IP Address 192.168.1.10

When configuring IP networks the information must be entered in "dotted decimal". The devices will translate each of the four decimal numbers into eight bits or an octet. Since eight bit can represent decimal numbers from 0 to 255, this range holds the only allowed values in "dotted decimal". For example, the IP address 10.1.0.128 is really binary 00001010.00000001.00000000.10000000 with the "." not really there but included here to help humans count the digits. The only exception to this is the slash notation method of giving the IP subnet mask as the number of leading "1"s. For example, the Class B natural mask is 255.255.0.0 or "/16" in the shorter notation[7].

5.6.2 IP Class

The IP Class is useful when documenting or discussing a Layer 3 network and subnetwork (subnet), but is not needed by the devices themselves. When one speaks of a Class B network, other network savvy engineers will know quite a number of things about the network such as the general size of the network. Devices will get this information from the first octet of the address.

---

[7] I have been unable to find a trustworthy citation for this notation being a Cisco Systems invention.

5.6.3 First Octet

The first octet of any IP address reveals critical information to the device such as the default or "natural" subnet mask. From the first octet, the device can determine how many bits of the address are used for the network address and how many are used for the host address. Further advances in devices, specifically routers, have led to the ability to choose a longer subnet mask than the natural one for the class as we will see in the discussion below.

5.6.4 Natural Subnet Mask

The subnet mask is used to determine the network and host part of the address. Where the subnet mask is binary "1"s, the address is the network part. Where the subnet mask is all binary "0"s, the address is the host part. Notice that this means the subnet mask must start with at least eight "1"s and once a "0" is encountered the subnet mask must contain all "0"s from that point on as in Figure 5.2. When an IP address is configured, the device will assume a subnet mask from the first octet (or the prefix) which is the "natural" subnet mask. **Note:** Class E does not have a defined natural subnet mask. Class E networks are reserved for experimental uses and must be avoided. This is why the subnet mask and number of hosts in Table 5.3 are not given.

5.6.5 Number of Hosts

The number of hosts allowed on any given network can easily be calculated once the number of network bits in the address and the number of host bits in the address are determined from the subnet mask. The maximum number of hosts, h_{max}, if n bits are used for the host part of the address is given by Equation 5.1 below:

$$h_{max} = 2^n - 2. \tag{5.1}$$

There are two reserved host addresses for any IP network. The host all "0"s is reserved to refer to the network as a whole as we will see when we discuss routing. Likewise, the host all "1"s is reserved for IP–based Layer 3 Broadcasts. Most devices will not allow a configuration that assigns them either of these reserved host addresses.

5.6.6 Usage

The Classes were originally designed to be used by different sizes of networks as measured by the number of hosts on the network. Class A was to be used by a limited number of international organizations or large network providers. Class B was to be used by moderate sized organizations such as a university. Class C was to be used for small to moderate size networks such as a business office. If you examine the table closely, you will find there are very few Class A networks and many Class C networks.

5.6.7 Prefix

IP networks are organized based upon a well–known technique called a prefix code. Prefix codes do not have a fixed length but are organized so that no two values have the same starting binary digits. Only Class A networks start with a binary "0" all others start with a binary "1". Likewise, all Class B networks start with a binary "10" and Class C are the only networks starting with a binary "110". This follows the same technique as Huffman Codes [25] for compressing binary encoding.

Table 5.5: Reserved IP Networks

	Usage
0	Routing such as the default network 0.0.0.0
10	Private Class A network, cannot connect to the Internet
127	Hardware usage such as loopback 127.0.0.1
169.254	Auto–configuration
172.16 to 172.31	Private Class B networks, cannot connect to the Internet
192.168.	Private Class C networks, cannot connect to the Internet
255	subnet masks

5.7 Reserved IPv4 Networks

A close examination of Tables 5.3 and 5.4 reveals that some possible IP networks are missing. The values of the first octet range from 0 to 255, yet three values are not accounted for: 0, 127, and 255. These networks are reserved for the uses given in Table 5.5 and explained below:

0   The most commonly encountered network with a first octet of 0 is the default route which is designated by 0.0.0.0 and will be explained more fully in the chapters on routing and routers.

10   The Class A network associated with the first octet of 10 is reserved for private networks such as an internal Intranet for a large company, and as such it may *not* be connected to the Internet. As a matter of fact, routers on the Internet normally drop all packets sent to a private network to avoid embarrassing route failures.

127   This network falls between Class A and Class B, but most devices assume a natural subnet mask of 255.0.0.0 for this network. The entire network is reserved for hardware use by a NIC with the exception of the loopback address of 127.0.0.1 which is paired with the host name of "localhost". This address is very useful for troubleshooting or to point a web browser to a web server on the same device.

169.254   This Class B network is reserved for auto–configuration of the IP address of a NIC [200] and should not be connected to the Internet[8].

172.16 to 172.31   Private Class B networks for Intranets and must not be connected to the Internet.

192.168   Any network starting with the first 16 bits (2 octets) of 192.168 are private Class C networks and must not be connected to the Internet.

255   All subnet masks must start with 255 or a minimum of eight binary "1"s.

5.8 Private IPv4 Networks

There are a number of networks reserved for private use to discourage people from simply picking some random network numbers for their own Intranets as has been done many times in the past. ICANN[9] will assign any networks not reserved in Table 5.5 to various organizations or individuals for their exclusive use. The assignment of IPv4 networks on the Internet is chaotic at best and to have someone squatting on an assigned network can cause local failures of the Internet that are difficult to identify and resolve, so the private networks are there to minimize accidental theft of IP addresses and networks. Private networks can be used by anyone for any purpose except to connect to the Internet which allows for thousands of Intranets to use 192.168.1.1 as the address of one of their routers without any problems because these addresses cannot be accessed from outside the Intranet.

When someone unknowingly connects a private address network to a public network, routing becomes very chaotic as local routers attempt to deal with the private networks. This is such a drastic problem and happens so often that we will address this possibility when we discuss configuring a router. We are all responsible for the protection of our own networks and the Internet at large. This is why routers should drop all incoming private traffic by sending it to the "bit bucket"; i.e., trashcan or null interface.

---

[8] On Windows machines this nice feature can lead to some difficult to resolve problems. In my opinion, this network should be explicitly "tanked" or routed to nowhere. How this is done will be discussed in Section 15.4.2

[9] ICANN[10]

This is the greatest strength of private networks and their greatest weakness: they cannot be not connected to the Internet at all.

5.9 Public IPv4 Networks

At this time, all possible public networks have been assigned by the ICANN to various entities. The Class A networks were used up first and followed very quickly by the Class B and Class C networks. Any device on a public network can be contacted by any other device on any public network. This leads to some interesting security issues, especially in places such as schools and colleges.

Public networks are assigned by giving the entity full rights to use the network address and all the addresses that start with that prefix. In order to be assigned a network address, certain minimum requirements must be met such as contact information and signatures of those responsible for maintaining the network[11]. For example, a university might be assigned a Class B address of 151.152.0.0 which entitles it to the full usage of all the IP addresses from 151.152.0.0 (the whole network) through 151.152.255.255 for the IP–based multicast.

Originally each IP network was restricted to using the natural subnet mask which was too rigid a constraint. Many organizations found it would be useful to break their assigned address space into multiple smaller networks. To allow this to happen, Classful IP gave way to Classless IP.

5.10 Classless IPv4 (CIDR)

In order to more closely follow the structure of the organization and to avoid wasting large numbers of IPv4 addresses it is possible, and usually desirable, to design a network with a VLSM[12] as in Section 5.13. The chosen subnet mask must be the same for all NICs on the network and *must* be at least as long as the "natural" subnet mask for the IP class. This allows a network address space to be broken up into smaller networks which can be interconnected at Layer 3. The only issue with Classless IP, or CIDR[13], is that some of the easiest to use routing protocols[14] only work with Classful IP. This is solved by using RIPv2 instead of RIPv1 as we will see later in Part II when routers are discussed. In fact, protocols that require Classful IP should be considered as deprecated and to be avoided.

---

[11] This can get to be quite tense if someone uses the assigned network to break the law. An investigation by the U.S. Secret Service or FBI is not something to take lightly.

[12] Variable Length Subnet Mask

[13] Classless Inter–Domain Routing

[14] RIPv1[15] and IGRP[16] for example

5.11 Sending a Unicast

There is one small problem that must be addressed: How does a NIC contact a specific IP address? Either the IP address has the same network part as assigned to the NIC or it doesn't. If the network parts are different, the packet is forwarded to the "default gateway" for the NIC (a router) or the NIC resolves the IP address to the correct MAC address using the ARP[17][18].

Regardless of whether the destination IP address is in the same network or not, the NIC must still send the actual frame to the correct destination MAC address. To facilitate this process, each NIC keeps a table, called the ARP Table as in Table 5.7 and Figure 5.4, of all the known IP–MAC address pairs. If the NIC has contacted this IP address before, it simply looks up the correct MAC address in the table, forms a frame with the IP packet as the data payload, and sends it out the wire[19]. If not, the NIC updates the ARP Table, as shown in Figure 5.3, using the steps in Algorithm 5.6. An actual Pi ARP Table is given Figure 5.4. Note that two IP addresses, 192.168.1.24 and 192.168.1.56, were not found so the information is incomplete for those addresses. The actual presentation of the ARP Table is up to the programmer who decided the details of how the table would be displayed. The output from the Windows `arp -a` is quite different although the same general information is given, see Figure 5.5. This is common with the output of commands in the networking world[20].

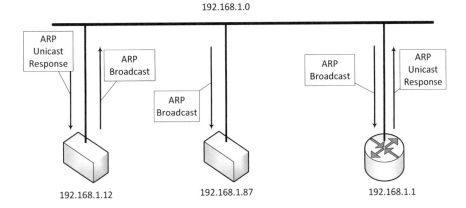

Fig. 5.3: An ARP Request from 192.168.1.12 for MAC of 192.168.1.1

---

[17] Address Resolution Protocol

[18] Q: How does a lost baby seal find its mother on the Internet?
A: It ARPs ... "arp! arp! arp!"

[19] The entries in the ARP Table are allowed to "age out" after a period of inactivity so that bad entries eventually disappear.

[20] This will not be pointed out again. Expect the output of similar commands to differ between different OSs. As time goes on, it seems some programmers are trying to minimize this.

Table 5.6 IPv4–Ethernet Address Resolution Protocol (ARP)

1: **procedure** ARP
2:      The source NICBroadcasts an ARP Request (Layer 2)
3:      Each NICon the LAN processes the ARP Request
4:      **if** IP address matches NIC's IP address **then**
5:          Destination NICupdates its ARP Table with the original NIC's information[21]
6:          NICforms an ARP Response with its IP address
7:          NICsend ARP Response to requester NIC
8:      **else**
9:          All other NICs ignore ARP Request
10:     **end if**
11:     Source NICupdates its ARP Table
12:     Source NICunicasts message to destination MAC
13: **end procedure**

Table 5.7: The ARP Table

Age	IP Address	MAC Address	Interface
1	192.168.1.12	f8:16:54:00:72:ab	eth0
10	192.168.1.15	b0:10:41:a9:c3:85	eth0
5	192.168.1.1	20:aa:4b:0d:93:8d	eth0

```
Raspberry Pi Model B Plus Rev 1.2
pi@howserPi1:~$ arp -a
? (192.168.1.222) at 24:5e:be:20:ec:d9 [ether] on eth1
? (192.168.1.200) at cc:af:78:67:af:55 [ether] on eth1
? (192.168.1.14) at f8:16:54:00:72:ab [ether] on eth1
? (192.168.1.24) at <incomplete> on eth1
? (192.168.1.10) at 28:39:5e:23:7e:6d [ether] on eth1
? (192.168.1.1) at 20:aa:4b:0d:93:8d [ether] on eth1
? (192.168.1.56) at <incomplete> on eth1
? (192.168.1.12) at 14:20:5e:55:3d:74 [ether] on eth1
pi@howserPi1:~$
```

Fig. 5.4: The Output From the Pi "arp -a" command

```
C:\Windows\system32>arp -a
Interface: 192.168.1.109 --- 0x5
  Internet Address      Physical Address        Type
  192.168.1.1           20-aa-4b-0d-93-8d       dynamic
  192.168.1.11          14-91-82-76-fb-f1       dynamic
  192.168.1.28          b8-81-98-24-7b-42       dynamic
  192.168.1.29          28-39-5e-ba-0a-7f       dynamic
  192.168.1.30          28-39-5e-ba-0a-7f       dynamic
  192.168.1.31          b8-27-eb-1d-e1-a4       dynamic
  192.168.1.200         cc-af-78-67-af-55       dynamic
  192.168.1.222         24-5e-be-20-ec-d9       dynamic
  192.168.1.255         ff-ff-ff-ff-ff-ff       static
  224.0.0.2             01-00-5e-00-00-02       static
  224.0.0.22            01-00-5e-00-00-16       static
  224.0.0.251           01-00-5e-00-00-fb       static
  224.0.0.252           01-00-5e-00-00-fc       static
  224.0.1.60            01-00-5e-00-01-3c       static
  239.255.255.250       01-00-5e-7f-ff-fa       static
  255.255.255.255       ff-ff-ff-ff-ff-ff       static

Interface: 192.168.56.1 --- 0x8
  Internet Address      Physical Address        Type
  192.168.56.255        ff-ff-ff-ff-ff-ff       static
  224.0.0.2             01-00-5e-00-00-02       static
  224.0.0.22            01-00-5e-00-00-16       static
  224.0.0.251           01-00-5e-00-00-fb       static
  224.0.0.252           01-00-5e-00-00-fc       static
  224.0.1.60            01-00-5e-00-01-3c       static
  239.255.255.250       01-00-5e-7f-ff-fa       static
  255.255.255.255       ff-ff-ff-ff-ff-ff       static

Interface: 192.168.222.25 --- 0xc
  Internet Address      Physical Address        Type
  192.168.222.1         24-5e-be-20-ec-d8       dynamic
  192.168.222.255       ff-ff-ff-ff-ff-ff       static
  224.0.0.2             01-00-5e-00-00-02       static
  224.0.0.22            01-00-5e-00-00-16       static
  224.0.0.251           01-00-5e-00-00-fb       static
  224.0.0.252           01-00-5e-00-00-fc       static
  239.255.255.250       01-00-5e-7f-ff-fa       static
  255.255.255.255       ff-ff-ff-ff-ff-ff       static
```

Fig. 5.5: The Output From the Windows "arp -a" command

Once the NIC has the MAC address for the IP address, the packets are simply wrapped in a Layer 2 frame and sent. There is also a protocol, RARP[22] to find the MAC address of the NIC that has been assigned a particular IP address.

---

[22] Reverse Address Resolution Protocol

ARPAWOCKY
(with apologies to Lewis Carrol)[23]

Twas brillig, and the Protocols
    Did USER-SERVER in the wabe.
All mimsey was the FTP,
    And the RJE outgrabe,

Beware the ARPANET, my son;
    The bits that byte, the heads that scratch;
Beware the NCP, and shun
    the frumious system patch,

He took his coding pad in hand;
    Long time the Echo–plex he sought.
When his HOST–to–IMP began to limp
    he stood a while in thought,

And while he stood, in uffish thought,
    The ARPANET, with IMPish bent,
Sent packets through conditioned lines,
    And checked them as they went,

One–two, one–two, and through and through
    The IMP–to–IMP went ACK and NACK,
When the RFNM came, he said "I'm game",
    And sent the answer back,

Then hast thou joined the ARPANET?
    Oh come to me, my bankrupt boy!
Quick, call the NIC! Send RFCs!
    He chortled in his joy.

Twas brillig, and the Protocols
    Did USER-SERVER in the wabe.
All mimsey was the FTP,
    And the RJE outgrabe.

D.L. COVILL
May 1973

---

[23] Apparently, some networking people have a sense of humor [61].

5.12 Layer 3 Devices

There are three devices that are used to connect Layer 3 networks: IP forwarders, IP switches (or Layer 3 switches), and routers. These devices can connect networks of different physical types and speeds as long as the same Layer 3 protocols are running on all the networks to be connected. This is one of the major strengths of the communications layer in that each layer is completely independent of the layers above or below.

While a number of different protocols that operate at Layer 3 can be interconnected, in practice the most common is IPv4 with IPv6 not far behind. In fact, a network engineer might go their whole career without encountering any other Layer 3 protocols besides IPv4 and IPv6. This is not an issue as routing any routable protocol works the same way.

Table 5.8: Layer 3 Devices and Configuration

Device	Configuration
IP Forwarder	Static: only networks present at power on
Router	Dynamic: Constantly learning best route to known networks
Layer 3 Switch	Dynamic: Constantly learning best route to known networks

5.12.1 Characteristics of Layer 3 Devices

All three of the devices we are interested in have some common characteristics as in Figure 5.6. They all must have multiple NICs: they must have one NIC on each network to be connected; they must have a table in memory of the network addresses and NIC for each network; and they must have a software/hardware process to move incoming packets to the correct next device (or "hop")[24]. For simplicity, we will call the software/hardware process the routing engine.

Because the connective device need only send a packet to the next "hop" along the way to its destination, the device has no need to learn the details of the entire network. This means the device can be studied without any concern for the tiny details of the networks involved which is why Layer 3 networks are often drawn as clouds[25]. This is a good thing because when large networks are owned by different corporations, the network details are a corporate resource and kept secret.

---

[24] Each time a packet is moved to a device on a different network is a "hop". Some routing protocols have a limit as to how many "hops" a packet can make before it is dropped to help prevent routing a packet in an endless loop.

[25] Do not confuse this with the **Cloud**.

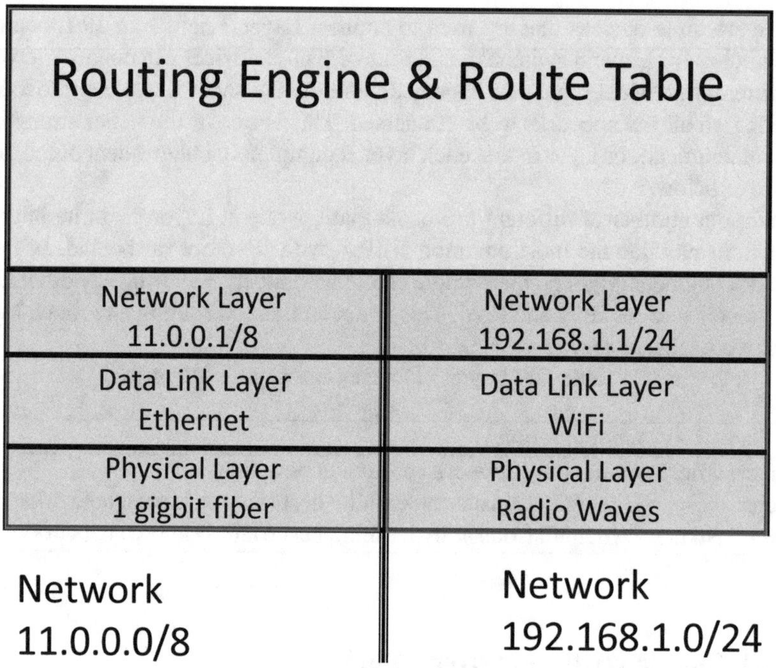

Routing Engine & Route Table

Network Layer 11.0.0.1/8	Network Layer 192.168.1.1/24
Data Link Layer Ethernet	Data Link Layer WiFi
Physical Layer 1 gigbit fiber	Physical Layer Radio Waves

Network
11.0.0.0/8

Network
192.168.1.0/24

Fig. 5.6: Typical Routing Device

5.12.2 IP Forwarder

The simplest device at Layer 3 is an IP Forwarder which can only move packets between its own interfaces. Typically, these devices move packets between two
interfaces: a private LAN NIC and a public WAN[26] NIC. Most home routers are
really IP Forwarders and do not have any ability to "learn" the network or respond
to changes in the network.

---

[26] Wide Area Network

5.12.3 Router

A true router has the ability to "learn" the network and respond to changes in the network automagically[27]. To do this, the router must have a route engine that is capable of running a routing protocol and updating the route table on the fly. A side–effect of this ability is that almost all of the route table can be built by the routing engine with little, or no, human input. The router must have more processing power than an IP forwarder, but the main function of the routing engine is still to move packets between networks quickly.

In order to maintain the route table, especially if a route becomes unavailable, all true routers also maintain a memory file of all the known routes in the network called the route cache. All changes to the route table and route cache must happen quickly because during the time the route table is being updated, it is possible to lose packets. Packets might also be corrupted as they pass through various networks. Corrupted packets are ignored by the router and the router is not usually required to notify any other devices of the packets it drops. Remember, at the communication layers, Layer 1 through Layer 3, devices do not have any responsibilities other than move messages as quickly as possible toward their destination. The upper layers are responsible for insuring the messages arrive, not Layer 3.

5.12.4 Layer 3 Switch

A Layer 3 switch operates on the same basic principles as a Layer 2 switch but switches packets instead of frames. For all intents and purposes, a Layer 3 switch acts like a router but uses different internal hardware and software. In almost all cases, the distinction between a Layer 3 switch and a router are all internal and do not change the way routing protocols work. Unless the difference is important, both will be referred to as "routers".

5.13 IPv4 Subnet Planner

In order to assign IPv4 addresses to an interface, one must first determine what network number and host numbers are available. The following Subnet Planner has been created to assist in determining the best values for the given parameters of a network based upon the IP Class, desired number of hosts, and the IP address range available.

---

[27] automagically is a pet term for things that happen "automatically[28]" without human intervention and keep us from messing the whole thing up.

There is a rule of thumb to follow when assigning CIDR IPv4 addresses or IPv6 addresses:

> Assign network bits from the left and host bits from the right.

IPv4 Subnet Planner

Allowed IP range: Start   End
IP Class: A   B   C
Natural Subnet Mask:
Required number of nodes[1,2]$(h$ or $h_m)$:
Number of host bits[3] $(h_m = 2^n - 2)$:
Best subnet mask length $(32 - n)$:

	OCTET 1	OCTET 2	OCTET 3	OCTET 4
IP ADDRESS (Binary)				
SUBNET Mask (Binary)				
NETWORK NUMBER (Bin)				
NETWORK NUMBER (Dec)				
BROADCAST IP (Binary)				
BROADCAST IP (Decimal)				
Lowest host IP (Binary)				
Lowest host IP (Decimal)				
Highest host IP (Binary)				
Highest host IP (Decimal)				

1. h is the minimum required number of hosts.

2. h_m is the maximum number of hosts expected on the network and is 2 *less* than the next power of 2 that is greater than h.

3. n is the number of host bits in the IPv4 address and is found by solving $(h_m = 2^n - 2)$ for n. $(32 - n)$ gives the number of network bits.

5.14 IPv4 Subnet Planner Example 1

Most home networks are in the range 192.168.1.0 to 192.168.1.255. How many hosts does that allow for and what is the range of allowed host IP addresses? To find out, we will fill out the Subnet Planner line–by–line.

1. Allowed IP Range Start: 192.168.1.0   End: 192.168.1.255
2. IP Class: This is a private IP address in Class C
3. The natural subnet mask for a Class C IP address is 255.255.255.0. When this is converted to binary, or 11111111.11111111.11111111.00000000, we find the

number of host bits, $n =$ the number of binary zeroes, is eight and the number of network bits is $32 - 8$ or 24. This is a valid subnet mask because in binary it is a string of 1s followed by a string of 0s. The subnet mask is also at least as long as the natural subnet mask, as is required by IPv4. If there was a binary 1 after the first binary 0 in the subnet mask, the subnet mask would have been invalid.

4. Required number of nodes is given by $h_m = 2^8 - 2$ or 254. $(256 - 2 = 254)$
5. Number of host bits we have already calculated to be 8.
6. The best subnet mask has 24 network bits. $(32 - 8 = 24)$
7. The starting IP Address in binary is: 11000000.10101000.00000001.00000000 which we enter as the IP ADDRESS.
8. The subnet mask in binary we found to be 11111111.11111111.11111111.00000000 which is entered in the SUBNET Mask.
9. The Network Number in binary is found by taking a logical *AND* of the bits in the starting IP address and the subnet mask for a binary value of 11000000.10101000.00000001.00000000.
10. The Network Number in decimal is found by converting 11000000.10101000.00000001.00000000 to 192.168.1.0.
11. The Broadcast IP is found by doing a logical *AND* with the starting IP address and the network mask. We then set the host bits of the starting IP address to all binary 1 to get the Broadcast IP in binary which in this example gives 11000000.10101000.00000001.11111111.
12. The Broadcast IP in decimal is found by converting the binary address to dotted decimal or 192.168.1.255.
13. The Lowest host IP in binary is found by adding 1 to the Network Number (Binary) which was found above to yield 11111111.11111111.11111111.00000001. The reason for this is left as an exercise.
14. The Lowest host IP in decimal is found by converting the Lowest host IP in binary to decimal which is 192.168.1.1.
15. The Highest host IP in binary is found by subtracting 1 from the Broadcast IP in binary or 11000000.10101000.00000001.11111110.
16. The Highest host IP in decimal is found by converting 11000000.10101000.00000001.11111110 to 192.168.1.254.

IPv4 Subnet Planner

Allowed IP range: Start192.168.1.0   End: 192.168.1.255
IP Class: A   B   **C**
Natural Subnet Mask: 255.255.255.0
Required number of nodes[1,2](h or h_m): 254
Number of host bits[3] $(h_m = 2^n - 2)$: 254
Best subnet mask length $(32 - n)$: 24

	OCTET 1	OCTET 2	OCTET 3	OCTET 4
IP ADDRESS (Binary)	11000000	10101000	00000001	00000000
SUBNET Mask (Binary)	11111111	11111111	11111111	00000000
NETWORK NUMBER (Bin)	11000000	10101000	00000001	00000000
NETWORK NUMBER (Dec)	192	168	1	0
BROADCAST IP (Binary)	11000000	10101000	00000001	11111111
BROADCAST IP (Decimal)	192	168	1	255
Lowest host IP (Binary)	11000000	10101000	00000001	00000001
Lowest host IP (Decimal)	192	168	1	1
Highest host IP (Binary)	11000000	10101000	00000001	11111110
Highest host IP (Decimal)	192	168	1	254

1. h is the minimum required number of hosts.

2. h_m is the maximum number of hosts expected on the network and is 2 *less* than the next power of 2 that is greater than h.

3. n is the number of host bits in the IPv4 address and is found by solving $(h_m = 2^n - 2)$ for n. $(32 - n)$ gives the number of network bits.

5.15 IPv4 Subnet Planner Example 2

Suppose you wish to use part of the private Class A address to assign 5 hosts in the smallest possible range starting with 10.0.25.0. First we must determine the correct number of host bits for this network from the formula $h_m = 2^n - 2$ where $h_m \geq 5$ and $h - m + 2$ is a power of 2. The next higher power of two is eight, so $H_m = 6$ and $n = 3$. We now know the starting IP is 10.0.25.0 and the number of host bits is 3 which leads to an IP subnet mask of 11111111.11111111.11111111.11111000 or 255.255.255.248 which is longer than the Natural Mask (255.0.0.0) for Class A and is a string of binary 1s followed by a string of binary 0s.

1. Allowed IP Range Start: 10.0.25.0   End: (not yet calculated)
2. IP Class: This is a private IP address in Class A
3. The natural subnet mask for a Class A IP address is 255.0.0.0. When this is converted to binary, or 11111111.00000000.00000000.00000000, we find the number of host bits, $n =$ the number of binary zeroes, is 3 and the number of network bits is $32 - 3$ or 29.
4. Required number of nodes is given by $h_m = 2^3 - 2$ or 66.
5. Number of host bits we have already calculated to be 3.
6. The best subnet mask has 29 network bits. $(32 - 3 = 29)$
7. The starting IP Address in binary is: 00001010.00000000.1010001.00000000 which we enter as the IP ADDRESS.
8. The subnet mask in binary we find to be
   11111111.11111111.11111111.11111000 which is entered in the SUBNET Mask.
9. The Network Number in binary is found by taking a logical *AND* of the bits in the starting IP address and the subnet mask for a binary value of 00001010.00000000.1010001.00000000.
10. The Network Number in decimal is found by converting
    00001010.00000000.1010001.00000000 to 10.0.25.0.
11. The Broadcast IP is found by doing a logical *AND* with the starting IP address and the network mask. We then set the host bits of the starting IP address to all binary 1 to get the Broadcast IP in binary which is
    0001010.00000000.10100001.00000111.
12. The Broadcast IP in decimal is found by converting the binary address to dotted decimal or 10.0.25.7, which means the highest IP address used is 10.0.26.7 or the IP Broadcast.
13. The Lowest host IP in binary is found by adding 1 to the Network Number (binary) which is found above to yield 00001010.00000000.1010001.00000001. The reason for this is left as an exercise.
14. The Lowest host IP in decimal is found by converting the Lowest host IP in binary to decimal which is 10.0.25.1.
15. The Highest host IP in binary is found by subtracting 1 from the Broadcast IP in binary or 0001010.00000000.1010001.00000110.
16. The Highest host IP is decimal is found by converting
    0001010.00000000.1010001.00000110 to 10.0.25.6.

IPv4 Subnet Planner

Allowed IP range: Start10.0.25.0   End: 10.0.25.7

IP Class: **A**   B   C

Natural Subnet Mask: 255.0.0.0

Required number of nodes[1,2](h or h_m): 6

Number of host bits[3] $(h_m = 2^n - 2)$: 3

Best subnet mask length $(32 - n)$: 29

	OCTET 1	OCTET 2	OCTET 3	OCTET 4
IP ADDRESS (Binary)	0000101	00000000	10100001	00000000
SUBNET Mask (Binary)	11111111	11111111	11111111	11111000
NETWORK NUMBER (Bin)	0000101	00000000	10100001	00000000
NETWORK NUMBER (Dec)	10	0	25	0
BROADCAST IP (Binary)	0000101	00000000	10100001	00000111
BROADCAST IP (Decimal)	10	0	25	7
Lowest host IP (Binary)	0000101	00000000	10100001	00000001
Lowest host IP (Decimal)	10	0	25	1
Highest host IP (Binary)	0000101	00000000	10100001	00000110
Highest host IP (Decimal)	10	0	25	6

1. h is the minimum required number of hosts.

2. h_m is the maximum number of hosts expected on the network and is 2 *less* than the next power of 2 that is greater than h.

3. n is the number of host bits in the IPv4 address and is found by solving $(h_m = 2^n - 2)$ for n. $(32 - n)$ gives the number of network bits.

5.16 IPv6 Addressing

IPv6 uses a much different addressing scheme than IPv4 and the two are *not* compatible. First, IPv6 addresses are 128 bits long versus only 32 bits for IPv4 which allows for 2^{128} IPv6 addresses versus 2^{32} IPv4 addresses[29]. The network part of an IPv6 address is fixed at 64 bits of which the first 48 are assigned by the IANA, but the final 16 bits are used to allow for the organization to create subnetworks as they

---

[29] IPv4 has only 4.2 billion possible addresses versus 340 trillion trillion trillion for IPv6.

need. It is important to keep in mind that IPv6 protocols will make use of route summarization whenever possible so care must be taken as to how those 16 bits are assigned.

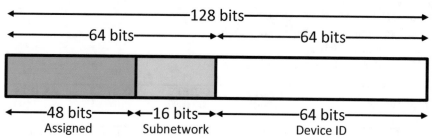

Fig. 5.7: The 128 bit IPv6 Address

5.16.1 Human Readable IPv6 Addresses

Table 5.9: Binary to Hexadecimal

Binary Pattern	Hexadecimal	Binary Pattern	Hexadecimal
0000	0	1000	8
0001	1	1001	9
0010	2	1010	a
0011	3	1011	b
0100	4	1100	c
0101	5	1101	d
0110	6	1110	e
0111	7	1111	f

Trying to write IPv4 addresses in binary is not easy and that is only 32 bits to get right and remember. For IPv6 even dotted decimal is too awkward and unwieldy, so IPv6 addresses are written in hexadecimal with some useful shortcuts possible. For example, an IPv6 address of:

00100000000000010000110110111000101011000001000011111111000000010000000000000000000000000000000000000000000000000000000000000000[30]

would never get copied or entered correctly by hand. However, it can be converted to an easier–to–handle format by the following steps [40].

1. First break the binary address into groups of 16 bits:
   0010000000000001 : 0000110110111000 : 1010110000010000 : 1111111000000001 :
   0000000000000000 : 0000000000000000 : 0000000000000000 : 0000000000000000

2. Break each group of 16 bits into 4 groups of 4 bits (4 bits = one nibble = one hex digit):

---

[30] In order to get all 128 digits on a single line of this page, they have to be reduced to a font that is too small to read. Computers can deal with this address in binary...we cannot.

16 bits	N 1	N 2	N 3	N 4
0010000000000001 :	0010	0000	0000	0001
0000110110111000 :	0000	1101	1011	1000
1010110000010000 :	1010	1100	0001	0000
1111111000000001 :	1111	1110	0000	0001
0000000000000000 :	0000	0000	0000	0000
0000000000000000 :	0000	0000	0000	0000
0000000000000000 :	0000	0000	0000	0000
0000000000000000	0000	0000	0000	0000

3. Using Table 5.9, convert each nibble to a hex digit to get the group of 16 bits as 4 hex digits:

16 bits	N 1	N 2	N 3	N 4
0010000000000001 :	0010	0000	0000	0001
2001 :	2	0	0	1
0000110110111000 :	0000	1101	1011	1000
0db8 :	0	d	b	8
1010110000010000 :	1010	1100	0001	0000
ac10 :	a	c	1	0
1111111000000001 :	1111	1110	0000	0001
fe01 :	f	e	0	1
0000000000000000 :	0000	0000	0000	0000
0000 :	0	0	0	0
0000000000000000 :	0000	0000	0000	0000
0000 :	0	0	0	0
0000000000000000 :	0000	0000	0000	0000
0000 :	0	0	0	0
0000000000000000	0000	0000	0000	0001
0000 :	0	0	0	0

4. Now collect the converted groups together to yield the IPv6 address:
   $2001 : 0db8 : ac10 : fe01 : 0000 : 0000 : 0000 : 0000$

5.16.2 Zero Compression

The address 2001:0db8:ac10:fe01:0000:0000:0000:0000 is still hard to deal with because of the length. When there are long strings of zeroes, they can sometimes be compressed, but only once per IPv6 address (the reason is left for an exercise.) When an entire consecutive groups of four hex digits are all zeroes, they can be replaced by "::". For example, the address we have been looking at can be compressed to yield 2001:0db8:ac10:fe01:: which tells us the required missing digits are all zeroes. This is a much more manageable notation.

5.16.3 Zero Suppression

Even after zero compression has been applied, there are still some zeroes that do not give us any information. As in with the numbers we are used to using on a daily basis, leading zeroes can be applied or removed from hexadecimal without changing the value of the field. After removing leading zeroes from the groups of four hex digits the final address is 2001:db8:ac10:fe01:: which is *still* interpreted as 2001:0db8:ac10:fe01:0000:0000:0000:0000. Any of the forms we have derived are acceptable and they are all the same address. [40]

5.17 IPv6 Header

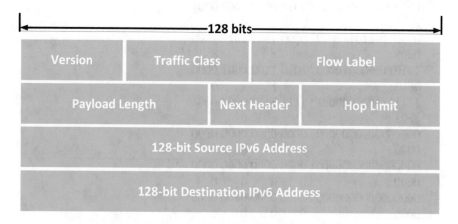

Fig. 5.8: The IPv6 Header

As is to be expected, the IPv6 header as shown in Fig. 5.7 is quite different from the IPv4 header shown in Fig. 2.6. The header has fewer fields that have only limited use and therefore has less overhead that the IPv4 header. Fortunately, the NICs and protocols take care of most of this for us.

5.18 IPv6 Route Summarization

Whenever possible, IPv6 summarizes subnetworks bit–by–bit from left to right, in other words, whenever possible networks are referred to in the route table by a variable length binary string that matches for the greatest possible length. To facilitate route summarization remember to follow the rule of thumb for assigning the network part of the IPv6 address. For most networks, this means the Subnet ID.

Assign network bits from the left and host bits from the right.

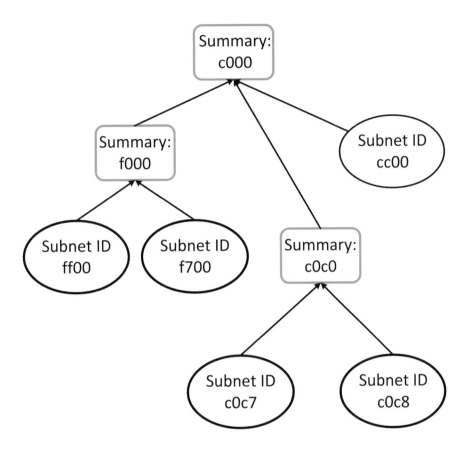

Fig. 5.9: IPv6 Subnet ID Summarization

For example, the subnetworks in Figure 5.9 can be summarized at various levels and the *entire* network can be summarized on the Subnet ID level to ff00. Remember that the first 48 bits of the 64 bit network part must match for the devices to be in the same network.

Projects

5.1 Connect a PC to the Internet and **ping** Yahoo.

    a. Does Yahoo answer?

    b. Find a website you can reach that does not answer a **ping**?

    c. Attempt to discover how your **ping** was routed using the trace route utility. This is **tracert www.yahoo.com** for Windows and **sudo traceroute www.yahoo.com** for Linux. Take a screen capture of the results.

    d. Why would some devices not answer a **ping** or trace route?

    e. How would trace route be useful when experiencing network issues?

Exercises

5.1 Why do you add 1 to the binary Network Number to get the lowest possible host IPv4 address?

5.2 Why do you subtract 2 from the maximum number of hosts to get the allowed number of hosts on an IPv4 network?

5.3 Why do we calculate all the values in both binary and dotted decimal? Why not just in decimal?

5.4 IPv4 Subnet Planner practice

    a. Fill out the IPv4 Subnet Planner for the maximum number of hosts for the 10.0.0.0 network.

    b. Fill out the IPv4 Subnet Planner for 2 hosts within the 172.17.1.0 network.

5.5 Expand the following IPv6 addresses to their full hex equivalents.

    a. 2001:ffd2:f00d::1:1

    b. c999::eda1:1003:12:7711:a

    c. ::1

5.6 What would be the effect of changing the Subnet ID of c0c8 in Figure 5.9 to 00a0?

5.7 Using the IPv6 prefix for your Group create a network with three subnets that can be summarized to one IPv6 network address such as fd86:9b29:e5e1:1000: for Group 1.

Further Reading

For more information on IPv6 addressing and use, there is a wonderful tutorial by the Internet Society at https://www.internetsociety.org/tutorials/introduction-to-ipv6/ [40].

The RFC below provide further information about the Network Layer. This is a fairly exhaustive list and most RFC are typically dense and hard to read. Normally RFC are most useful when writing a process to implement a specific protocol.

RFCs Directly Related to This chapter

IPX	Title
RFC 1132	Standard for the transmission of 802.2 packets over IPX networks [83]
RFC 1234	Tunneling IPX traffic through IP networks [88]

Layer3	Title
RFC 4778	Operational Security Current Practices in Internet Service Provider Environments [248]

RFCs Directly Related to IPv4 Addressing

IPv4	Title
RFC 0826	An Ethernet Address Resolution Protocol: Or Converting Network Protocol Addresses to 48.bit Ethernet Address for Transmission on Ethernet Hardware [67]
RFC 0894	A Standard for the Transmission of IP Datagrams over Ethernet Networks [71]
RFC 0895	Standard for the transmission of IP datagrams over experimental Ethernet networks [72]
RFC 1234	Tunneling IPX traffic through IP networks [88]
RFC 1577	Classical IP and ARP over ATM [115]
RFC 1700	Assigned Numbers [124]
RFC 1812	Requirements for IP Version 4 Routers [130]
RFC 1917	An Appeal to the Internet Community to Return Unused IP Networks (Prefixes) to the IANA [143]
RFC 1933	Transition Mechanisms for IPv6 Hosts and Routers [145]
RFC 2030	Simple Network Time Protocol (SNTP) Version 4 for IPv4, IPv6 and OSI [153]
RFC 3330	Special-Use IPv4 Addresses [200]

IPv4	Title
RFC 3378	EtherIP: Tunneling Ethernet Frames in IP Datagrams [202]
RFC 3787	Recommendations for Interoperable IP Networks using Intermediate System to Intermediate System (IS-IS) [222]
RFC 3974	SMTP Operational Experience in Mixed IPv4/v6 Environments [228]
RFC 4361	Node-specific Client Identifiers for Dynamic Host Configuration Protocol Version Four (DHCPv4) [235]
RFC 5692	Transmission of IP over Ethernet over IEEE 802.16 Networks [261]
RFC 5994	Application of Ethernet Pseudowires to MPLS Transport Networks [264]
RFC 7393	Using the Port Control Protocol (PCP) to Update Dynamic DNS [279]
RFC 7608	IPv6 Prefix Length Recommendation for Forwarding [282]
RFC 7775	IS-IS Route Preference for Extended IP and IPv6 Reachability [284]
RFC 8115	DHCPv6 Option for IPv4-Embedded Multicast and Unicast IPv6 Prefixes [290]
RFC 8468	IPv4, IPv6, and IPv4-IPv6 Coexistence: Updates for the IP Performance Metrics (IPPM) Framework [299]
RFC 8539	Softwire Provisioning Using DHCPv4 over DHCPv6 [304]

RFCs Directly Related to CIDR

CIDR	Title
RFC 1517	Applicability Statement for the Implementation of Classless Inter-Domain Routing (CIDR) [108]
RFC 1518	An Architecture for IP Address Allocation with CIDR [109]
RFC 1519	Classless Inter-Domain Routing (CIDR): an Address Assignment and Aggregation Strategy [110]
RFC 1520	Exchanging Routing Information Across Provider Boundaries in the CIDR Environment [111]
RFC 1812	Requirements for IP Version 4 Routers [130]
RFC 1817	CIDR and Classful Routing [131]
RFC 1917	An Appeal to the Internet Community to Return Unused IP Networks (Prefixes) to the IANA [143]
RFC 2036	Observations on the use of Components of the Class A Address Space within the Internet [155]
RFC 4632	Classless Inter-domain Routing (CIDR): The Internet Address Assignment and Aggregation Plan [242]
RFC 7608	IPv6 Prefix Length Recommendation for Forwarding [282]

RFCs Directly Related to IPv6 Addressing

IPv6	Title
RFC 1680	IPng Support for ATM Services [123]
RFC 1881	IPv6 Address Allocation Management [138]
RFC 1888	OSI NSAPs and IPv6 [139]
RFC 1933	Transition Mechanisms for IPv6 Hosts and Routers [145]
RFC 1972	A Method for the Transmission of IPv6 Packets over Ethernet Networks [148]
RFC 2030	Simple Network Time Protocol (SNTP) Version 4 for IPv4, IPv6 and OSI [153]
RFC 2080	RIPng for IPv6 [158]
RFC 2081	RIPng Protocol Applicability Statement [159]
RFC 2461	Neighbor Discovery for IP Version 6 (IPv6) [172]
RFC 2464	Transmission of IPv6 Packets over Ethernet Networks [173]
RFC 3646	DNS Configuration options for Dynamic Host Configuration Protocol for IPv6 (DHCPv6) [220]
RFC 3974	SMTP Operational Experience in Mixed IPv4/v6 Environments [228]
RFC 4191	Default Router Preferences and More-Specific Routes [232]
RFC 4193	Unique Local IPv6 Unicast Addresses [233]
RFC 4361	Node-specific Client Identifiers for Dynamic Host Configuration Protocol Version Four (DHCPv4) [235]
RFC 5006	IPv6 Router Advertisement Option for DNS Configuration [254]
RFC 5340	OSPF for IPv6 [259]
RFC 5692	Transmission of IP over Ethernet over IEEE 802.16 Networks [261]
RFC 5902	IAB Thoughts on IPv6 Network Address Translation [263]
RFC 5994	Application of Ethernet Pseudowires to MPLS Transport Networks [264]
RFC 6085	Address Mapping of IPv6 Multicast Packets on Ethernet [268]
RFC 6106	IPv6 Router Advertisement Options for DNS Configuration [269]
RFC 7393	Using the Port Control Protocol (PCP) to Update Dynamic DNS [279]
RFC 7503	OSPFv3 Autoconfiguration [280]
RFC 7608	IPv6 Prefix Length Recommendation for Forwarding [282]
RFC 7775	IS-IS Route Preference for Extended IP and IPv6 Reachability [284]
RFC 8064	Recommendation on Stable IPv6 Interface Identifiers [289]
RFC 8115	DHCPv6 Option for IPv4-Embedded Multicast and Unicast IPv6 Prefixes [290]
RFC 8362	OSPFv3 Link State Advertisement (LSA) Extensibility [294]

IPv6	Title
RFC 8415	Dynamic Host Configuration Protocol for IPv6 (DHCPv6) [296]
RFC 8468	IPv4, IPv6, and IPv4-IPv6 Coexistence: Updates for the IP Performance Metrics (IPPM) Framework [299]
RFC 8501	Reverse DNS in IPv6 for Internet Service Providers [302]
RFC 8539	Softwire Provisioning Using DHCPv4 over DHCPv6 [304]

Other RFCs Related to Layer 3

For a list of other RFCs related to the Network Layer but not closely referenced in this chapter, please see Appendices B, B, and B.

Chapter 6
The OSI Upper Layers

6.1 Overview of the Upper Layers

Until now the concentration has been on the OSI layers responsible for building the network. The end–points of the data conversation are not responsible for the actual movement of messages but are responsible for the higher level functions to insure that the conversation meets the requirements of the final application. When the conversation is initiated, the details of the functioning of the upper levels are negotiated before data begins to flow.

6.2 The Transport Layer, Layer 4

The Transport Layer runs on both the end–points of the data conversation and is responsible for either guaranteed delivery over TCP or best–effort delivery over UDP depending upon the requirements of the application.

6.2.1 Connectionless vs Connection Oriented

In networking, conversations between devices can be carried out as connectionless or connection oriented. This has a significant impact on the function of the Transport Layer.

6.2.1.1 Connection Oriented Conversations

A good example of a connection oriented conversation is a plain old telephone call. The person who places the call must know the correct phone number (the address)

© Springer Nature Switzerland AG 2020
G. Howser, *Computer Networks and the Internet*,
https://doi.org/10.1007/978-3-030-34496-2_6

and then dial it. The telephone system sets up a connection between the two telephones and then causes the other telephone to ring. If the person on the other end answers, the call is placed and data can flow. If the call is not completed no data can flow and the call setup has failed.

In a packet network such as the Internet, this means that a pathway must be found from the IP address starting the conversation to the IP address of the other end–point. Once this pathway is set up, all the packets that are part of the conversation flow over this exact same pathway. All packets take roughly the same time to transit the pathway and they are not able to pass one another. A packet may get lost or corrupted, but the packets that arrive at the end–point arrive in the order they were sent[1]. When the conversation ends, the pathway must be taken down and those resources freed for other conversations. A good analogy would be packages placed in the sequential cars of a train.

6.2.2 Connectionless Conversations

There is a problem with connection–oriented conversations. The Internet is extremely volatile and the best path from one end–point to another changes millisecond by millisecond. Physical media goes down. Congestion occurs on lines and changes constantly. The solution to this is connectionless conversations via packet switching. As before, the messages are broken down into packets each with information as to the destination. At each interconnection between small networks (routers), the packet is sent along the best path to the next hop. As the local situation changes, a router might send two packets for the same destination to different next hops. Packets no longer arrive in order since it is quite likely that packet 12 might go a longer route than packet 13 because a some router local conditions changed. Connectionless conversations present an extra problem for the destination, but react to changing conditions whereas connection oriented conversations cannot do this. The Internet is a jungle of shifting connectionless conversations with packets taking all kinds of routes to go from A to B.

A good analogy for connectionless conversations is the US Postal Service delivering the chapters of this book each in its own envelope. Each envelope has a destination and can all go in the same truck, but it is likely that some envelopes could go in different trucks. At each intersection or hop, a truck takes what looks like the best route to the Post Office. As traffic changes, the best route changes so the envelope with chapter 10 could arrive before the one with chapter 2 and an envelope could break open or get lost (which is not likely as the Post Office is not careless).

---

[1] Asynchronous Transfer Mode (ATM) works this way. This makes things much simpler.

6.2.3 Sending a Message

Most messages traveling over the Internet are larger than the data payload of an IP packet and must be broken down into smaller segments ("slicing and dicing") which can fit into a packet. The Transport Layer accepts a message from the Session Layer and then breaks the message into smaller parts and pre–pends a header with information as the number of segments and the serial number of the current segment. The Transport Layer at the destination must then hold these segments until all the pieces of the message have arrived and can be reassembled into the original message or S-PDU[2].

6.2.4 Receiving a Message

The destination Transport Layer must receive the packets that hold the segments of the original message. The header in the segment is used to determine the order in which the pieces are to be reassembled and the number of pieces to expect. Connections over the Internet are connectionless, so packets may arrive out of order or not at all. When all have arrived, the Transport Layer is able to reconstruct the original message and release it to the Session Layer.

6.2.5 Guaranteed Delivery

Many applications depend upon message being received at the destination without losses or corruption. For example, a credit card customer would be upset if the amount charged against their account was posted three or four times. Likewise, the credit card company would not be happy if the charge was authorized but the message to debit was lost in transit. The Transport Layer, TCP on the Internet, deals with these and other problems.

   In order to provide guaranteed delivery, the two end–points must trade extra messages which means extra overhead for the network. In order to accomplish guaranteed delivery, each message, called a T-PDU[3], has a header which denotes the number of message segments and the sequence number of the message segment in the data payload of this packet. When a message segment is received, the destination must verify the integrity of the message segment and then reply with an acknowledgment (**ACK**) if the segment is correct or a negative acknowledgment (**NAK**) if

---

[2] Session Layer PDU

[3] Transport Layer PDU

there are any problems. If a **NAK** is sent, the sender re-sends *only* the message segment with the error[4].

6.2.6 Best–Effort Delivery

If the loss of a message can be tolerated or cannot be corrected, best–effort delivery
sends messages with less overhead than guaranteed delivery. The message is "sliced
and diced" into message segments that will fit into the packet payload. The destination collects these segments the same way as before and checks each message
segment for errors. If any errors occur, the destination drops the entire message and
continues on to the next message. *No effort is made to notify the sender.*

Streaming services such as streaming video, IP telephony, or streaming radio
are not able to resend packets that do not arrive or are corrupted; therefore, these
services do not benefit from the overhead required for guaranteed delivery. For these
services best–effort, UDP on the Internet, makes more sense. Control messages such
as those between switches are sent over different best–effort ICMP[5].

6.2.7 Flow Control

The transport layer is also responsible for flow control, which we will discuss in
more detail in Chapter 7. There are many different methods available for flow control but all have the goal of insuring that the sender does not overwhelm the receiver.
If the sender has nothing to send there is no problem at Layer 4, but the assumption
is that at any given time the receiver can only store a limited number of messages
for later processing. If the receiver processes messages slower than the sender produces them, the unprocessed messages are stored in memory called a buffer. When
the buffers are full any further messages are dropped which causes extra overhead
in the case of guaranteed delivery. For obvious reasons flow control is not usually
used in best–effort delivery.

6.3 The Session Layer, Layer 5

The Session Layer, Layer 5, is responsible for sessions: initialization of sessions,
maintenance of sessions, and the graceful termination of sessions. It should be noted
that the *graceful* termination of a session is probably more important that any other
function of the Session Layer. During initialization of a session, resources are al-

---

[4] The receiver has the option of **ACK**ing multiple segments with one message. This is similar to
"lock step" or "handshake" communications.

[5] Internet Control Message Protocol

located to a session and if the session is not gracefully terminated these resources are not returned by the processes involved. Over time, this leads to complete loss of resources, commonly known as a "memory leak", which will lead to all processes "freezing" for lack of resources. This is not a good thing.

6.3.1 Session Initialization

The first responsibility of the Session Layer is to allow or deny new sessions. A session could be denied for a number of reasons. The endpoint might only be able to support a limited number of sessions and when this number is reached all new sessions requests are denied. Each end–point has limited memory and processing resources and when theses limits are reached all further sessions must be denied. A more interesting possibility is that certain requests may be denied for security reasons such as an untrusted IP subnet or MAC address. Whatever the reason, the end–point receiving the session request is allowed to deny any and all requests.

Allocation of Resources

The Session Layer is responsible for allocating a number of resources such as memory, buffer space, session IDs, and possibly other resources. Once a resource is allocated to a particular session between two end–points it cannot be allocated to another session. This helps prevent over subscription of the limited resources available and resources are *always* limited.

Accepting or Refusing Sessions

If the requested resources are not available, the Session Layer denies the session and no messages will pass between the end–points. A session is usually refused due to lack of resources, but it is conceivable that a NIC could be designed to refuse sessions based upon MAC address or some other security issue.

Session Maintenance

After a session is initialized, data messages can be exchanged between two NICs as S-PDUs which allows the Session Layer to inter–operate with any variety of lower Layers. Unfortunately, things happen which could cause the session to be paused or dropped. In order to detect a dropped session, each end–point keeps a timer that is reset to zero anytime a message is sent or received. When this clock reaches some agreed upon value, the session is dropped and restarted. Preventing dropped sessions and dealing with the session when it is dropped is called Session Maintenance.

6.3.2 Keep–Alive and Heartbeats

It is entirely possible that one end–point may not need to communicate with the other for some period of time. If there are no messages for an agreed upon time–out time, the session is considered to be dropped and must be restarted. This can lead to problems with re-initializing the session such as wasted overhead, disallowed sessions, or memory leaks. Therefore, the session layer must have a method to keep a quiet session from becoming an unnecessarily dropped session. The two most common methods for doing this are known as "keep–alives" and "heartbeats".

Keep–Alive

A session may be negotiated to use keep–alives to deal with quiet times when one end–point or the other has nothing to send. When this happens, the quiet end–point sends a special message to inform the other Session Layer that the session is still active but there is nothing to process at this time. Both ends then reset their time–out clocks back to zero and start timing the session again. When a data message is sent, both ends also reset their time–out clocks. Since the keep–alive is a message, no additional code is needed to trigger this reset of the clocks, which is actually a clever way to handle this problem[6].

Heartbeats

A second way to deal with sessions that go quiet is called "heartbeats" and is similar to "keep–alive". The end–points keep time–out clocks in the same way, but each end–point sends a special "heartbeat" message at an agreed upon interval. These messages are ignored, but the end–points reset the time–out clock each time a "heartbeat" is sent or received[7].

6.3.3 Pausing and Resuming a Session

Another possibility for dealing with a quiet time in a conversation is for the Session Layer to send a special "pause" message which is ignored by the other end–point except to mark the session as paused. For example, the device might be busy with OS maintenance such as memory defragmentation and pause the process involved in this session. Suppose the device is a printer and is out of paper. It makes no

---

[6] A good analogy for "keep alives" is asking if the other person is still there when a cellphone conversation is quiet for too long. Is it really quiet or has the connection dropped?

[7] A good analogy for "heartbeats" is the way many people talk on the phone. Every so often, even if they are not listening, they say "uh huh" or "really?" so the other person knows they are still there.

sense to continue the session until there is paper to print on, so the Session Layer on the printer might be able to send a message back to your laptop to wait to send more pages. This should not be confused with flow control because it takes more resources to pause a session than for a sending end–point to wait to send a message.

If a session can be paused, there must be some graceful way to resume it and this is also handled by the Session Layer. A special message is sent by the end–point that paused the session to resume it.

6.3.4 Dropped Sessions

When an end–point time–out timer reaches a certain value, the session is assumed to be dropped. Rather than start a new session, the Session Layer must try to resume the session. If this does not work then the session is terminated.

6.3.5 Session Termination

Sessions that are started by the Session Layer must eventually be terminated. Dropped sessions present more of a problem than sessions that end properly.

Graceful Termination vs Dropped Session

When an end–point is ready to stop communicating, a special termination request is sent by the Session Layer. The two end–points then agree to end the session *gracefully*. If this is not done, the other end–point will eventually determine the session is dropped and must attempt to restart the session which takes a fair amount of messaging and other resources. Also dropped sessions are not handled as cleanly as ones that are ended properly.

De–Allocation of Resources

When a session is terminated, both end–points still have resources allocated to that session. The Session Layer is responsible for marking these resources as available for reuse. If this does not happen, those resources remain allocated and eventually the end–point will run out of resources to allocate to new sessions and no new sessions can occur. This is the cause of the infamous "memory leak." Early in the deployment of the web, a certain browser was famous for not returning resources at the end of sessions and was known to slow down PCs[8] running Windows. Some web servers did not properly end sessions handled by child processes and left those

---

[8] Personal Computers

processes running while new child processes were spawned. Eventually the device would physically halt for lack of resources.

6.4 The Presentation Layer, Layer 6

The Presentation Layer has nothing to do with how the user or a process views the data but has everything to do with how the data is presented to the end–points of the conversation. The Presentation Layer is responsible for encoding/decoding, compression/decompression, and encryption/decryption of the data messages sent back and forth as P-PDUs[9]. This is one of the few times when the difference between encoding and encrypting data is important.

6.4.1 Encoding

In order for a process to deal with an outside message, there must be a prior agreement as to how the data is to be encoded or represented. One of the most common encoding schemes is ASCII which was developed to solve a very real problem in the mainframe days. There are many different methods to represent characters as eight bit patterns and mainframes regularly used multiple different schemes to store data. To facilitate electronic communications between devices a common encoding scheme must be agreed upon and one of the main tasks of Layer 6 is to take messages from the native encoding of the device and translate them into whatever encoding the two end–points have agreed to use during the current session. Unless the devices both use the same native encoding method, such as ASCII, this step means the difference between a message exchange and all messages being dropped as error messages because some header information in the message may need to be encoded as well.

6.4.2 Compression

Sending messages over media that is owned by someone else is rarely free and often charged on a usage rate such as the number of gigabytes used[10]. One way to reduce the message size and therefore the time and cost is to compress the message. The other end–point must know the compression scheme and how to decompress the message. The compression scheme is typically negotiated during session initialization, see Section 6.3.1. Compressing a message is much like zipping a file; it

---

[9] Presentation Layer PDUs

[10] This problem is well understood by anyone with a cellphone that does not have an unlimited plan.

will normally result in a smaller message but sometimes it may not result in enough savings to make it useful.

6.4.3 Encryption

If privacy is a concern, the messages may be encrypted by Layer 6 once the encryption scheme has been negotiated at session initialization. Bear in mind that the encryption is most likely well known and not very powerful as it is being done by a NIC and not a powerful computer. Encryption at the Presentation Layer will at least annoy an eavesdropping device and make troubleshooting at least that much more difficult as well.

6.5 The Application Layer, Layer 7

The highest layer of the OSI Model is the Application Layer which has two main tasks:

1. To deal with Client and Server announcements.
2. To "map" a set of messages to the correct API[11] or process.

To some extent, an administrator can control how announcements are handled and likewise how mapping is done, but care must be taken to insure that all devices and processes involved understand these choices. The author is of the opinion that it is safest to take the defaults whenever possible but has violated the rules when that was more convenient or desirable[12].

6.5.1 Services and Processes

Services are resources provided to the network by processes running on some device, usually called a Server, and are used by processes called Clients. Although it may not be obvious from this chapter, keep in mind that a device can be both a client and server for different services at the same time. Although a device can host many services at the same time, we will find when we discuss Internet services in Part IV it is best to think of these services as being independent and possibly on devices scattered throughout the network.

---

[11] Application Program Interface

[12] Always understand the rules and the reasoning behind them before you break any rules.

6.5.2 Announcements

Service announcements present a difficult, but typical, problem for the designer and administrators of the network. To be effective at all, service announcements must be seen by all the devices that might make use of the service and yet announcements could easily overwhelm the network. It is a balancing act to limit the number and scope of announcements without a device missing an important announcement.

Client announcements present exactly the same problems as service announcements. For simplicity, the four classes of announcements (see Figure 6.1) will be handled from the most problematic to the least.

	Active Announcements	Passive Announcements
Client	Client Broadcasts	Client Posts
Service	Service Broadcasts	Service Posts

Fig. 6.1: The Four Classes of Announcements

Active Service Announcements

Active Service Announcements happen when a service is available. This is much like the annoying ice cream trucks that circle the neighborhood playing some childrens song to draw a crowd[13]. The service uses an IP Broadcast to the entire IP network to announce that it is available. Clients that need that service respond to the IP address that originated the Broadcast and a session is started between the two. The problem is that if a service is idle it must keep Broadcasting the announcement over and over until it is not idle. A heavily used network with many idle services making Active Service Announcements wastes a large amount of the available bandwidth on these announcements. Limiting the Active Service Announcements could keep clients that need the service from connecting to the service even though it is idle.

---

[13] I lived in a neighborhood where the ice cream truck played "Pop Goes the Weasel" with a hiccup in the same place every time through the song. Thankfully, I was not armed or dangerous.

Active Client Announcements

An Active Client Announcement is similar to an Active Service Announcement except the announcement is Broadcast by the client, not the server. An analogy would be a swimmer screaming for help as he or she does not care which of the bystanders comes to help as long someone comes to help. Again, there is the same problem with balancing a potentially large number of Broadcasts against a desired service remaining idle.

Passive Service Announcements

Passive Service Announcements are posted to a special location instead of Broadcast and clients go to this location to determine if a desired service is available some-where on the network. A real world example would be posting a flyer for guitar lessons on a bulletin board in a coffee shop. Potential students (clients) must know where to look for such flyers or they will not know a guitar teacher has openings. On the network, available services must be posted to a location, be it a file or database of services, known to all clients which wish to use that service. The client looks up the available services and then contacts the service it wishes to use. The service accepts the request by replying to the client and marks itself as unavailable where services are posted. This is much more complex than active announcements, but much easier to manage and less of an impact on the network.

Passive Client Announcements

Passive Client Announcements are posted by clients requiring a service to a known location where services know to find these announcements. A real world example is Craig's List help wanted ads. People post the type of help they need and those who can provide that service contact them. A network example most people are familiar with is printing to a pool of printers. A heavily used lab might have multiple printers that can be used by anyone in the lab. The lab machines are configured to print to a print queue (posting a passive announcement) and the first available printer picks up the job from the queue and prints it. While the printer is busy, it does not look for more jobs and is effectively unavailable for other users. Once this is set up, it is very simple and effective. Indeed, the users may not even understand anything at all about how this all works behind the scenes.

Passive Announcements by Sockets or Ports

A low impact but effective way to manage connections between clients and services is the use of sockets or ports, see Table 6.1. These act like a two byte suffix to the device Layer 3 address and facilitate one–to–one and other mappings as discussed in Section 2.16.

Table 6.1: Port/Socket Numbers

Port Range
0 - 1024 Well Known Port Numbers
1025 - 4096 Registered Port Numbers
4096 - 65,536 User Port Numbers

Table 6.2: Well–Known TCP and UDP Ports

Port Protocol
15 Netstat
21 File Transfer Protocol (FTP)
23 Telnet
25 Simple Mail Transfer Protocol (SMTP)
53 Domain Name Service (DNS)
68 Dynamic Host Configuration Protocol (DHCP) client side
69 Trivial File Transfer Protocol (TFTP)
80 Hypertext Transfer Protocol (HTTP)
88 Kerberos (Security Server)
110 Post Office Protocol, Version 3(POP3)
119 Network News Transfer Protocol (NNTP)
123 Network Time Protocol (NTP)
137 Windows Internet Naming Service (WINS)
139 NetBIOS over TCP/IP (NBT)
143 Internet Message Access Protocol (IMAP)
161 Simple Network Management Protocol (SNMP)
443 Secure Sockets Layer (SSL)
515 Line Printer (LPR)
1701 Layer 2 Tunneling Protocol (L2TP)
1723 Point–to–Point Tunneling Protocol (PPTP)
8080 HTTP Proxy (Commonly used, not reserved)

Port numbers less than about 4096 [125] should be avoided as they may be assigned to services already on the network, see Table 6.2 for some of the more commonly used ports. Typically, ports above 4096 are safe, but it might be best to use only ports above 10,000. If a user should accidentally use an assigned port, the service that normally uses that port will no longer be available.

Fig. 6.2: Ports and Bi-Directional Communications

On many Linux boxes, such as a Raspberry Pi, it is easy to get a list of the active ports used for sending and receiving by entering `sudo netstat` or just the ports being "listened to" by entering `sudo netstat -lptun4` to get an output such as in Figure 6.3. Notice the port numbers are given as a suffix to the IP address such as "0.0.0.0:22" for `sshd`.

```
pi@router1-1:~$ sudo netstat -lptun4
Active Internet connections (only servers)
Proto Recv-Q Send-Q Local Address            Foreign Address
             State        PID/Program name
tcp    0      0 127.0.0.1:2608              0.0.0.0:*
             LISTEN       640/isisd
tcp    0      0 192.168.1.49:53            0.0.0.0:*
             LISTEN       530/named
tcp    0      0 192.168.1.201:53           0.0.0.0:*
             LISTEN       530/named
tcp    0      0 127.0.0.1:53               0.0.0.0:*
             LISTEN       530/named
tcp    0      0 0.0.0.0:22                 0.0.0.0:*
             LISTEN       644/sshd
tcp    0      0 0.0.0.0:25                 0.0.0.0:*
             LISTEN       844/sendmail: MTA:
tcp    0      0 127.0.0.1:953              0.0.0.0:*
             LISTEN       530/named
tcp    0      0 127.0.0.1:2601             0.0.0.0:*
             LISTEN       607/zebra
tcp    0      0 127.0.0.1:3306             0.0.0.0:*
             LISTEN       842/mysqld
tcp    0      0 127.0.0.1:2602             0.0.0.0:*
             LISTEN       642/ripd
tcp    0      0 127.0.0.1:587              0.0.0.0:*
             LISTEN       844/sendmail: MTA:
tcp    0      0 127.0.0.1:2604             0.0.0.0:*
             LISTEN       653/ospfd
udp    0      0 0.0.0.0:520                0.0.0.0:*
                          642/ripd
udp    0      0 0.0.0.0:54284              0.0.0.0:*
                          374/avahi-daemon: r
udp    0      0 192.168.1.49:53            0.0.0.0:*
                          530/named
udp    0      0 192.168.1.201:53           0.0.0.0:*
                          530/named
udp    0      0 127.0.0.1:53               0.0.0.0:*
                          530/named
udp    0      0 0.0.0.0:68                 0.0.0.0:*
                          529/dhcpcd
udp    0      0 0.0.0.0:5353               0.0.0.0:*
                          374/avahi-daemon: r
pi@router1-1:~$
```

Fig. 6.3: Output From the Command **netstat -lptun4**

6.5.3 Receiver Ports

Services are processes that "listen" for messages on a specific port. For example, a web server on a Linux box is a copy of httpd waiting for the OS to forward a message received on port 80. If some other process is "listening" to port 80, then this box cannot be a normal web server; that port is taken and a normal web browser cannot contact this box correctly.

It is entirely possible to create any IP–based service simply by having the client and service communicate on ports known to both. A private messaging service could be created by a JAVA program sending messages on port 12345 and listening on port

54321 with another copy of the same program running on a different machine with sending and receiving ports reversed, as in Figure 6.2. Such a program does not present any major difficulties and could easily be used to send encrypted messages over the Internet.

6.5.4 Sender Ports

In order to send a message to a process on a different IP address, the NIC must use the well–known port number or the port number negotiated as part of the session[14]. For example, a message sent with an outgoing port of 80 will be directed to a web server if there is a web server "listening" on port 80, if not the message will be dropped.

---

[14] FTP is notoriously difficult to manage at times because transfers are moved to a semi-random port number for downloads.

Exercises

6.1 How does the use "best–effort" delivery affect the actions at:

    a. Layer 6
    b. Layer 5
    c. Layer 4
    d. Layer 3
    e. Layer 2
    f. Layer 1

6.2 Why isn't flow control used in "best–effort" delivery such as streaming video?

6.3 Many mainframes use a standard called EBCDIC to encode data but use AC-SII for data transmissions. Which layer of the OSI Model does the required translation between EBCDIC and ASCII?

6.4 Why might Layer 5 refuse a new connection?

6.5 What happens at Layer 5 when you unplug a cable between two endpoints of an on going conversation?

6.6 In networking we typically ignore the upper layers. Is this a good idea or a bad one?

Chapter 7
Flow Control

Overview

Flow control is needed anytime two endpoints communicate; especially when one endpoint is slower than the other or has limited resources[1]. Thus we will concentrate on controlling the flow of messages with the understanding the message can be PDU at any layer of the OSI Model.

For the following discussions, one endpoint is assumed to be the "sender" and the other the "receiver", but the roles could be constantly switching during a conversation. This is not a problem as we can look at a single set of exchanges and understand the next set could have the roles reversed.

7.1 No Flow Control

In many cases there is no need for flow control. For example, there is usually no flow control in the case of streaming live media. When the receiver is busy, the sender cannot stop sending because the data cannot be stopped. In this case, the receiver simply misses the message and continues on. Good examples of this are radio broadcasts, telephone conversations, and television. The receiver often has no transmitter and cannot send any messages back to the transmitting tower to slow down or stop.

Many UDP conversations do not use flow control because delivery of the messages is not guaranteed.

---

[1] In the real world resources are always limited. If you add resources to a bottleneck, some other part of the system becomes the new bottleneck.

7.2 Start–Stop Flow Control

Figure 7.1 is a simple example of Start–Stop flow control. The receiver accepts messages until the memory allocated for incoming messages is full. At this point the receiver sends a "STOP" message to the sender. The sender then waits for a "Start" message before sending any more messages. This is a trivial answer to flow control and is too simplistic for most applications. As an added issue, a message could be lost if the receiver's buffer fills up too quickly for it to send the "STOP" message in a timely fashion.

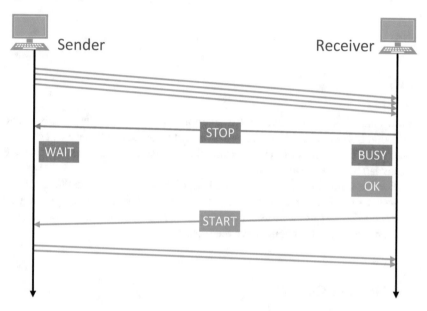

Fig. 7.1: Start–Stop Flow Control

Table 7.1: Start–Stop Flow Control Detail

Sender State		Receiver State
1. Idle		Idle
2. Transmit message 1	$\longrightarrow$	Receive
3. Transmit message 2	$\longrightarrow$	Receive
4. Transmit message 3	$\longrightarrow$	Receive
5. Transmit message 4	$\longrightarrow$	Receive
6.	$\longleftarrow$ STOP	Busy
7. Wait		Busy
8. Wait		Busy
9. Wait	$\longleftarrow$ START	Receive
10. Transmit message 5	$\longrightarrow$	Receive
$\vdots\ \vdots$	$\vdots$	$\vdots$

The steps in Figure 7.1 are given in detail below.

1. When there is no data to send, both sides are idle.
2. The sender sends message 1 to the other endpoint. The receiver places the message in memory (a buffer) for processing.
3. The sender sends message 2 to the other endpoint. The receiver places message 2 in memory (a buffer) for processing.
4. The sender sends message 3 to the other endpoint. The receiver places message 3 in memory (a buffer) for processing.
5. The sender sends message 1 to the other endpoint. The receiver places the message in memory (a buffer) for processing. The receiver's buffer is now full.
6. The receiver sends a "STOP" message to the sender and no longer can receive messages until the buffers are empty.
7. The sender goes into a wait state and no longer sends any messages. The receiver continues to process messages to clear out its receive buffers.
8. The sender continues to wait while the receiver continues to process message.
9. At some point the receiver will catch up enough to be ready to receive messages again. The receiver sends a "START" message to inform the sender to begin sending again.
10. The sender leaves the wait state and begins to send messages again.

7.3 Lock Step or Handshake Flow Control

In networking there are many cases where the loss or corruption of a message would cause catastrophic failures. If one endpoint is sending a long encryption key in a number of small messages, the loss of a message would lead to an invalid key. The receiving endpoint would have no way of knowing whether the key is correct or was corrupted during transmission. If a bad key is found, the entire key may need to be resent. On early networks with high error rate, this was unacceptable. In order to ensure the correct exchange of messages, a lock step, sometimes called "handshake", mechanism can be used as shown in Figure 7.2.

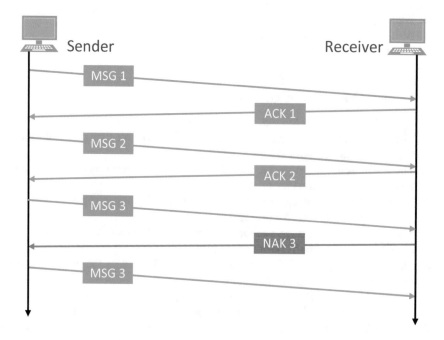

Fig. 7.2: Lock–Step Flow Control

Table 7.2: Lock–Step Flow Detail

Sender State		Receiver State
1. Idle		Idle
2. Transmit message 1	$\longrightarrow$	Receive
3. Wait	$\longleftarrow$ ACK 1	Transmit
4. Transmit message 2	$\longrightarrow$	Receive
5. Wait	$\longleftarrow$ ACK 2	Transmit
6. Transmit message 3	$\longrightarrow$	Receive
7. Wait	$\longleftarrow$ NAK 3	Transmit
8. Transmit message 3	$\longrightarrow$	Receive
9. Wait	$\longleftarrow$ ACK 3	Transmit
10. Transmit message 4	$\longrightarrow$	Receive
⋮ ⋮	⋮	⋮

Lock step flow control is fairly straight forward. Each message *must* be acknowledged as correctly received before the next message can be sent. Lock step flow control is common where errors would be catastrophic and common. For example, this type of flow control is used in ATM[2] when negotiating the characteristics of a newly discovered connection such as an ATM switch.

1. Both sides are idle when there are no messages to be sent.
2. The sending side sends the first message with some simple label such as ″1″ and then goes into a wait state.
3. The receiving side successfully processes the message and sends an acknowledgment.
4. When the acknowledgment for message 1 is received, the sender can then send message 2.
5. The receiving side acknowledges message 2 once it is successfully processed.
6. Upon receiving the acknowledgment for message 2, the sending side sends message 3.
7. Suppose the third message is corrupted or lost. The receiving side can send a negative acknowledgment to request the message be resent.
8. If the sending side receives a negative acknowledgment, it sends the damaged message again. This continues until the message is correctly received or the conversation is dropped because of too many errors.
9. This goes on until there are no more messages to send.

---

[2] Asynchronous Transfer Mode

7.4 Fixed Window Flow Control

An obvious extension to Lock Step flow control is for the two endpoints of the conversation to agree upon an allowed number of unacknowledged message the sender can have in process. When the conversation is started, the sender has a fixed window of messages that can be sent before any acknowledgment from the receiver and simply sends when needed as long as the window is not filled. When the window is filled, the sender must stop until messages are acknowledged before sending again, as shown in Figure 7.3

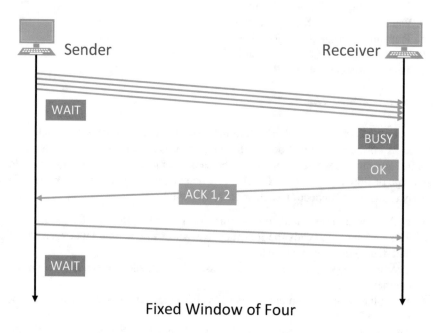

Fig. 7.3: Fixed Window Flow Control

Table 7.3: Fixed Window Flow Control Detail

Sender	In process		Receiver
1. Idle	0		Idle
2. Transmit	1	message 1 $\longrightarrow$	Receive
3. Transmit	2	message 2 $\longrightarrow$	Receive
4. Transmit	3	message 3 $\longrightarrow$	Receive
5. Transmit	4	message 4 $\longrightarrow$	Receive
6. Wait	4		Busy
9. Wait	4	$\longleftarrow$ ACK 1,2	Transmit
10. Transmit	4-2+1	message 5 $\longrightarrow$	Receive
10. Transmit	4	message 6 $\longrightarrow$	Receive
11. Wait	4		Busy
$\vdots$	$\vdots$	$\vdots$	$\vdots$

Fixed window flow control is much like Start–Stop and Lock Step flow control except the two endpoints negotiate a fixed number of messages that can be in a received but not acknowledged state.

1. Both negotiate the maximum number of messages that can be in process between the sender and receiver[3]. In this example the size of the fixed window is 4.
2. The sending side sends the first message with some simple label such as "1" and and continues to send until four unacknowledged messages have been sent.
3. The receiving side successfully processes the first two messages and sends an acknowledgment noting which messages are confirmed.
4. The acknowledgment for messages 1 and 2 is received, and the sender calculates it can send another message because the fixed window will allow two more messages.
5. The sending side then sends messages 5 and 6 which fills the fixed window size of four messages unacknowledged.
6. The sending side then waits for acknowledgments from the receiving side.
7. This goes on until there are no more messages to send.

7.5 Sliding Window Flow Control

Fixed window flow control leads directly to Sliding Window Flow Control where the receiving endpoint can change the size of the window as needed. When the receiver is busy, it has the option to send a message to the sender to change the size

---

[3] The window size can be different from A to B and B to A depending upon the characteristics of the connection.

of the window to a smaller number of messages. When the receiver is not busy, it can send a message to increase the size of the window. In all other respects, Sliding Window Flow Control behaves exactly like Fixed Window Flow Control.

Table 7.4: Sliding Window Flow Control Detail

	Sender	In process		Receiver
1.	Idle	0		Idle
2.	Transmit	1	message 1 $\longrightarrow$	Receive
3.	Transmit	2	message 2 $\longrightarrow$	Receive
4.	Transmit	3	message 3 $\longrightarrow$	Receive
5.	Transmit	4	message 4 $\longrightarrow$	Receive
6.	Wait	4		Busy
9.	Wait	4	$\longleftarrow$ ACK 1,2	Transmit
10.	Wait	4	$\longleftarrow$ Window size 2	Transmit
11.	Wait	$2-2+2$		Busy
12	Wait	4	$\longleftarrow$ ACK 3	Transmit
13.	Wait	$2-2+1$		Busy
14.	Transmit	2	message 5 $\longrightarrow$	Receive
15.	Wait	$2-2+2$		Busy
	$\vdots$	$\vdots$	$\vdots$	$\vdots$

1. Both endpoints are idle and have an agreed upon window size of 4.
2. The sender sends message 1 which leaves a window of 3.
3. The sender sends message 2 which leaves a window of 2.
4. The sender sends message 3 which leaves a window of 1.
5. The sender sends message 4 which leaves a window of 0.
6. The sender must wait until the window allows at least 1 message while the receiver continues to process messages.
7. After the receiver determines messages 1 and 2 were received correctly and processed to the point at which there is space for another message, the receiver sends an acknowledgment for messages 1 and 2.
8. Because it is very busy, the receiver determines it cannot process the messages this quickly and slows down the conversation by sending a window size of 2. The Sender updates the window size to 2.
9. The window is now textbf$2-2+2$, or zero, so the sender must still wait.
10. After the receiver finishes processing message 3 it sends an acknowledgment to the sender.
11. The sender calculates a window of 1.
12. With a window of 1 the sender can send message 5.
13. The window is again zero so the sender must wait.

7.6 Poll–Select Flow Control

In some networks the devices only communicate with a central master node. All data flows between the client nodes and the master node under the complete control of the master node and absolutely *no* messages are sent client–to–client. In this case a good choice for flow control would be either Lock Step, see Section 7.3, or a technique called Poll–Select[4]. This technique consists of two phases called POLL if the master was idle and expecting to receive a message and called SELECT when the master has data to send to a client station.

7.6.1 Poll

When the master is prepared to receive messages, it sends a special POLL message to each station requesting data. If a station does not have data to send, it sends a NAK[5] to alert the master to the fact that it has no data to send. The master then sends a POLL to the next station. If it has data to send it replies with a data message which the master acknowledges with an ACK[6]. It is important that the master poll the stations in a round–robin fashion to avoid starving a station that has data to send.

---

[4] Poll–Select was used extensively on Burroughs and IBM mainframes that supported "dumb" terminals.

[5] Negative Acknowledgment

[6] Acknowledge transmission

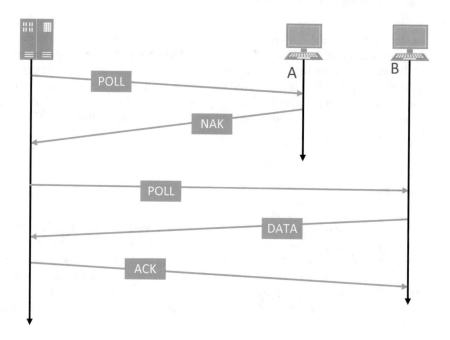

Fig. 7.4: Poll Example

Master		Station
1. Idle		Idle
2. Transmit	Poll $A \longrightarrow$	Receive
3. Wait	$\longleftarrow$ NAK	Transmit
4. Transmit	Poll $B \longrightarrow$	Receive
5. Wait	$\longleftarrow$ message (data)	Transmit
6. Transmit	ACK to $B \longrightarrow$	Receive
$\vdots$	$\vdots$	$\vdots$

The process of the master polling each station for data is very straight forward and works very well when the stations are on shared media.

1. The master and all stations are idle.
2. The master begins to poll the stations on the line using a round–robin technique, starting with A.
3. Station A has no data to send at this time and so replies with a NAK message[7].
4. Since A has no need to send, the master polls the next station which happens to be "B".
5. Station B has data to send and transmits it.
6. If the master receives the data correctly it replies with an ACK.

---

[7] This is often nothing more than the station identifier A and the NAK character.

7.6.2 BNA Group POLL

BNA[8] introduced a faster polling method known as group polling. The master could send a group poll to a list of stations all at the same time. Each station would respond with an ACK if it had data to transmit and a NAK otherwise. The master would then poll the stations that replied with an ACK in order. This is much more efficient than single polling.

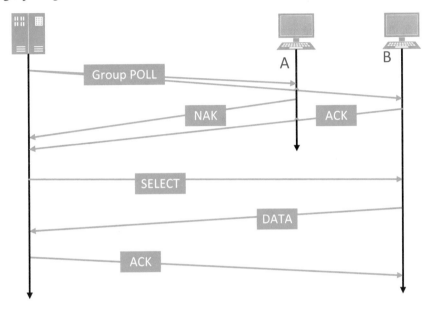

Fig. 7.5: BNA Group POLL Example

Master		Station
1. Idle		Idle
2. Transmit	Group POLL $\longrightarrow$	Receive
3. Wait	$\longleftarrow$ NAK from station *A*	Transmit
4. Wait	$\longleftarrow$ ACK from station *B*	Transmit
5. Transmit	POLL *B* $\longrightarrow$	Receive
6. Wait	$\longleftarrow$ message (data) *B*	Transmit
7. Transmit	ACK to *B* $\longrightarrow$	Receive
$\vdots$	$\vdots$	$\vdots$

The process of the master polling each station for data is very straight forward and works very well when the stations are on shared media. Notice how the group poll only allows the master to learn which stations have data to send. The master must still send a normal POLL to get the information but even so this saves network traffic when some of the stations are idle for periods of time. One useful side–effect

---

[8] Burroughs Network Architecture

is that group polling defaults to normal polling for stations that do not use group polling.

1. The master and all stations are idle.
2. The master begins to poll the stations on the line using a round–robin technique, starting with A.
3. Station A has no data to send at this time and so replies with a NAK message[9].
4. Since A has no need to send, the master polls the next station which happens to be "B".
5. Station B has data to send and transmits it.
6. If the master receives the data correctly it replies with an ACK.

7.6.3 SELECT

When the master has a message for a station, it notifies the station via the SELECT protocol as in Figure 7.6.

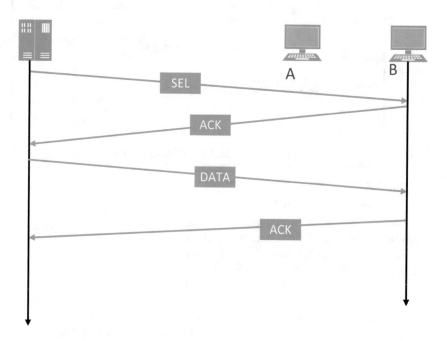

Fig. 7.6: Select Example

---

[9] This is often nothing more than the station identifier A and the NAK character.

Master		Station
1. Idle		Idle
2. Transmit	SELECT $B \longrightarrow$	Receive
3. Wait	$\longleftarrow$ ACK from station B	Transmit
5. Transmit	message (data) $B \longrightarrow$	Receive
6. Wait	$\longleftarrow$ ACK B	Transmit
$\vdots$	$\vdots$	$\vdots$

1. Master and all stations are idle.
2. To send a message to B the master sends a SELECT to B.
3. When B is read to receive the message it replies with an ACK.
4. The master sends the data with B's address.
5. When B has correctly received the data it sends the master an ACK

Exercises

7.1 Give a non–electronic example of each of these types of flow control.

    a. No flow control.
    b. Start–Stop flow control.
    c. Lock Step flow control.
    d. Fixed Window flow control.
    e. Sliding Window flow control.

7.2 Why would the receiver acknowledge more than one message with a single ACK?

7.3 Poll–Select flow control is not common in peer–to–peer networks. Why?

Chapter 8
Raspberry Pi Operating System

Overview

Like most modern operating systems Raspbian is controlled by a large number of configuration files. When Raspbian boots it reads these files at various times during initialization. As the files are read Raspbian sets internal flags and sometimes even creates new configuration files to reflect the desired behavior of the OS and therefore the Raspberry Pi.

Many times the configuration files must be manually edited to select various optional actions to be taken by the OS. This is to be expected as Raspbian is based upon the Linux distribution called Debian which is designed to be configured either manually or via the graphic desktop GUI. Either way, it is a good idea to become familiar with directly editing files using the Linux editor **vi**.

In this chapter the student, or hobbyist, will have the opportunity to gain some familiarity with **vi**. There is another very easy to learn editor included with most Linux distribution called **nano**. Either editor will do what needs to be done fairly painlessly.

8.1 Creating and Loading a Custom Pi OS

This section gives the instructions to create a custom Pi OS with the required packages to provide all Internet services. A 4 gigabit microSD card should be used. Current versions of Raspbian automatically expand the file system to use the full microSD card which means the custom image will be as large as the card can hold. This means transfers will be slower and the custom image will take a huge amount of space for backup. The custom image used for this book was created on a 4 gigabit microSD.

© Springer Nature Switzerland AG 2020
G. Howser, *Computer Networks and the Internet*,
https://doi.org/10.1007/978-3-030-34496-2_8

8.1.1 Transferring the Image to a microSD Card

Before the Pi can boot on the OS, the OS must be transferred to a microSD[1] card. This is done differently on Windows and UNIX/Linux.

Windows Transfer

An easy way to transfer images to a microSD card is by using two free programs, **SDFormatter** [310] from the SDA[2] to format the microSD card (see Section 8.3) and **Balena Etcher** [4].[3]. The transfer should be to a cleanly formatted microSD using the SDFormatter program[4].

One of the nice features of Balena Etcher is an automatic verification of the transfer to the microSD. It is really helpful to know the transfer has completed and it has been verified.

Fig. 8.1: Balena Etcher on Windows

---

[1] Older Pi Microcomputers have a normal SD slot and can boot on a microSD using an adapter. Many microSD cards come with an SD adapter. This adapter may be needed to read/write a microSD on a laptop or desktop computer.

[2] SD Association

[3] **Win32DiskImager** [316] can write an SD card image to a hard disk and is a great way to backup an SD card.

[4] Using the Windows `format` command is usually unsuccessful in my experience.

Linux Transfer

If the Unix (Apple MAC) version of SD Formatter does not work on your distribution of Linus, then simple Linux format should correctly format the microSD card.

Fortunately, Balena Etcher works *exactly* the same way on Windows, Linux, and UNIX as shown in Fig. 8.2.

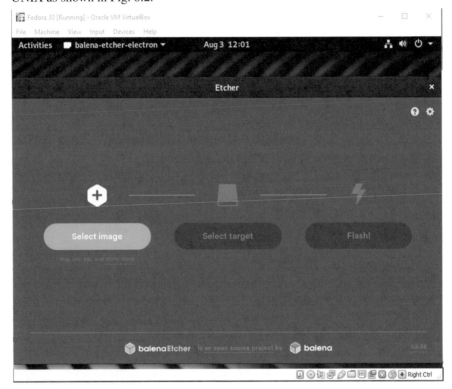

Fig. 8.2: Balena Etcher on Linux

Apple and Linus users who have trouble with Balena can use **sudoddif=/path/ to/Raspbian-image.imgof=/dev/name-of-sd-card-disk** to transfer to transfer the image.

8.1.2 Enabling SSH

For security reasons, the Pi image has ssh[5] turned off by default. If you have a monitor and keyboard attached to your Pi, you can turn on ssh as part of the initial configuration of the Pi, see Section 8.2. However, if you don't have a monitor and keyboard you can easily enable ssh by the following steps.

---

[5] Secure Shell (ssh)

Windows

1. After you have copied the image to a new microSD, open the device from Windows[6].
2. If the folder is not **boot**, then navigate to the **boot** folder
3. Right click anywhere on the directory window that is blank and create a new text file.
4. Save this file as **ssh** with *no extension*.
5. Close the folder and eject the device.

Linux

1. After you have copied the image to a new microSD, open the directory**/media/boot**[7].
2. If the folder is not **boot**, then navigate to the **boot** folder
3. Create an empty file by entering **sudo touch ssh**.
4. Close the folder and eject the device.

8.1.3 Boot the Pi on the Custom Image

1. Power off the Pi.
2. Place the microSD card in the slot and power up the Pi.
3. Contact the Pi via **ssh**. If the Pi does not display a screen, type **Pi** and press enter.
4. Log on using the default user, **Pi**, and password **raspberry**
5. Each time you reboot the Pi, you should correct the time with a command such as: **sudo date MMDDhhmmYY** where:

   a. **MM** is the two–digit month
   b. **hh** is the two–digit hour in 24–hour format
   c. **mm** is the two–digit minute
   d. **YY** is the two–digit year or **YYYY** is the four–digit year

8.1.4 Raspberry Pi First Log–on

The first time the Pi is booted, the following screen is presented over a serial link:
```
Raspbian GNU/Linux 9 raspberrypi ttyAMA0
raspberrypi login: pi
Password:
```

---

[6] Windows does not understand the image and will tell you to format the drive before you use it **Do not format the drive!** Ignore this Windows warning.

[7] How you get to the **boot** directory may be different on some distributions.

```
Last login: Fri Mar 15 23:05:07 GMT 2019 on ttyAMA0
Linux raspberrypi 4.14.79+ #1159 Sun Nov 4 17:28:08 GMT 2018
    armv6l
The programs included with the Debian GNU/Linux system are
    free software;
the exact distribution terms for each program are described in
    the
individual files in /usr/share/doc/*/copyright.
Debian GNU/Linux comes with ABSOLUTELY NO WARRANTY, to the
    extent
permitted by applicable law.
SSH is enabled and the default password for the 'pi' user has
    not been changed.
This is a security risk - please login as the 'pi' user and
    type 'passwd' to set a new password.
pi@raspberrypi:~\$
```

8.1.5 Install Required Packages

The Pi OS does not come with all the required packages installed and there is no guarantee that all the pre-installed packages are up to date. Connect the Pi to the Internet and contact the Pi via **ssh** or the console cable[8] and log in. The following commands will update the OS and install the required packages. These commands[9] must be issued with the prefix command sudo[10] and the final step is optional. For our purposes, the software can be slightly out of date without presenting a security risk but production Pi Microcomputers must be updated regularly as security issues and bugs are fixed fairly often.

- sudo apt-get update[11]
- sudo apt-get upgrade[12]
- sudo apt-get install telnet
- sudo apt-get install quagga
- sudo apt-get install apache2
- sudo apt-get install bind9 bind9utils bind9-doc dnsutils
- sudo apt-get install isc-dhcp-server
- sudo apt-get install mysql-server[13]
- sudo apt-get install mariadb-server
- sudo apt-get install mailman

---

[8] If you do not use a console cable, you must know the IP address of the Pi for **ssh**

[9] If a command stops with a question, answer **Y** or take the default. Anything else is beyond the scope of this text.

[10] Sudo

[11] Always, always, always do this before intalling any packages on the Raspberry Pi. I learned this the hard way.

[12] See the note about **sudo apt-get update**. Ada Fruit changes their mirror sites frequently.

[13] This may fail. There is a newer, freeSQL[14]database that is installed in the next step.

- sudo apt-get install Alpine
- One of the two email MTAs

    1. sudo apt-get install sendmail ssmtp[15]
    2. sudo apt-get install postfix postfix-mysql dovecot-core dovecot-imapd dovecot-pop3d dovecot-lmtpd dovecot-mysql
    3. NOTE: postfix will ask you a number of questions while being installed. These settings can be changed later, so take your best guess.

- sudo apt-get update

    Shut down and power off the Pi before removing the microSD card.

8.2 Setting Up the Pi

The Raspberry Pi was designed as an inexpensive platform to teach computer science by way of hands-on experience. With a little effort, you should be able to locate a kit from Ada Fruit [56] with everything needed except the USB/Ethernet dongles to provide additional interfaces for the Raspberry Pi Hobby Computer or an instructor may give you specific directions on how to obtain your hardware.

In this Section you will set up the Pi hardware and a custom Pi OS.

8.2.1 Equipment Lists

This equipment list assumes a maximum of 32 students in eight groups of four students each. For each additional set of 32 students, another set of the class equipment *may* be needed. As with the Internet, there is no theoretical limit to the size of the class network. In reality there will be limits on physical space for the students, electrical outlets, network wiring (Ethernet), and other normal classroom limitations[16] that will limit the number of groups before any network limitations are reached.

8.2.2 Class Equipment

The instructor will need a Pi to act as the Top Level Domain name server and to allow the instructor to monitor, and possibly display via a projector, the status of the class network. If a star network is planned for the labs, an additional Ethernet switch with at least one port per Group will be required.

---

[15] **ssmtp** is an extension to make sendmail easier to configure.

[16] Safety concerns for the number of people in a room will most likely be the deciding factor, but multiple rooms can be connected together. The last time this class was taught by the author the groups were in two adjacent room.

8.2.3 Group Equipment

Each Group should have access to a switch with at least four ports. Optionally each Group should have up to four spare USB/Ethernet dongles.

8.2.4 Indiviual Equipment

Each individual in the group will need a Raspberry Pi computer with some additional equipment. It is also assumed that each person will have access to a computer with at least one USB port and an RJ45 Ethernet port.

8.2.5 The Raspberry Pi Hobby Computer

The Raspberry Pi computer was created to provide a cheap, but stable, computer for use in the classroom from Elementary Schools through University. The Pi is cheap, rugged, and extraordinarily flexible. On the Raspberry Pi home site [58] and numerous hobby sites such as CanaKit.com [8], you can find step–by–step instructions to build a powerful desktop computer, a hand–held touch–screen calculator, and even a controller for a lighted Halloween costume. Many of these projects can also be found as YouTube [317] demonstrations.

Our goal is to use the Raspberry Pi to build routers, web–servers, and the other services required to build a tabletop intranet with all the same services as the Internet.

Details of the Available Raspberry Pi microcomputers

The best place to purchase a Raspberry Pi is from www.raspberry.org or the class could purchase them through a bookstore as a course requirement. Keep in mind that the Pi is a surprisingly versatile and powerful computer. A Pi can be used for any number of interesting projects, even if some are a little silly[17]. Each member of the group will need the same equipment.

It is best to purchase a kit that contains, at a minimum:

- The Pi computer version 2B or better. Version 3 makes a great home router/access point because of the built in wireless NIC.
- A case[19]

---

[17] For example, a Pi makes a great wireless access point or TOR[18] access point. It also makes a great controller for wearable lighting.

[19] A case is not strictly required unless you plan to carry the Pi around, accidentally put something heavy on top of it, or spill something. Buy a clear case in order to observe the on–board LEDs.

- A power supply. The best ones are a USB cable and separate adapter so that the Pi can be powered from a laptop USB port.
- Ethernet cable (CAT5 or better)
- A microSD card with at least a 4 gigabit capacity. 4 gigabits should be all that is needed and larger cards will take more room for backups. Most kits will have a microSD card with Raspbian pre–installed which is perfect for our needs.

Additional Required Equipment

- A microSD–to–SD reader or USB microSD reader. Often one will come with the microSD card.
- A USB/Ethernet dongle[20]. If purchased from Ada Fruit [56] or CanaKit [8], it will work with the Raspberry Pi.
- A Raspberry Pi serial or console cable.

  – These cables allow you to contact the Pi directly without requiring an IP address.
  – A serial cable connected to a laptop will also power the Pi.
  – This allows you to observe what happens when the Pi reboots.

- A laptop computer with an Ethernet NIC and an open USB port.

Optional Equipment

- One or two additional USB/Ethernet dongles.

8.3 Raspbian and Debian

The OS of choice for the Pi is Raspbian which is a stripped–down version of the Debian distribution of Linux. The best place to obtain Raspbian is from www.raspberry.org or to install a custom version created for this class. It is possible to download a version of Raspbian that supports a graphic desktop, but the OS would no longer fit on a 4 gigabit microSD card.

Installing the Class Image (Windows)

Before loading the image on a microSD, it is a good idea to erase and format the card. The Windows and DOS format commands are not very reliable when used on any SD card, so it is best to use a SD Format utility (free is best) such as SD Card

---

[20] A spare Ethernet dongle is nice to have and allows for more experimenting with routing.

Formatter [310], see Figure 8.3. Please note that this is a very slow process and some microSD card readers are faster than others.

Formatting a device should never be interrupted as the device, the microSD card in this case, could be left in a corrupted state ("bricked") and be unusable. typically the worst that happens is that the microSD card must be formatted again. The other issue to watch for is that the SD Card Formatter defaults to the first device it can format. If this is not the proper SD device, the wrong device might be formatted which can lead to a bit of embarrassment or loss of data. Lastly, the default is a quick format which writes a new FAT[21] index to the card without erasing data. If a full format has been performed there is less chance of corrupted data remaining behind to cause errors that are extremely difficult to track down. When time permits, always perform an over–write format to write good, blank data to the microSD card. This is much safer than a quick format but takes a significant amount of time.

Fig. 8.3: Formatting a microSD Card

---

[21] File Allocation Table (16 bit version)

The next step is to transfer the image to the microSD card. **This cannot be done with any built–in Windows copy function.** The easiest way to transfer the image using Windows or Linux is to use Balena Etcher.

1. Insert the microSD card into an adapter and insert into the desktop computer or laptop.
2. Ignore any and all Windows error messages about the device being corrupt or needing to be formatted. Simply close those message windows.
3. Start Balena Etcher
4. Click on Select Image and use the folder icon to navigate to the correct image file.
5. Click on Select Target and select the correct device.
6. Double–check the settings.
7. Click Flash! when ready.
8. When done, navigate to the **/boot** directory and create an empty file named **ssh**.

   - Linux: `touch ssh` or `sudo touch ssh` depending upon your user permissions.
   - Windows: Right–click in the **boot** window and select New then Text document and create a file **ssh** with *no* extension. **ssh.txt** will not work.

9. Eject the device from Windows or unmount it from Linux.

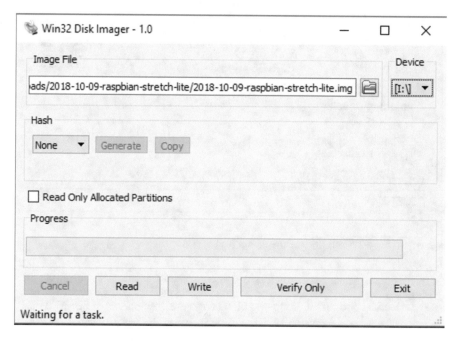

Fig. 8.4: The Win32 Disk Imager Utility

Downloading the Most Current Pi OS

As an alternative to using a custom Pi OS created for you by the instructor, you can download Raspbian and create your own custom image. This is especially useful if you happen to have an HDMI[22] monitor, USB Mouse, and USB keyboard lying around so that you can make use of the Pi Desktop[23].

Simply download the most current version of Raspbian from the web–site and follow the instructions for building the custom image given in Section 8.1. Once a custom image has been created, it can be loaded onto multiple Pi Microcomputers by following the steps above in this section.

8.4 Configuring A New Raspberry Pi File System

The first thing that should be done after booting a new file system on an SD card is to assign a **hostname** [86] and *possibly* expand the file system to use the entire card. **NOTE:** Do not expand the file system if you plan to multi–boot on this SD card or if you cannot afford approximately 32 to 50 gigabits of storage for backups. Backups will consist of exact, full copies of the microSD card which is typically 16 gigabits. It might be wise to have enough off–line storage to keep 3 to 4 copies of the microSD card. For example, you can back up your Pi to a laptop or desktop computer[24].

1. pick a unique **hostname** to identify this Pi.

   a. Confer with your group before you choose a name for your Pi as it *must* be unique within your group. If you have no better idea, use "router$g - n$" where g is your group number and n is the number of the Pi ($n = 1, 2, 3, 4$).

   b. The **hostname** is comprised of uppercase, lowercase letters, and numbers[25]. Avoid spaces and special characters.

   c. Choose a mnemonic name and consider labeling the physical case and/or SD. This is a matter of personal choice.

   d. Avoid generic names that might be related to Internet services such as **www, ns, dns, ftp, mail, or email**.

   e. Later we will find it is easy to assign aliases to the Pi, so the choice of a hostname is not critical as long as it is unique in your group.

   f. This procedure can be used later to change the hostname if needed.

---

[22] High Definition Multimedia Interface

[23] All of the work done on the Pi can be done from the desktop as well as from the command line. This is beyond the scope of this book.

[24] I do not recommend trying to keep multiple microSD cards as version 1, version 2, and so on. They are too small to label easily.

[25] In my opinion, it is best to pick a 6-10 letter name all lowercase. Some installations like to choose names that go with their business such as: crust, cheese, and anchovy for a pizza company.

2. Insert the SD card into the proper slot on the Pi
3. Connect the console cable [57], see Figure 8.5.

> WARNING: Either connect the red lead as shown *or* connect the USB power supply *but not both*! Connecting both could damage the Pi and render it completely useless.
>
> This is known as "bricking" the Pi because a dead Pi is as useless as a brick.

Fig. 8.5: Connecting the Console Cable to the Pi
Note: Your Pi may look different from the one shown, but the console cable
connects the same way to all Pi's.

4. Open **putty** and connect to the Pi using the correct COM port where the serial cable is connected.
5. Many commands require extra permissions and must be preceded by **sudo**.
6. When the login prompt appears, login as **pi** with a password of **raspberry** (both lowercase[26]). Later on you should change this password for security purposes.
7. Set the date and time. If you do not do this each time you login, you will run into file creation date problems and problems with name service.
   sudo date mmddhhmmyyyy
8. Visually verify the date

---

[26] Linux is case sensitive, so many users tend to use all lowercase to keep things simple. Personally, I find this to be a good idea although many use "camel case".

Warning: If the **root** password is lost, the OS will need to be re-installed.

[**Optional**]Normally, it is best not to administer the Pi as root but by using the **sudo** command. However, if the Pi is to be used in a production setting it is best to set the password of **root** at this point.[27]

1. **sudo passwd root**
2. When prompted, enter a password for **root**. Your group should either have a common **root** password or know each others passwords. Later on, you can change this for security if you wish.
3. A forgotten **root** password is a problem. To reset it, you must reload the OS and all your changes will be lost.
4. Change to **root** by entering **su** – ("sudo" space minus sign) and the password for **root**

8.4.1 Raspbian Configuration Utility: raspi-config

When you first log on, you should see a screen similar to:

```
pi@raspberrypi:~ $
Linux raspberrypi 4.19.57-v7+ #1244 SMP Thu Jul 4 18:45:25 BST
    2019 armv7l

The programs included with the Debian GNU/Linux system are
    free software;
the exact distribution terms for each program are described in
    the
individual files in /usr/share/doc/*/copyright.

Debian GNU/Linux comes with ABSOLUTELY NO WARRANTY, to the
    extent
permitted by applicable law.
Last login: Sat Sep  7 18:17:18 2019 from 192.168.1.12

SSH is enabled and the default password for the 'pi' user has
    not been changed.
This is a security risk - please login as the 'pi' user and
    type 'passwd' to set a new password.

pi@raspberrypi:~ $ sudo raspi-config
```

Fig. 8.6: Run **sudo raspi-config**

Expand the file system and set the hostname using **sudo raspi-config** at the command prompt, see Figures 8.6, 8.7, 8.8, and 8.9.

1. Set the date and time *each* time you log on by **sudo date** if it is wrong.

---

[27] It might be best not to change this password for a classroom situation, especially if you are new to Linux.

2. Set the **hostname**.

   a. If the screen is not displayed, restart the program.
   b. Use the arrow keys to move the highlight to "2 Network Options" and press enter.

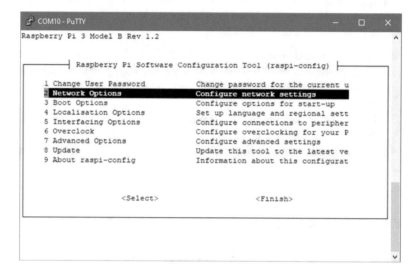

Fig. 8.7: Network Options Menu

   c. Choose option " N1 Hostname Set the visible name for this Pi on a network " and press enter.

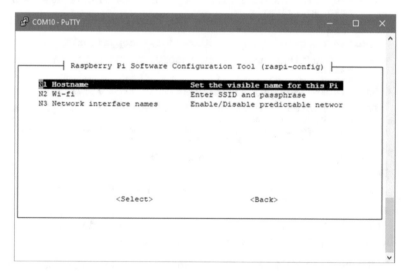

Fig. 8.8: Network Options (to change **hostname**)

d. Read the message from Raspbian and press enter.

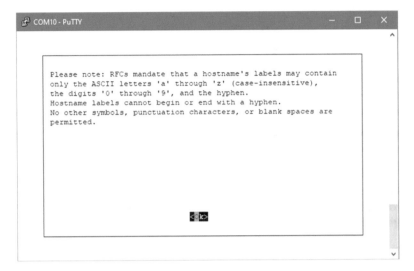

Fig. 8.9: Hostname Rules

e. Backspace to delete the current name and enter the **hostname** without leading or trailing spaces.

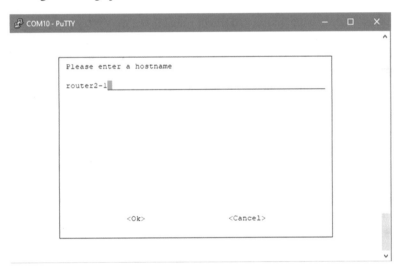

Fig. 8.10: New Hostname

f. Press enter to accept the new name

3. Exit **raspi-config** by moving the highlight to "Finish" and pressing enter.
4. The file system will automatically be expanded to fill the entire SD card with the next reboot.

5. The Raspberry Pi should reboot autmatically. If not, reboot the Pi by entering either:

    a. **sudo init 6**[28]

    b. **sudo reboot**[29]

    c. Alternatively, the system may be halted by entering either **sudo init 0** or **sudo halt**.[30]

8.4.2 Network Interfaces on the Pi

The standard Pi has multiple network interfaces and can be adapted to have additional interfaces if needed. It is this feature which makes the Pi an excellent choice for a router. Because Raspbian is a Linux distribution, the network interfaces are given the standard Linux names.

Loopback or **lo**

All devices that have IP installed must support the Loopback address (127.0.0.1) for hardware and software support. Linux users and manuals may call this the **localhost** which is essentially the same thing. This address points back to the device. Any messages sent to **lo** will travel down the OSI stack to Layer 1 where they will then be sent back up the OSI stack as if the messages had been received from the physical media. This is a critical interface for many reasons. First, if a **ping** is sent to the loopback address (127.0.0.1) and the IP hardware and software are properly installed, the **ping** will be answered. This should be the first step in checking the function of a NIC. If the loopback address does not answer, there is a problem with the installation, or with the hardware NIC, and this must be corrected before there is any hope of the interface working.

    Secondly, this allows Internet services on the device to be tested in isolation from other potential issues. For example, a web server installed on this device should answer if a browser on this device is pointed at the URL[31] **http://127.0.0.1**. Many devices take advantage of the loopback address for configuration purposes and for web support. A programmer might design a program to run on a device using the loopback address and a web server to provide a seamless local and remote interface.

---

[28] This command will work exactly the same way on any UNIX/Linux system.

[29] On some systems, this command will notify all users and then reboot after a system defined delay. On the Pi, there does not seem to be any difference between **init 6** and **reboot**

[30] If the Pi is not gracefully shut down with a **halt** message the file system on the Pi may be corrupted. Fortunately, these problems can *usually* be corrected on the next boot. If you simply remove power from a running Pi enough times the file system will eventually be corrupted to the point that the OS will need to be reloaded and all your data and configurations will be lost.

[31] Universal Resource Locator

Thirdly, the loopback address provides a sneaky place to dump unwanted network traffic. If a remote host is pointed at the loopback address, any messages for that host are sent out but dumped at the loopback instead of being received at some remote host. Certain anti–spyware products have used this to allow installed software to continue to function but not report back to a remote host; thus keeping information private.

Ethernet

The standard Raspberry Pi comes equipped with an Ethernet interface (RJ45 jack) mounted on the board. This interface is always named **eth0** and is the primary Ethernet connection. The Pi also can support multiple USB Ethernet connections by attaching USB/Ethernet dongles. These additional NICs are named **eth1, eth2, eth3** and so on[32].

```
Raspberry Pi Model B Plus Rev 1.2

pi@howserPi1:~$ sudo ifconfig -a
eth0: flags=4099<UP,BROADCAST,MULTICAST>  mtu 1500
        ether b8:27:eb:15:f1:54  txqueuelen 1000   (Ethernet)
        RX packets 0  bytes 0 (0.0 B)
        RX errors 0  dropped 0  overruns 0  frame 0
        TX packets 0  bytes 0 (0.0 B)
        TX errors 0  dropped 0 overruns 0  carrier 0  collisions 0

lo: flags=73<UP,LOOPBACK,RUNNING>  mtu 65536
        inet 127.0.0.1  netmask 255.0.0.0
        inet6 ::1  prefixlen 128  scopeid 0x10<host>
        loop  txqueuelen 1000  (Local Loopback)
        RX packets 6  bytes 278 (278.0 B)
        RX errors 0  dropped 0  overruns 0  frame 0
        TX packets 6  bytes 278 (278.0 B)
        TX errors 0  dropped 0 overruns 0  carrier 0  collisions 0

pi@howserPi1:~$
```

Fig. 8.11: Interface Status Without Any Connections

The Pi will attempt to make all interfaces functional. Ethernet NICs will *not* be fully functional until they are connected to another Ethernet NIC on some other device, see Figures 8.11 and 8.12. Notice the interface now has an IPv4 address that we did not choose[33]. This is the auto–configuration address and is semi–randomly assigned by the NIC to itself. In general, this is not a good outcome but it is easily

---

[32] If the Pi is powered via a Console cable or from a USB port there is a limit to how many USB devices can be supported. If too many are attached, the Pi Console will get a message that an under voltage is detected. In this case, the Pi should be powered using the supplied AC adapter.

[33] An interface running IPv4 is not fully functional, or "up", unless **ifconfig** displays an IP address, a netmask, and an IP Broadcast address. This means the NIC is not only powered and configured correctly, but that it has sensed carrier or activity from another NIC on the LAN.

corrected as we will see in Section 15.4.2. One method to correct this is discussed
in Exercise 1.

```
Raspberry Pi Model B Plus Rev 1.2

pi@howserPi1:~$ sudo ifconfig -a
eth0: flags=4163<UP,BROADCAST,RUNNING,MULTICAST>  mtu 1500
        inet 169.254.209.209  netmask 255.255.0.0  broadcast 169.254.255.255
        inet6 fe80::7500:c8d4:f013:c93b  prefixlen 64  scopeid 0x20<link>
        ether b8:27:eb:15:f1:54  txqueuelen 1000  (Ethernet)
        RX packets 39  bytes 4177 (4.0 KiB)
        RX errors 0  dropped 0  overruns 0  frame 0
        TX packets 43  bytes 6240 (6.0 KiB)
        TX errors 0  dropped 0 overruns 0  carrier 0  collisions 0

lo: flags=73<UP,LOOPBACK,RUNNING>  mtu 65536
        inet 127.0.0.1  netmask 255.0.0.0
        inet6 ::1  prefixlen 128  scopeid 0x10<host>
        loop  txqueuelen 1000  (Local Loopback)
        RX packets 12  bytes 848 (848.0 B)
        RX errors 0  dropped 0  overruns 0  frame 0
        TX packets 12  bytes 848 (848.0 B)
        TX errors 0  dropped 0 overruns 0  carrier 0  collisions 0

pi@howserPi1:~$
```

Fig. 8.12: Interface Status After Connecting **eth0**

Wireless

All newer Pi Microcomputers have an on–board wireless adapter that Raspbian denotes as **wlan0**. As with Ethernet interfaces, the Pi can support multiple wireless
interfaces via the use of USB wireless adapters. As with all OSs, Raspbian and the
Pi hardware do not support all USB adapters but adapters purchased through Ada
Fruit will be supported. If the Pi does not have an on–board wireless adapter, **wlan0**
will not show up.

8.4.3 The Test–bed Network

The network will be built of inter–connected groups of Raspberry Pi microcomputers as shown in Figure 8.13. Each group will be required to apply to the local ISP
for a range of IP addresses and to register a domain name (Exercise 8.9). Each group
should come up with a fake organization that would need to provide information to
the public via a web–site. For example, your group might wish to be an NGO[34] that
provides winter sweaters for small dogs and register a domain name

Note 8.1 Domain names are *not* case sensitive and can contain only the letters A–Z,
a–z, numbers 0–9, and the hyphen. The hyphen cannot be in the first or last position

---

[34] Non-Governmental Organization

of a part of the name and cannot be in the first (hostname) part of the name. Avoid two hyphens together. "coldpuppies.org" (with the instructor's Pi). This would be a good time to start designing a set of web–pages and collecting a few images and/or text files for your web–site.

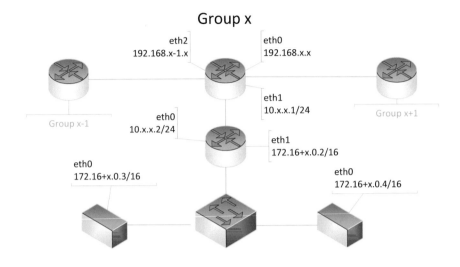

Fig. 8.13: The Group Network Diagram (Ring)

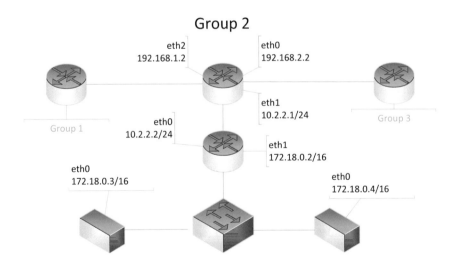

Fig. 8.14: The Group 2 Network Diagram (Ring)

8.4.4 Backing Up the Pi OS

Warning: This method can be used to back up the Pi before dangerous changes are made. This could be critical on a production system if Raspbian is being updated, but normally backups are made to make it possible to return to an earlier configuration of the Pi.

Backups Using Windows and WIN32 Disk Imager

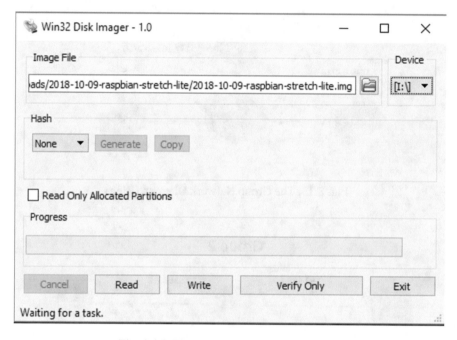

Fig. 8.15: The Win32 Disk Imager Utility

1. Gracefully power down the Pi.
2. Insert the microSD card into an adapter and insert into the desktop computer or laptop.
3. Ignore any and all Windows error messages about the device being corrupt or needing to be formatted. Simply close those message windows.
4. Use the small folder icon to navigate to the correct backup folder and enter a unique image file name.
5. Use the device drop–down box to select the correct device.
6. Double–check the settings.
7. To create an image file from the device, or to back–up the microSD card, click on the "Read" button.

8. Eject the device from Windows.

8.5 Manipulating Configuration Files

Like most modern operating systems Raspbian is controlled by a large number of configuration files. When Raspbian boots it reads these files at various times during initialization[35]. As the files are read Raspbian sets internal flags and sometimes even creates new configuration files to reflect the desired behavior of the OS and therefore the Raspberry Pi.

Many times the configuration files must be manually edited to select various optional actions to be taken by the OS. This is to be expected as Raspbian is based upon the Linux distribution called Debian [28] which is designed to be configured either manually or via the graphic desktop GUI[36]. Either way, it is a good idea to become familiar with directly editing files using the Linux editor **vi** or its more colorful version **vim**[37].

In this chapter the student, or hobbyist, will have the opportunity to gain some familiarity with **vi**. There is another editor included with most Linux distribution called **nano** that is also easy to learn and use. Either editor will do what needs to be done fairly painlessly.

8.6 Creating and Editing a Simple File

It is entirely possible to work on a UNIX/Linux[38] computer without ever opening a terminal window. However, the Raspberry Pi does not have a keyboard, mouse, or display so all operations must be done in a terminal window (also known as "command line"). For the configurations required in this course, a command line editor must be used. The editor of choice is most often **vi**[39][40].

---

[35] Most of these operations are displayed when the Pi is rebooted. There are many, many of these and I doubt many people have any idea what they all do.

[36] Graphical User Interface

[37] For some reason **vim** is not usually a part of Raspbian distributions. This is a very minor loss that could be overcome by installing the package as part of the custom image. This is not really worth the trouble.

[38] For this course, UNIX and Linux can be used interchangeably unless specifically noted.

[39] If **vim** is not a valid command, use **vi** instead as **vim** is an enhanced version with all the same commands.

[40] Even if you use **nano**, you should still learn **vi** as some specialized distributions do not include **nano**.

8.7 Brief Introduction to the **vi** Editor

vi has two distinct modes of operation which leads to two separate sets of commands.

8.7.1 Command Mode

When it first executes, **vi** is in command mode. Part of the text of the file will be shown and can be edited, but command mode is more useful for moving around in the file to make small edits than to make large changes. Making large changes is easiest in edit mode. To return to command mode from edit mode press **<Esc>**.

8.7.2 Edit Mode

Edit mode can be entered by pressing the **(INSERT)** key or **i**. See the above section for other ways to enter edit mode.

There are fewer commands in the edit mode. Edit mode is designed to make large changes by simply typing the text. The cursor can be moved via the arrow keys on the keyboard or by entering command mode by pressing **(esc)**.

8.8 Example: Edit **dummy.config**

It is assumed that you can open a terminal window on the Pi and log in. There are some basic steps that should be used before a configuration file is edited.

Assume the configuration file **dummy.config** must be edited. Before any changes are made, the file should be backed up:

cp dummy.config dummy.config-org

or some similar command should be used. If the configuration file is changed to the point where it is advisable to start over, the command

cp dummy.config-org dummy.config

will restore the file.

If the configuration file does not exist and must be created, this can be done by "touch–ing" the file: **touch dummy.config**.

When all else fails or you get extremely lost using **vi**, issue the following commands:

  (esc) key (places the session in command mode)

    :   sets cursor to the bottom of the screen for a command

    q!   quits the session leaving the file on disk unchanged

Open up a terminal window, connect to the Pi, and log in. Here are some things to try:

1. Change to your home directory "**cd ~**"
2. Verify the current directory "**pwd**" (it should end in **/Pi/home**)
3. Create a dummy configuration file "**touch dummy.config**"
4. Type in the following text **with the proper line breaks**:

```
#This is a comment line in most of the configurations.
#If you remove the ``#'', the line is no longer a comment
.
Name of group:  mygroup name
Domain applied for:  something.com
Another member of my group:  Jane Doe
Some random line of configuration nonsense goes here
#Configuration updated on:  today
```

5. Write the file to the Pi by **(esc):wq**
6. Verify the current path by **pwd**
7. View the files in the current directory by **ls -l**
8. List the configuration to the standard out (screen or terminal window) via either:

- **cat dummy.config**
- **more dummy.config**
- **less dummy.config**

9. You should see something like the screen below.

```
pi@howserPi1:~\$ cat dummy.conf
s is a comment line in most of the configurations.
#If you remove the ``#'', the line is no longer a
   comment.
Name of group: mygroup name
Domain applied for: something.com
Another member of my group: Jane Doe
Some random line of configuration nonsense goes here
#Configuration updated
```

8.9 **vi** Helpful Hints

This is a quick guide to **vi**[41] commands that you will find useful[42].

Invoking **vi**:   **vi** *filename*
Format of **vi** commands: *[count][command]*
                          (count repeats the effect of the command multiple times)

---

[41] Vi text editor

[42] These hints closely follow **ACNS Bulletin ED–03** (February 1995) which can be found on the web. There are many, many **vi** and **vim** cheat sheets on the web.

Command Mode versus Input Mode

vi starts in command mode. The positioning commands operate only while vi is in command mode. You switch vi to input mode by entering any one of several vi input commands. Once in input mode, any character you type is taken to be text and is added to the file. You cannot execute any commands until you exit input mode. To exit input mode, press the escape (**<Esc>**) key.

Input Commands (end with <Esc>)

a   Append after cursor
i   Insert before cursor
o   Open line below
O   Open line above
:**r** *filename* Insert *filename* after the current line
     Any of these commands leaves **vi** in input mode until you press **<Esc>**. Pressing the **<RETURN>**/**<Enter>** key will not take you out of input mode.

Change Commands (input mode only)

cw Change word (Esc)
cc Change line (Esc) - blanks line
c$ Change to end of line
rc Replace character with *c*
R   Replace (Esc) - typeover
s   Substitute (Esc) - 1 char with string
S   Substitute (Esc) - Rest of line with text
.   Repeat last change

Changes During Insert Mode

<ctrl>h Back one character
<ctrl>w Back one word
<ctrl>u Back to beginning of insert

File Management Commands

`:w name`	Write edit buffer to file name
`:wq`	Write to file and quit
`:q!`	Quit without saving changes
`ZZ`	Same as `:wq`
`:sh`	Execute shell commands (`<ctrl>d`)

One of the most important **vim** commands is `:q!` which saves the file *without changes*. If you get confused or make changes you don't want, this command can be a life–saver.

Windows Motions

`<ctrl>d`	Scroll down (half a screen)
`<ctrl>u`	Scroll up (half a screen)
`<ctrl>f`	Page forward
`<ctrl>b`	Page backward
`/string`	Search forward
`?string`	Search backward
`<ctrl>l`	Redraw screen
`<ctrl>g`	Display current line number and file information
`n`	Repeat search
`N`	Repeat search reverse
`G`	Go to last line
`nG`	Go to line *n*
`:n`	Go to line *n*
`z<Enter>`	Reposition window: cursor at top
`z.`	Reposition window: cursor in middle
`z-`	Reposition window: cursor at bottom

Cursor Motions

H	Upper left corner (home)
M	Middle line
L	Lower left corner
h	Back a character
j	Down a line
k	Up a line
— **^**	Beginning of line
$	End of line
l	Forward a character
w	One word forward
b	Back one word
f*c*	Find *c*
;	Repeat find (find next *c*)

Deletion Commands

dd or *n***dd**	Delete *n* lines to general buffer
dw	Delete word to general buffer
d*n***w**	Delete *n* words
d)	Delete to end of sentence
db	Delete previous word
D	Delete to end of line
x	Delete character

Recovering Deletions

p Put general buffer after cursor
P Put general buffer before cursor

Undo Commands

u Undo last change
U Undo all changes on line

Rearrangement Commands

yy or **Y**	Yank (copy) line to general buffer
\z6yy	Yank 6 lines to buffer *z*
yw	Yank word to general buffer
\a9dd	Delete 9 lines to buffer *a*
\A9dd	Delete 9 lines; Append to buffer *A*
\ap	Put text from buffer *a* after cursor
p	Put general buffer after cursor
P	Put general buffer before cursor
J	Join lines

Parameters or Options

:set list	Show invisible characters
:set nolist	Don't show invisible characters
:set number	Show line numbers
:set nonumber	Don't show line numbers
:set autoindent	Indent after carriage return
:set noautoindent	Turn off autoindent
:set showmatch	Show matching sets of parentheses as they are typed
:set noshowmatch	Turn off showmatch
:set showmode	Display mode on last line of screen
:set noshowmode	Turn off showmode
:set all	Show values of all possible parameters

Move text from file *old* to file *new*

vi *old*	
*a*10yy	yank 10 lines to buffer *a*
:w	write work buffer
:e *new*	edit new file
*a*p	put text from *a* after cursor
:30,60w *new*	Write lines 30 to 60 in file new

Regular Expressions (search strings)

^   Matches beginning of line
$ Matches end of line
.   Matches any single character
*   Matches any previous character
.*   Matches any character

Search and Replace Commands

Syntax:

```
:[address]s/old--text/new--text/
```

Address Components

.	Current line
n	Line number *n*
.+*m*	Current line plus *m* lines
$	Last line
/*string*/	A line that contains "*string*"
%	Entire file
[addr1], [addr2]	Specifies a range

Search Examples with the vi Editor

The following example replaces only the **first** occurrence of Banana with Kumquat in each of 11 lines starting with the current line (.) and continuing for the 10 that follow (.+10).

```
:.,.+10s/Banana/Kumquat
```

The following example replaces **every** occurrence (caused by the **g** at the end of the command) of **apple** with **pear**.

```
:%s/apple/pear/g
```

The following example removes the last character from every line in the file. Use it if every line in the file ends with M̂ as the result of a file transfer. Execute it when the cursor is on the first line of the file.

`:%s/.$//`

Projects

1. Testing network interfaces on the Pi

   a. Connect the Pi Microcomputers in your group in pairs by connecting the RJ45 Ethernet ports with a cable and issue the command **sudo ifconfig -a** on the command line of each Pi.
   b. Record the settings for **eth0**. Does the interface have an IP address?
   c. If both interfaces are up, try to **ping** the interface from the same Pi. Does it respond?
   d. Try to **ping** the **eth0** interface on the other Pi. Does it respond?
   e. While the **ping** is running, reboot the other Pi by the command **sudo init 6** or **sudo reboot**. What happens as one Pi reboots? What happens when both Pi Microcomputers are back up and running?
   f. What impact would this have on a network should a Pi reboot for some reason? Is this a serious problem when we build a network of Pi devices?

2. Testing network interfaces on the Pi
3. Connect the Pi Microcomputers in your group in pairs by connecting the RJ45 Ethernet ports with a cable and issue the command **sudo ifconfig -a** on the command line of each Pi.
4. Pick numbers for each of the Pi Microcomputers in your group if you have not already done so. In the following steps, "g" is your group number and "x" is the number of your Pi.
5. Issue the commands:

   a. **sudo ifconfig eth0 down**
   b. **sudo ifconfig eth0 192.168.g.x**
   c. **sudo ifconfig eth0 up**
   d. **sudo ifconfig** or **sudo ifconfig -a**

6. Attempt to **ping** the other Pi Microcomputers in your group. What are the results for each and *why*?
7. Reboot your Pi. How does the result of **sudo ifconfig eth0** compare to what you had before?
8. Can you **ping** the other Pi Microcomputers? Why or why not?
9. What are some possible advantages and disadvantages of the behavior you observed?

Exercises

Fill in a form with the information below and turn it into your instructor to register
your group domain name. This domain name can be changed later by the instructor.

Table 8.1: Domain Registration Form

Domain Registration Form

Group number:

Requested domain name:

Information to be found on this domain's website:

Names of People in Group: •

Chapter 9
The Laboratory Network

Overview

It is important to have a consistent physical network for all of the protocols examined in this text for three reasons:

1. Device addresses are somewhat dependent upon the topology of the physical network. If the physical network changes for each set of protocols, the addressing might need to be changed which would lead to errors in the configuration of some of the services on the network. If the physical topology does not change, these services should be located at predicable addresses. This should lead to a significantly shorter time to implement protocols and services. In short, keeping the topology consistent should prevent unnecessary changes to the network.
2. Improvements in network services and resilience can attributed to more efficient operations if the physical network remains the same.
3. With a very few exceptions other than routing, protocols and services are independent of the underlying physical networks. Protocols and services depend only upon connectivity.

All of the protocols and services implemented on the Raspberry Pi's will be presented as two options: a ring topology or a star topology. Bear in mind that the physical topology of the Laboratory Network is immaterial with the exception of the total number of interfaces that must be supported by each group. Each group must support *one* additional interface on Pi $g.1$ for a ring topology which requires an additional USB/Ethernet dongle. It is important that Pi $g.1$ is powered by a one–piece power supply rather than a USB cable connected to a laptop USB port[1]. If this might be an issue, the star topology network should be used instead of the ring.

---

[1] The two piece power supplies that consist of a cable and converter should be avoided as well. Either might work but there is a good possibility that Raspbian will report an under–voltage situation. This could shorten the lifetime of the Pi. One of the Pi's used to test the lab networks reports under–voltages constantly. This is annoying but it has not failed in the first year of operation.

© Springer Nature Switzerland AG 2020
G. Howser, *Computer Networks and the Internet*,
https://doi.org/10.1007/978-3-030-34496-2_9

In all diagrams, figures, and tables the various Pi's will be designated as *g.n* where *g* is the group and *n* is Pi 1, 2, 3, or 5. For example, the Pi that connects Group 3 to the other groups would be Pi 3.1 in all figures.

9.1 IPv4 Ring Network Backbone

Ring Laboratory Network Backbone

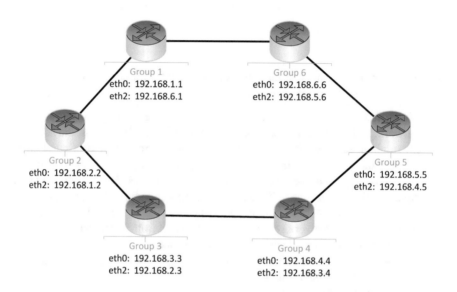

Fig. 9.1: Ring Topology Backbone

The backbone of the Ring Network requires an additional interface, **eth2** on Pi *g*.1[2], to build a ring topology as in Figure 9.1. In reality, the complete network will have two rings for redundancy, but the main topology is a ring of each group's Pi *g*.1. In order to build a ring of groups, each Pi *g*.1 will require an additional USB/Ethernet dongle and cable in order to connect to the group to the "left" and the group to the "right". While the ring version of the Laboratory Network does take additional USB/Ethernet dongles, it is more interesting than the Star Network because of the additional subnetworks involved and it presents more failure possibilities to explore.

---

[2] If there are more than 14 groups, it might be a good idea to use two digit group numbers instead of one digit. This is not a problem with the suggested IPv4 numbering, but it can cause a problem with the suggested IPv6 numbering.

Ring topologies can be used to connect a number of autonomous networks, such as large ISPs, or for *compact* backbone networks[3]. Some of the technologies that use rings include FDDI, Token Ring, and extremely high–speed SONET. In fact many Telcos lease bandwidth carried over fiber optic SONET rings extending over large areas.

For example, in the following examples we will look at the addressing and connections for Group 2 out of five groups[4].

Table 9.1: Group Equipment for a Ring Lab Network

Quantity	Description
8	Ethernet Cables
6	USB/Ethernet Dongles
4	Raspberry Pi's
1	Ethernet Switch or Hub

9.1.1 Ring IPv4 With a Group of Four Pi's

Table 9.2: Ring IPv4 for a Group g with Four Pi's

Pi (g.n)	Interface	IPv4 Address	Connects to Pi	Interface
g.1	eth0	$192.168.g.g/24$	g+1.1	eth1
g.1	eth1	$10.g.0.1/16$	g.2	eth0
g.1	eth2	$192.168.g-1.g/24$	g-1.1	eth1
g.2	eth0	$10.g.0.2/16$	g.1	eth1
g.2	eth1	$10.g.g.2/24$	**Group g switch**	N/A
g.2	eth2	$10.g+1.g+1.g*10/24$	**Group g+1 switch**	N/A
g.3	eth0	$10.g.g.3/24$	**Group g switch**	N/A
g.3	eth1	$172.g+16.0.3/16$	g.4	eth1
g.4	eth0	$10.g.g.4/24$	**Group g switch**	N/A
g.4	eth1	$172.g+16.0.4/16$	g.2	eth1

---

[3] If the backbone requires WAN connections, the cost becomes a concern due to the cost of extra links to complete the ring.

[4] If you are still not comfortable with IPv4 addressing, see Section 5.5

Table 9.3: Ring IPv4 For Group 2 with Four Pi's

Pi (g.n)	Interface	IPv4 Address	Connects to Pi	Interface
2.1	eth0	192.168.2.2/24	3.1	eth1
2.1	eth1	10.2.0.1/16	2.2	eth0
2.1	eth2	192.168.1.2/24	1.1	eth1
2.2	eth0	10.2.0.2/16	2.1	eth1
2.2	eth1	10.2.2.2/24	**Group 2 switch**	N/A
2.2	eth2	10.3.3.20/24	**Group 3 switch**	N/A
2.3	eth0	10.2.2.3/24	**Group 2 switch**	N/A
2.3	eth1	172.18.0.3/16	2.4	eth1
2.4	eth0	10.2.2.4/24	**Group 2 switch**	N/A
2.4	eth1	172.18.0.4/16	2.2	eth1

Ring Groups of Four

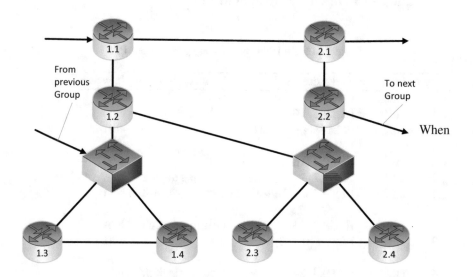

Fig. 9.2: Ring Two Groups of Four

two groups of four Pi's interconnect, the connections are made as shown in Table 9.2, Table 9.3, and and Figure 9.2. Notice that quite a few subnetworks are created and the topology of the Group network is rather complicated. This is done to show the capabilities of routers to build and operate complex networks without much input from the Network Administrator. This network will be used to demonstrate the ability of routing protocols to "learn" the network and how these protocols can easily handle networks that would present problems if humans had to configure all the possible characteristics of even small Layer 3 networks.

9.1.2 Ring IPv4 With a Group of Three Pi's

Table 9.4: Ring IPv4 For Group 2 with Three Pi's

Pi (g.n)	Interface	IPv4 Address	Connects to Pi	Interface
2.1	eth0	192.168.2.2/24	3.1	eth1
2.1	eth1	10.2.2.1/24	2.2	eth0
2.1	eth2	192.168.1.2/24	1.1	eth1
2.2	eth0	10.2.2.3/24	2.1	eth1
2.2	eth1	172.18.0.3/16	2.3	eth1
2.3	eth0	10.2.2.4/24	**Group 2 switch**	N/A
2.3	eth1	172.18.0.4/16	2.2	eth1
2.3	eth2	10.3.3.20/24	**Group 3 switch**	N/A

Ring Group of Four and of Three

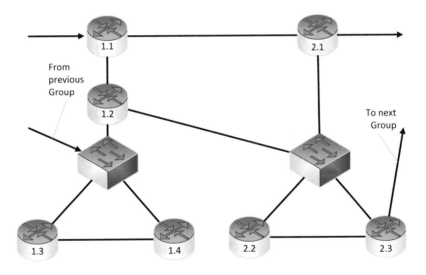

Fig. 9.3: Ring Group of Four Connected to a Group of Three

Table 9.4 and Figure 9.3 give the details to connect a Group with 4 Pi's to a Group with 3 Pi's. Most of the Group's interior networks are still created, but not all.

9.1.3 Ring IPv4 With a Group of Two Pi's

Table 9.5: Ring IPv4 For Group 2 with Two Pi's

Pi (g.n)	Interface	IPv4 Address	Connects to Pi	Interface
2.1	eth0	192.168.2.2/24	3.1	eth1
2.1	eth1	10.2.2.1/24	**Group 2 switch**	N/A
2.1	eth2	192.168.1.2/24	1.1	eth1
2.2	eth0	10.2.2.3/24	**Group 2 switch**	N/A
2.2	eth1	10.3.3.20/24	**Group 3 switch**	N/A

Ring Group of Four and of Two

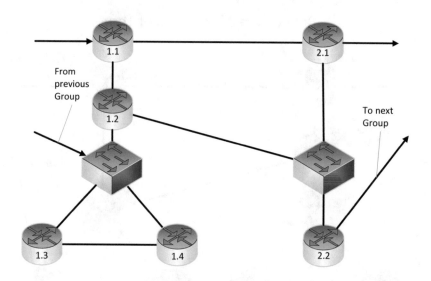

Fig. 9.4: Ring Group of Four Connected to a Group of Two

Table 9.5 and Figure 9.4 show a Group with 2 Pi's connected to a Group with 4 Pi's. While this is a completely legitimate choice, it might be better to move one Pi to the smaller group and have two groups of three. Either way will work. It is even possible to have a Group with only one Pi running all the services, but that is not very instructive.

9.2 IPv4 Star Network Backbone

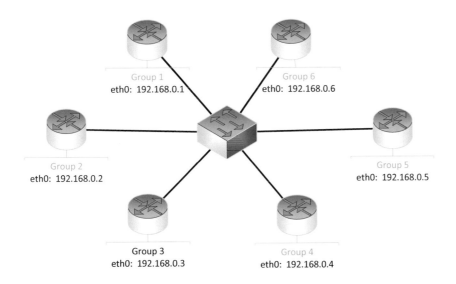

Star Laboratory Network Backbone

Group 1
eth0: 192.168.0.1

Group 6
eth0: 192.168.0.6

Group 2
eth0: 192.168.0.2

Group 5
eth0: 192.168.0.5

Group 3
eth0: 192.168.0.3

Group 4
eth0: 192.168.0.4

Fig. 9.5: Star Topology Backbone

Table 9.6: Group Equipment for a Star Lab Network

Quantity	Description
7	Ethernet Cables
5	USB/Ethernet Dongles
4	Raspberry Pi's
1	Ethernet Switch or Hub

The backbone of the Star[5] Network is built as a virtual network[6] inside an Ethernet switch as in Figure 9.5.

In addition to the equipment required for each group (see Table 9.6) the center of the star topology requires an Ethernet switch with at least one port per group. If there are more groups than available switch ports, additional switches will be needed and they must be concatenated to appear as a single switch for the network to function

---

[5] This topology is sometime called "hub and spoke" but "star" seems to be more common.

[6] This network is virtual in the sense that it exists *only* as a logical network built by the configuration of the component devices rather than by the physical wiring. This is actually a VLAN and has no existence if the switch is powered off.

properly. It is relatively easy to replace the center of the star with a network, but this is beyond the scope of this book[7].

For example, in the following examples we will look at the addressing and connections for Group 2 out of five groups[8].

9.2.1 Star IPv4 With a Group of Four Pi's

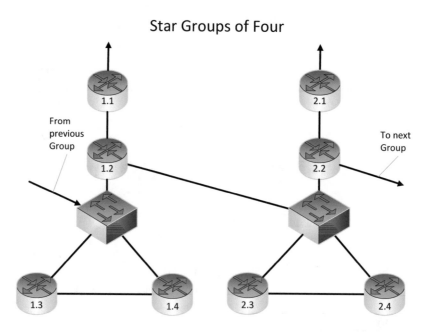

Fig. 9.6: Star Two Groups of Four

---

[7] For example, multiple switches could be interconnected with repeaters. With a change to the IP addressing of the switch networks, routers could be used to connect the switches instead.

[8] If you are still not comfortable with IPv4 addressing, see Section 5.5

Table 9.7: Star IPv4 For Group g with Four Pi's

Pi (g.n)	Interface	IPv4 Address	Connects to Pi	Interface
$g.1$	eth0	192.168.0.g/24	**Star switch**	N/A
$g.1$	eth1	10.g.0.1/16	$g.2$	eth0
$g.2$	eth0	10.g.0.g/16	$g.1$	eth1
$g.2$	eth1	10.g.g.2/24	**Group g switch**	N/A
$g.2$	eth2	10.$g+1.g+1$.20/24	**Group g+1 switch**	N/A
$g.3$	eth0	10.g.g.3/24	**Group g switch**	N/A
$g.3$	eth1	172.$g+16$.0.3/16	$g.4$	eth1
$g.4$	eth0	10.g.g.4/24	**Group g switch**	N/A
$g.4$	eth1	172.$g+16$.0.4/16	$g.2$	eth1

Table 9.8: Star IPv4 For Group 2 with Four Pi's

Pi (g.n)	Interface	IPv4 Address	Connects to Pi	Interface
2.1	eth0	192.168.0.2/24	**Star switch**	N/A
2.1	eth1	10.2.0.1/16	2.2	eth0
2.2	eth0	10.2.0.2/16	2.1	eth1
2.2	eth1	10.2.2.2/24	**Group 2 switch**	N/A
2.2	eth2	10.3.3.20/24	**Group 3 switch**	N/A
2.3	eth0	10.2.2.3/24	**Group 2 switch**	N/A
2.3	eth1	172.18.0.3/16	2.4	eth1
2.4	eth0	10.2.2.4/24	**Group 2 switch**	N/A
2.4	eth1	172.18.0.4/16	2.2	eth1

The connections are made using Ethernet cables in the manner of Table 9.7 to create the partial network shown in Figure 9.6. Notice that each group creates networks 10.g.0.0/24 and 172.16 + g.0.0/16. Groups 1 and 5 will interconnect in a similar manner to form a ring of networks involving each groups switches for redundancy and fault tolerance as we will see in the Projects for this chapter.

9.2.2 Star IPv4 With a Group of Three Pi's

Table 9.9: Star IPv4 For Group 2 with Three Pi's

Pi (g.n)	Interface	IPv4 Address	Connects to Pi	Interface
2.1	eth0	192.168.0.2/24	**Star switch**	N/A
2.1	eth1	10.2.2.1/24	**Group 2 switch**	N/A
2.1	eth2	10.3.3.20/24	**Group 3 switch**	N/A
2.2	eth0	10.2.2.2/24	**Group 2 switch**	N/A
2.2	eth1	172.18.0.1/16	2.4	eth1
2.3	eth0	10.2.2.3/24	**Group 2 switch**	N/A
2.3	eth1	172.18.0.3/16	2.2	eth1

Star Group of Four and of Three

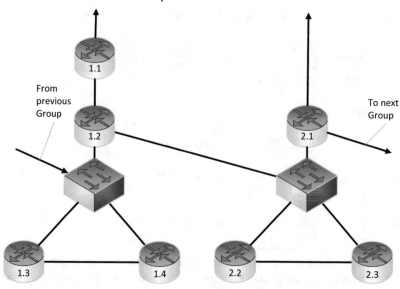

Fig. 9.7: Star Group of Four Connected to a Group of Three

The connections made as in Table 9.2.2 above should be used if Group 2 only has three Raspberry Pi's. Notice, the same networks are created, with the exception of network 10.g.0.0/16, and the inter–group connections are the same as for Group 2 with four Pi's.

9.2.3 Star IPv4 With a Group of Two Pi's

Table 9.10: Star IPv4 For Group 2 with Two Pi's

Pi (g.n)	Interface	IPv4 Address	Connects to Pi	Interface
2.1	eth0	192.168.0.2/24	**Star switch**	N/A
2.1	eth1	10.2.2.1/24	**Group 2 switch**	N/A
2.1	eth2	10.3.3.20/24	**Group 3 switch**	N/A
2.2	eth0	10.2.2.2/24	**Group 2 switch**	N/A

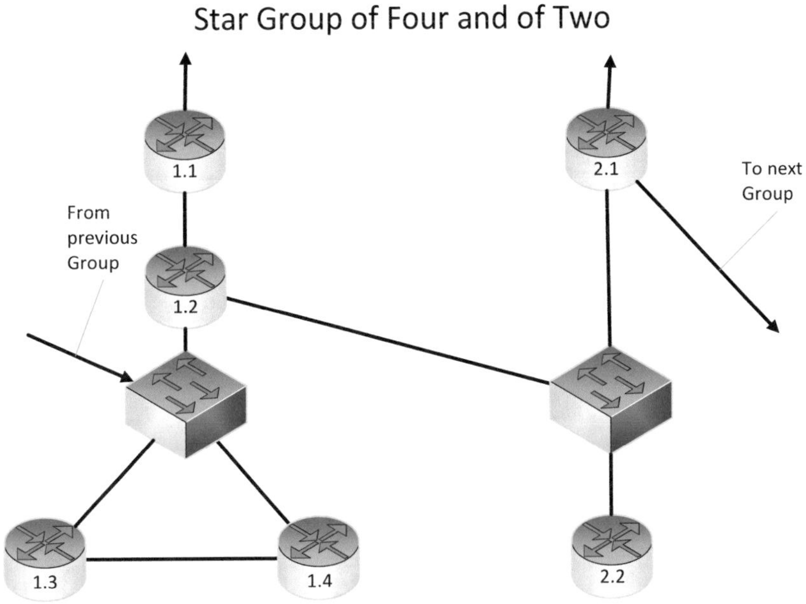

Fig. 9.8: Star Group of Four Connected to a Group of Two

The connections made as in Table 9.10 above should be used if Group 2 has only two Raspberry Pi's as in Figure 9.8. The only new network created is $10.g.g.0/16$, and the inter–group connections are the same as for Group 2 with four Pi's[9].

9.2.4 Star IPv4 With a Group of One Pi

If a group has only one Raspberry Pi, the group is wired exactly like Table 9.10 with Pi 2.2 deleted completely from the network. As with the Ring Laboratory Network,

---

[9] As noted for the Ring Laboratory Network, my personal opinion is that it would be better to make the 4-2 groups into two groups of three Pi's.

this involves running all the services for the Group on a single Pi which is not as instructive.

9.3 IPv6 Addressing for the Laboratory Network

Table 9.11: Private IPv6 Network Part (64 bits)

7 bits	1	40 bits	16 bits	64 bits
Prefix	L	Global ID	Subnet ID	Interface ID
fc00::/7	1	Random		
fd		$869b29e5e1^a$	$g000^b$	$::gni^c$

a – This is a randomly generated number. Do not use this one, generate your own.
b – Subnetwork chosen by the Group where g is the Group number in hex. The other digits are chosen by the group as needed.
c – Static host part where g is the Group number in hex, n is the Pi number (1, 2, 3, or 4), and i is the interface number (0 is **eth0**, 1 is **eth1**, and so on.)

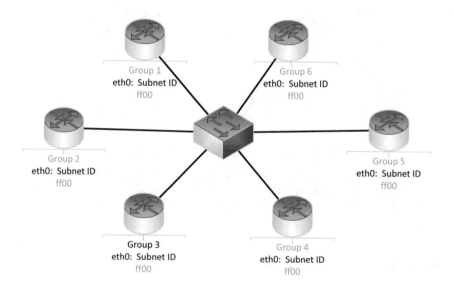

Fig. 9.9: IPv6 Subnet IDs for the Star Lab backbone

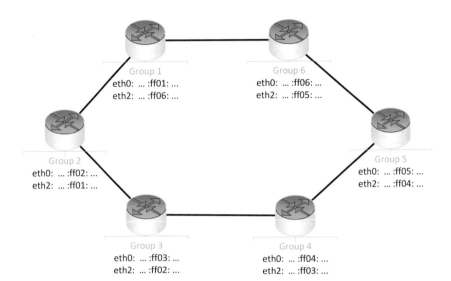

Fig. 9.10: IPv6 Subnet IDs for the Ring Lab backbone

The Raspberry Pi computers can be connected in either a star as in Figure 9.9 or a ring as in Figure 9.10. The entire network will share a single prefix and global ID while each group will have its own group Subnet ID space.

The address space in IPv6 is so large that it is expected that every device can have an assigned public address and there should be no need for any private addresses; but for the Laboratory Network private IPv6 addressing will be used. This is done so that if the Laboratory Network should accidentally get connected to the Internet the first public router encountered will send the traffic to a black hole network. To do this, a unique private network part needs to be generated following the guidelines of RFC 4193, Section 3.2 to minimize the probability of a non-unique network [233]. This involves some work or a web site, https://simpledns.com/private-ipv6, will do the heavy lifting for us [311]. Please do not use 869b29e5e1, as it is used in all IPv6 examples in this text.

9.3.1 IPv6 Laboratory Network Part

Table 9.12: Lab Network IPv6 Addresses

Network Part	
backbone	fd86:9b29:e5e1:ff00::/64
Group 1	fd86:9b29:e5e1:1000::/64
Group 2	fd86:9b29:e5e1:2000::/64
Group 3	fd86:9b29:e5e1:3000::/64
Group 4	fd86:9b29:e5e1:4000::/64
Group 5	fd86:9b29:e5e1:5000::/64
Group 6	fd86:9b29:e5e1:6000::/64

Table 9.13: Star backbone IPv6 Network Prefix

Pi #1	Interface	IPv6 Address
Group 1, Pi #1	eth0	fd86:9b29:e5e1:ff00::110/64
	eth1	fd86:9b29:e5e1:1xxx::111/64
Group 2, Pi #1	eth0	fd86:9b29:e5e1:ff00::210/64
	eth1	fd86:9b29:e5e1:2xxx::211/64
Group 3, Pi #1	eth0	fd86:9b29:e5e1:ff00::310/64
	eth1	fd86:9b29:e5e1:3xxx::311/64
Group 4, Pi #1	eth0	fd86:9b29:e5e1:ff00::410/64
	eth1	fd86:9b29:e5e1:4xxx::411/64
Group 5, Pi #1	eth0	fd86:9b29:e5e1:ff00::510/64
	eth1	fd86:9b29:e5e1:5xxx::511/64
Group 6, Pi #1	eth0	fd86:9b29:e5e1:ff00::610/64
	eth1	fd86:9b29:e5e1:6xxx::611/64

Table 9.14: Ring backbone IPv6 Network Prefix

Pi #1	Interface	IPv6 Address
Group 1, Pi #1	eth0	fd86:9b29:e5e1:ff01::110/64
	eth1	fd86:9b29:e5e1:1xxx::111/64
	eth2	fd86:9b29:e5e1:ff06::112/64
Group 2, Pi #1	eth0	fd86:9b29:e5e1:ff02::210/64
	eth1	fd86:9b29:e5e1:2xxx::211/64
	eth2	fd86:9b29:e5e1:ff01::212/64
Group 3, Pi #1	eth0	fd86:9b29:e5e1:ff03::310/64
	eth1	fd86:9b29:e5e1:3xxx::311/64
	eth2	fd86:9b29:e5e1:ff02::312/64
Group 4, Pi #1	eth0	fd86:9b29:e5e1:ff04::410/64
	eth1	fd86:9b29:e5e1:4xxx::411/64
	eth2	fd86:9b29:e5e1:ff03::412/64
Group 5, Pi #1	eth0	fd86:9b29:e5e1:ff05::510/64
	eth1	fd86:9b29:e5e1:5xxx::511/64
	eth2	fd86:9b29:e5e1:ff04::512/64
Group 6, Pi #1	eth0	fd86:9b29:e5e1:ff06::610/64
	eth1	fd86:9b29:e5e1:6xxx::611/64
	eth2	fd86:9b29:e5e1:ff05::612/64

The only critical concern when assigning the Subnet ID is future route summarization. Routing is more efficient the more the Subnet IDs can be summarized.

9.3.2 IPv6 Subnetting the backbone

The backbone of the Laboratory Network will have different addressing for a star configuration and a ring configuration as was done for IPv4 as well. For the star configuration the network part of the IPv6 address for all backbone interfaces will use a Subnet ID of ff00. For the example network using a Global ID of 869b29e5e1, the network part would be fd86:9b29:e5e1:ff00::/64 as in Table 9.12 and Table 9.13.

For a ring configuration the addressing is more complicated, but also more informative. Each of pair of backbone Pi computers is connected by a separate subnetwork to allow for routing to work while still allowing for route summarization on the backbone. This is done in case the Laboratory Network eventually connects to another Laboratory Network via BGP[10]. In that case each network would have its own Global ID and summarize to that ID. The two networks would then only need to exchange two routes (one summary route per network) in order for all subnetworks to be reachable from any subnetwork.

9.3.3 IPv6 Group Subnet IDs

In Table 9.12 each Group is assigned a Subnet ID of the form $g000$ which in this case is really a Subnet ID range. In order to have the proper subnetworks for routing IPv6, each group will need to assign Subnet IDs within the Group's range. If this is done correctly, some routing protocols will be able to summarize these subnetwork routes into a single route for the entire group. On the surface this does not seem very important, but think in terms of all the subnetworks that exist in a single large ISP or the whole Internet. Route summarization not only shrinks the size of the route table, but the size of the route table directly determines the speed at which a packet can be routed. The speed gained by route summarization on a large internet can be enormous.

Many lessons were learned from IPv4 and applied to IPv6. For example, IPv6 addresses are designed and allocated to improve route summarization. Another improvement was in automatically assigning host addresses. That leaves the Group with only the issue of assigning Subnet IDs to the various networks the Group creates. This is left as a project for the Group.

---

[10] Border Gateway Protocol

Projects

These are suggested Projects for this chapter but you are encouraged to explore any possibility that comes to mind. If time permits, do each project for both IPv4 and then IPv6.

Project 9.1 Build the appropriate Ring or Star network for the participating Groups[11].

Project 9.2 Find the IP addresses of the active interfaces on your Pi by entering **`sudo ifconfig -a`**. Can you find the addresses?

Project 9.3 Attempt to ping both the IPv4 and IPv6 loop–back address of each active interface, other than **`wlan0`**.

Project 9.4 Attempt to ping as many other interfaces as you can. Record what you find. Were you able to ping most of the interfaces or only a few? Why?

Project 9.5 Start a number of successful pings

> 9.1 pick a random connection (cable) in your group. Predict what will happen to a running pings if you disconnect that cable.
>
> 9.2 Disconnect the cable and observe the results. Were your predictions correct? If not, what actually happened? Why?
>
> 9.3 What will happen to your pings if another group disconnects one of their cables?

Project 9.6 Try to connect with another Pi via **`ssh`** or **`telnet`**.

Project 9.7 If the Ring network is used, predict what will happen if two cables connecting Pi#1s are removed. Were your predictions correct?

Project 9.8 If the Star network is used, predict what will happen if the center switch is turned off. Were your predictions correct?

Exercises

9.1 Draw a map of your Group network showing the following:

- Group number
- Clearly label each Pi as Pi #1, Pi #2, and so on.
- Clearly label each interface as eth0, eth1, and so on.
- Make a table of the IPv4 and IPv6 addresses of all the connected interfaces on each Pi.

9.2 Does the IPv4 addressing of Group 1 impact the IPv4 addressing of Group 2? Support your answer.

---

[11] For the hobbyist, I suggest using as many groups of at least two Pi's as you can. You might want one of the Groups to have at least 3 Pi's for later Projects such as Domain Name Service.

9.3 Show that the summarization in Figure 5.9 is correct.

9.4 Why was the backbone assigned Subnet IDs starting with f instead of 0?

9.5 What is your group number in decimal, hex, and binary.

9.6 Could you subnet your group to summarize to $gg00$ and still be able to route packets to the other groups?

9.7 Does subnetting change the IPv6 loop–back address of a Pi? Does it change the IPv4 address of your Pi?

9.8 Why are no host parts given for the IPv6 Laboratory Networks?

9.9 Can you give the same host number to an interface in both IPv4 and IPv6?

Part II
The Router

Overview

Two roads diverged in a wood, and I–
I took the one less traveled by,
And that has made all the difference."

Robert Frost

The Internet is built by connecting large numbers of small networks together and correctly moving information (packets) from device to device. In order to do this effectively, these networks are connected using devices known as routers. Simply put, a router connects to two or more Layer 3 networks and insures that packets move from network to network correctly. If networks were simple, or at least did not change, this would be a fairly easy thing to do.

The complexity of the Internet is beyond the ability of most of us to work with correctly and changes much too quickly for anyone to be able to keep up. Sites and machines go on the network and disappear from the network at an alarming speed. Entire networks connect to the Internet frequently and just as frequently change or completely disappear. The dynamic nature of the Internet insures that humans trying to configure devices to reflect the true status of the Internet are bound to fail. Fortunately, there are three main devices to handle these problems for us: IP Forwarders, Layer 3 Switches, and Routers.

It is important to remember that while routers are the main building blocks of the Internet and to a large extent determine the topology of the Internet, the main function of a router is to move packets from one network to another. It is easy to get caught up in the details of how routers "learn" the network and forget that this is a secondary function. Routers move packets at high speed and limit the scope of Layer 2 broadcasts.

IP Forwarders

IP Forwarders are rarely used on the Internet because they are too limited in their operations. These devices have multiple interfaces but can *only* move packets between those interfaces. Packets received on a given interface are quickly sent out a specified interface and here lies the problems. IP Forwarders are configured when installed to reflect the local networks connected to the device *and cannot automatically adjust to changes in those networks or any networks attached to them.* Even if the Internet were static in nature, IP Forwarders would be of limited use except as SOHO[12] routers and we will not expend much effort on them. However, most home networks are connected to the Internet by an IP Forwarder.

Routers and Layer 3 Switches

Unlike IP Forwarders, routers and Layer 3 switches can dynamically update their view of the Internet. While routers and Layer 3 switches have completely different hardware philosophies, from the NIC on out they appear to be the same. For our purposes, we will only make a distinction between the two when absolutely necessary and will refer to both as "routers."

Using Routers to Build Large Interconnected Networks (Internets)

Because routers are used to connect Layer 3 networks built upon Layer 2 networks, most of the work of actually building the Internet consists of properly configuring routers. All of the services we have come to know as the Internet (or World Wide Web, which is not exactly the same thing) are implemented as messages passed back and forth as Layer 3 packets. This means that insuring packets are transferred quickly and correctly results in most of the effort to connect a small network to the Internet at large. Indeed, routers of all kinds are the backbone of the Internet and determine the logical topology of the Internet.

---

[12] Small Office/Home Office

Chapter 10
Routing

Overview

The critical function of Layer 3 networks is to somehow find a way to get messages, or more correctly message fragments, from the local device to the target device regardless of where that target is located. Ideally this needs to be done without the device being aware of exactly where geographically the target is located and without a network administrator being charged with locating in "cyber space" all conceivable destinations or targets[1].

Fortunately, from the network viewpoint all routers are simply routers. IP Forwarders, which include most home routers, are somewhat different in that they cannot "learn" network changes on the home, or private network, side and do not need to know network changes on the Internet Service Provider side at all. The term "router" will be used for both true routers and Layer 3 switches unless a distinction must be made.

10.1 Introduction to Routing

To an extent, both end–points of a data conversation must be identified by Layer 3 address before messages can be exchanged, but the extent as to how much must be known about how to get a message from one end–point to the other depends upon the type of connection being made. For a connection–oriented conversation both end–points and a path between them must be identified in advance, but this is not true for a connectionless conversation. Both have their strong points and weaknesses.

---

[1] The author worked with a mainframe that required minimal prior information (name and IP address) of all devices on the Internet that it was expected to be able to communicate with. It was such a nightmare that a separate UNIX minicomputer was purchased just for email. Imagine being forced to look up every device and reconfigure the network *before* you could visit it with your web browser. Now extend that to each and every email server and other devices you do not even know you are contacting. It does not work.

© Springer Nature Switzerland AG 2020
G. Howser, *Computer Networks and the Internet*,
https://doi.org/10.1007/978-3-030-34496-2_10

10.1.1 Connection Oriented Conversation or Call Setup

Connection oriented conversations require that a path through the Layer 3 networks be found and set up end–to–end *before* any messages can be exchanged. typically one end (the caller) initiates the call setup and the other end simply accepts or denies the call. This is how the regular telephone system works and this type of connection has some important advantages. Data cannot flow unless a connection has been established which means no data can be lost during the connection process. Many networking protocols that use connection oriented conversations, such as ATM, also allow the caller to specify the minimum delay, bandwidth, and other characteristics of the call must be met or the connection is not completed. A good analogy is that of a freight train with a single delivery in multiple boxes in different cars. The boxes might be damaged or lost, but they arrive in the same order as the cars they were packed into. Packets all take the same route with connection oriented conversations and cannot get out of order. To an extent, the delay and other characteristics of the arrival of packets can be controlled.

The problem with connection oriented conversations is they cannot react to changes in the network and the Internet changes from millisecond to millisecond. The best route when the call is set up may not stay the best route. If one segment of the connection drops out for even a single packet, the entire connection is wiped out and the call must be rebuilt before more data can move. Call setup can take quite a bit of time in terms of the network, and data flow is delayed until the call set up is completed. For many data conversations this is unnecessary and maybe even unacceptable. Native IP does not work this way and routers do not normally work this way[2].

10.1.2 Connectionless Transport or Packet Switching

On the Internet most data conversations are connectionless. The sending end–point need only have the IP address of the target device and minimal information about the sender's local Layer 3 network to send data. If the destination is on the same Layer 3 network sending is trivial; however, if the destination is on any other Layer 3 network, the sender need not have any knowledge at all of how to find the destination. It simply sends the packets to the default gateway, or local router interface, and the router handles things from there.

With connectionless conversations packets are sent from router to router by the best path possible at the time using a simple, greedy algorithm: choose the best path to the next router that claims to know how to reach the destination network. If there is no next router that claims to know how to reach the destination network, the packet

---

[2] The most common exception I know of is MPLS[3] which is a transparent way to use IP over ATM to take advantage of the best characteristics of ATM call setup and IP ease of reacting to network changes.

is either sent to the router's default gateway ("network of last resort") or dropped as undeliverable. As the network changes, the new best next steps are learned and packets sent accordingly. This greedy choice, pick the router that claims to have to shortest path to the destination, means that routers need not know any details except those of the local network. This is why Layer 3 networks, such as the Internet, are drawn as clouds. The details are not known and really don't matter. As long as the packets go into the cloud and drop out at the right place everything works.

A good analogy, which we have used before, is sending the chapters of a book each in a separate FedEx package and each package in a different truck. Each time a truck comes to an intersection, literally a packet arriving at a different router, the driver takes the best street to the next intersection, or a packet sent to the current best next router. The best choice changes as the traffic changes or streets are opened or closed. The packages may take different routes and may arrive out of order or be lost completely. The truck drivers, and unfortunately the author who sent the chapters, do not know if the packages all arrive; they do not know if their packaged arrives in the proper order; they do not know if all the packages took the best route overall; and *they do not care!*

With connectionless conversations, reconstructing the message requires waiting until all the packets arrive and can be placed into correct order. It takes longer to determine if a packet is lost or delayed. It is possible to require more resources on each end–point than with connection oriented conversations and dedicating resources to a conversation is extremely difficult. Indeed, IP networks were not originally designed with the idea of dedicating bandwidth which is crucial to audio and visual data.

The overriding advantage is clear: connectionless conversations are able to react to changes in the network. The loads on the Internet change constantly and quickly.

10.2 Network Requirements

In order to provide connectionless conversations, the Internet must be built to meet certain requirements. Some device, a router, must be designed to quickly examine a packet, determine somehow the best next "hop" to reach the destination based upon current network status, and quickly send the packet along. This means that routers, with certain exceptions, must be able to report changes in the network to other routers and react quickly to changes reported by other routers. An assumed requirement is that packets must have a two–part destination address which is met by only allowing IP addressing on the Internet. Other protocols, such as AppleTalk and IPX, could be allowed but it requires fewer resources to hide other protocols inside the IP packet. Routing on the Internet is complicated enough without forcing all routers to maintain separate best next hops for each protocol[4].

---

[4] This is the case with an Intranet. Intranets are not as complex as the Internet and so it does not over–burden the routers to route multiple protocols. It still could take more resources and typically Intranets only allow IP for all the same reasons as the Internet bars all non–IP traffic.

How these requirements are met and enforced will be covered in detail in Chapter 12.

Exercises

10.1 Would you expect each of the following to be connection–less or connection oriented? Support your answers as some of these could be considered either one.

    a. Streaming a movie on your phone.

    b. Contacting a website.

    c. Calling your Aunt Bea.

    d. In the *The Return of the King* [314], the kingdom of Gondor had people located on a string of mountain peaks between Gondor and Rohan such that if a signal fire was lit on a mountain peak, the people on the next peak could see it and light their signal fire and so on until the kingdom of Rohan could see the signal. Gondor lighting signal fires to request help from Rohan.

    e. Sending a telegram.

    f. Registered mail.

    g. Listening to the radio[5]

10.2 Give an example of an electronic connection oriented conversation, other than those listed in question 1?

10.3 Give an example of an electronic connection–less conversation, other than those in question 1?

10.4 Explain in your own words why packets can arrive out of order if using connectionless communications.

10.5 Explain how packets can, or cannot, arrive out of order if using connection–oriented communications.

10.6 Why hasn't Microsoft moved folder and printer sharing from NetBIOS which is not routable to a routable set of protocols?

---

[5] Hopefully everyone still knows what a radio is.

Chapter 11
The Router

Overview

All routers, regardless of type or manufacturer, have many of the same common parts. Some parts are implemented as hardware, some are purely software, and some are virtual components but each plays a critical role in the functioning of the router. In order to avoid introducing bias into our discussion, this chapter will deal with a *generic* router running open–source protocols that can be implemented on a Raspberry Pi.

11.1 IP Forwarders

An IP Forwarder is a commonly used special case of a standard router. A typical IP Forwarder can *only* send packets between two networks, the local network and an ISP. The most common IP Forwarder is the SOHO router such as most home WiFi routers. These devices provide a local network via WiFi and LAN ports on the back of the router. A connection to the ISP via the special port on the back marked "WAN". Packets from the local network that must leave the network are forwarded to the ISP. All incoming packets are sent to the appropriate device on the LAN via the destination IP address. IP Forwarders typically cannot have the actual routing configuration changed and do not dynamically learn the network. There are IP Forwarders that can forward between multiple networks with minimal configuration, usually a "default route", and also do not support dynamic routing. All IP Forwarders are of limited use on the Internet except for connecting private networks to the Internet[1].

The router ports and WiFi are bridged together to form a Layer 2 LAN for local traffic. It is important to remember that Layer 3 networks, such as your home net-

---

[1] This is done using a technique known as NAT[2] as discussed in Chapter 23.1

© Springer Nature Switzerland AG 2020
G. Howser, *Computer Networks and the Internet*,
https://doi.org/10.1007/978-3-030-34496-2_11

work or the Internet, are logical Layer 3 networks built on top of Layer 2 networks that handles all local traffic such as unicasts and Broadcasts.

11.2 Parts of a Router

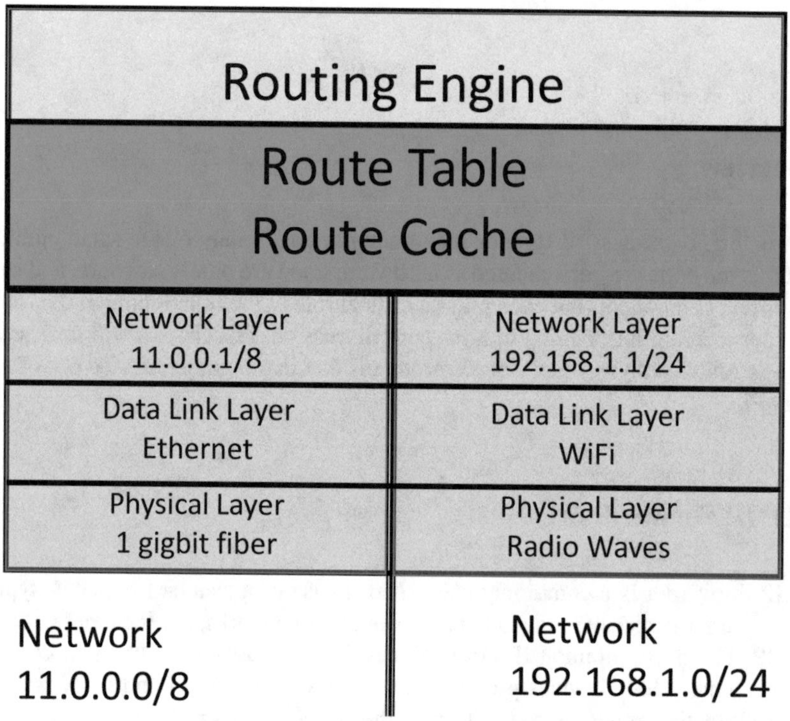

Routing Engine	
Route Table Route Cache	
Network Layer 11.0.0.1/8	Network Layer 192.168.1.1/24
Data Link Layer Ethernet	Data Link Layer WiFi
Physical Layer 1 gigbit fiber	Physical Layer Radio Waves

Network 11.0.0.0/8	Network 192.168.1.0/24

Fig. 11.1: A Router Connecting Two Layer 3 Networks

All routers, IP Forwarders, Layer 3 Switches, or typical routers, have the same logical structure depicted in Figure 11.1. Routers *must* have multiple network interfaces[3] otherwise the router can only connect to one network and such a router would be pointless. In order to be able to determine the best interface to use to reach a destination, the router *must* have a table of known networks/interfaces. This is the route table.

Lastly, every router needs some process to scan the received packets and route table to determine where to forward the packets. This process is usually dedicated

---

[3] A router can route between logical interfaces on the same physical connection, especially on fiber connections. Such a router is sometimes called "a router on a stick."

software but can be implemented in hardware/firmware as well[4]. We will call this process the routing engine. If the router is able to "learn" the network, in other words the router is able to respond to dynamic changes in the network, the routing engine software must be able to dynamically update the route table. Dynamic changes to the route table are usually done with the aid of a route cache.

11.3 Network Interfaces

Each NIC of a router can, and should, connect to an independent network. The incoming NIC handles all of the functions of the lower three layers of the OSI model or essentially the IP protocols associated with either TCP, UDP, or ICMP and simply delivers packets to the Routing Engine. Likewise, the out–going NIC handles the lower three layers for all out–going packets.

A router might implement the upper–layer protocols on multiple NICs for management and/or routing protocol purposes, but this is not really a router function. Many routers perform few, if any, of the upper layer functions because they are not managed remotely. In any case, routing does not directly require TCP or UDP[6].

Most modern routers have more than two NICs and many devices that have multiple NICs can be used as a router. A prime example of a device that can become a router is a PC or Raspberry Pi.

The first step in building a router is to assign the proper network addresses to the various interfaces of the router. This can be done dynamically, as is done with most SOHO routers, or it can be done statically. Home routers typically get an IPv4 address for the "up stream" or "WAN" interface from the ISP via a protocol known as DHCP[7], but when first constructed, the test–bed network does not have a service to provide this information[8]. For now, the Pi will be assigned a static IP address that will survive a reboot. Many times it is not desirable for a device change to *not* survive a reboot. The reasons are left to the student as an exercise, see Exercise 2.

Refer to the Group Network Diagram, Figure 8.13, for the IP address that goes to the group with the next highest number, $10.x.0.1/24$, and the next lower group number, $10.x - 1.0.x/24$, where x is your group number. The highest and lowest numbered groups will connect to the instructor's Pi and should determine those addresses using the instructor's group of 0 or one higher than the highest group number. For the following example we will use a group number of 3 which gives two interface addresses of $10.2.0.3/24$ (eth1) and $10.3.0.1/24$ (eth0). The members of the group should pair with a member of the next highest and next lowest group.

---

[4] Hardware routing is not common, but some Layer 3 switches implement the route engine in hardware (ASICs[5]) using dedicated processors instead of software for speed.

[6] Yes, there are many upper–layer functions that are performed by most routers but they do not concern us during the present discussion.

[7] Dynamic Host Configuration Protocol

[8] You will get a chance to configure a DHCP server later. For now we do not want one.

11.3.1 Temporary Assignment of Interface Addresses

There are a number of ways to temporarily assign a specific IP address to a device interface, or NIC, in Linux. Such an assignment is only valid until the device is rebooted but will survive the interface being taken down and brought back up or a temporary loss of connectivity such as a cable being unplugged or replaced.

1. Connect a dongle to a USB port of the Pi
2. Connect the RJ45 connector of the Pi to the dongle of the next highest group (4 in this case) with an Ethernet cable.
3. Connect to the Pi with the console cable. Obviously, you cannot connect to the Pi via **ssh** at this point.
4. Check the status of the Interfaces by **sudo ifconfig -a**.
5. Set the address 10.3.0.1/24 on the **eth0** interface.

   a. **sudo ifconfig eth0 down** (to shutdown the interface)
   b. **sudo ifconfig eth0 10.3.0.1 netmask 255.255.255.0**
   c. **sudo ifconfig eth0 up** (to bring the interface back up)

6. Check the status of the Interfaces by **sudo ifconfig -a**. **eth0** should now be up and running with the proper address as in Figure 11.2.
7. Set the address 10.2.0.3/24 on the **eth1** interface.

   a. **sudo ifconfig eth1 down** (to shutdown the interface)
   b. **sudo ifconfig eth1 10.2.0.3 netmask 255.255.255.0**
   c. **sudo ifconfig eth1 up** (to bring the interface back up)

8. Check the status of the Interfaces by **sudo ifconfig -a**. Both interfaces should be up and running at this point as in Figure 11.2.
9. Once both Pi Microcomputers have been configured, it should be possible to "ping" each other.

   a. First ping the interface on the local Pi by **ping 10.3.0.1**
   b. If this works, ping the other Pi by **ping 10.3.0.4**
   c. Do the same with the other group, in this case group 2 at 10.2.0.1

```
pi@mail:~$ sudo ifconfig -a
eth0: flags=4163<UP,BROADCAST,RUNNING,MULTICAST>  mtu 1500
      inet 10.3.0.1  netmask 255.255.255.0  broadcast 10.3.0.255
      inet6 fe80::4a70:8959:d8c8:511a  prefixlen 64  scopeid 0
         x20<link>
      ether b8:27:eb:d7:6b:bc  txqueuelen 1000  (Ethernet)
      RX packets 1599  bytes 128400 (125.3 KiB)
      RX errors 0  dropped 0  overruns 0  frame 0
      TX packets 1798  bytes 113769 (111.1 KiB)
      TX errors 0  dropped 0 overruns 0  carrier 0  collisions 0

eth1: flags=4163<UP,BROADCAST,RUNNING,MULTICAST>  mtu 1500
      inet 10.2.0.3  netmask 255.255.255.0  broadcast 10.2.0.255
      inet6 fe80::c480:1cf7:ee16:98a2  prefixlen 64  scopeid 0
         x20<link>
      ether 00:05:1b:24:48:8a  txqueuelen 1000  (Ethernet)
      RX packets 322  bytes 24708 (24.1 KiB)
      RX errors 0  dropped 0  overruns 0  frame 0
      TX packets 404  bytes 33199 (32.4 KiB)
      TX errors 0  dropped 0 overruns 0  carrier 0  collisions 0

lo: flags=73<UP,LOOPBACK,RUNNING>  mtu 65536
    inet 127.0.0.1  netmask 255.0.0.0
    inet6 ::1  prefixlen 128  scopeid 0x10<host>
    loop  txqueuelen 1000  (Local Loopback)
    RX packets 1090  bytes 104017 (101.5 KiB)
    RX errors 0  dropped 0  overruns 0  frame 0
    TX packets 1090  bytes 104017 (101.5 KiB)
    TX errors 0  dropped 0 overruns 0  carrier 0  collisions 0
pi@mail:~$
```

Fig. 11.2: Interfaces With IPv4 Addresses Assigned

After the groups are able to ping each other, start a ping while one of your neighbor groups reboots. The ping should return a message that the address is unreachable because this method of assigning addresses does not survive a reboot. For a network to work properly, the addresses of the interfaces of the routers should be predictable but this method of assigning addresses does not do that.

If a device provides *any* network service, it must be reachable at a predictable address. The Pi router must have static addresses for all interfaces.

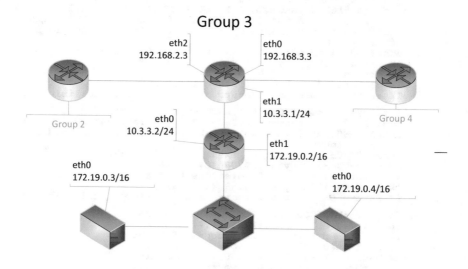

Fig. 11.3: Group Diagram for Group 3

11.3.2 Static Assignment of Interface Addresses

When Raspbian boots, it uses the information found in the file **/etc/dhcpcd.con**[9] to determine which interfaces will request address information from the network and which will be statically assigned addresses. If we change the information in that file, Raspbian will assign a static IP address to the interfaces we choose and will try and obtain a dynamic address for all other networks. Because networks change and devices move from network to network it is preferable to let the network assign addresses to client NICs[10], but for devices that will be providing network services we need to assign a static addresses.

There is a problem with the standard for assigning an IPv4 address to an interface. If an IP address for a NIC is not statically assigned, or assigned by a protocol known as DHCP, then a randomly generated address beginning with 169.254 is assigned to the interface, which usually causes problems. Indeed, sometimes it is extremely difficult to get a NIC to accept any other address once it has assigned itself a 169.254

---

[9] This is the DHCP Client daemon configuration file. Later in Chapter 14 we will configure Pi #3 in each group to provide this information to laptops automatically.

[10] For example, to laptops that move from network to network and often disappear forever.

address. This problem will have to be dealt with more than once with the Pi. The first step is to assign an address as the Pi is booting up[11].

The following steps will cause the Pi to automatically assign the desired IPv4 addresses to the proper interfaces at boot time[12].

1. Connect to the Pi and sign on
2. Back up the original configuration file by entering: **sudo cp -p /etc/dhcpcd. conf/etc/dhcpcd.conf-yymmdd** where yymmdd is today's date (or yyyym-mdd).
3. Edit the file using **vi**: **sudo vi /etc/dhcpcd.conf**
4. Add the following lines at the beginning of the file:

```
 1  interface eth0
 2  static ip_address=10.x.0.x/24
 3  static routers=10.x.0.x+1
 4  static domain_name_servers=172.16+x.0.3 172.17+x.0.3
 5
 6  interface eth1
 7  static ip_address=10.x-1.0.x/24
 8  static routers=10.x-1.0.1
 9  #static domain_name_servers=172.16+x.0.3 172.17+x.0.3
10
11  interface eth2
12  static ip_address=10.x.x.1/24
13  static routers=10.x.x.1
14  #static domain_name_servers=172.16+x.0.3 172.17+x.0.3
```

5. halt the Pi: **sudo halt**
6. Correctly connect all the cables as shown in Figure 8.13
7. Power up the Pi and check the interfaces: **sudo ifconfig -a**

For the router (Pi #1) for Group 2, the first few lines of the **/etc/dhcpcd.conf** file are:

```
 1  # A sample configuration for dhcpcd.
 2  # See dhcpcd.conf(5) for details.
 3
 4  interface eth0
 5  static ip_address=10.2.0.2/24
 6  static routers=10.2.0.3
 7  static domain_name_servers=172.18.0.3 172.19.0.3
 8
 9  interface eth1
10  static ip_address=10.1.0.2/24
11  static routers=10.1.0.1/24
12  #static domain_name_servers=172.18.0.3 172.19.0.3
13
14  interface eth2
15  static ip_address=10.2.2.1/24
16  static routers=10.2.2.1/24
17  #static domain_name_servers=172.18.0.3 172.19.0.3
```

---

[11] This can also be done with DHCP but at this point the Pi does not have access to a DHCP service and it is easier to use alternate methods to assign static addresses to a router. A SOHO router typically uses DHCP to assign an address to the upstream interface to the ISP.

[12] Remember: Each Raspberry Pi and each *interface* must have a unique IPv4 address that is correct for the network it attaches to.

11.4 The Routing Engine

Once a packet is detected it is passed to the routing engine for forwarding. The routing engine must quickly decide which IP address should be the next step in the delivery of the packet to the final destination. Notice that the router typically does not have a direct connection to the final destination or even a NIC that is part of the network where the final destination can be found. The router does however have a table, see Section 11.7, of all the networks known to the router.

11.5 Installing the Quagga Routing Engine

One of the more interesting free routing engines is an open source routing project named Quagga [43]. Quagga[13] uses a routing engine still called zebra routing daemon which functions in cooperation with the Linux kernel and a suite of routing protocols which includes most popular open source routing protocols such as RIP[14], OSPF[15], BGP, Babel, and others. All of these protocols are designed to implement dynamic networks that allow the network to react to changes in the physical network (Layer 1) as connections, devices, and routers are added or leave the network[16] Routing protocols allow this to happen *without* human intervention. We will explore these routing protocols in Part III.

11.6 Installing Quagga

Quagga is not pre-installed in Pi Raspbian but should be installed if you are using a custom image. If not, it can be installed following the procedure in Section 8.1. Quagga installs the **zebra** routing daemon to update the route table kept in the Linux kernel which implements the actual routing engine.

11.6.1 TCP and UDP Ports

Insure that the ports required to connect to the configuration processes are open:
```
pi@router2-2:~\$ sudo cat /etc/services | egrep -i
    [2-2]60[1-9]
```

---

[13] A quagga is an extinct(?) close relative of the zebra. The name was chosen with quagga became a "fork" of the very similar zebra project which is now defunct [44].

[14] Route Interchange Protocol

[15] Open Shortest Path First (IPv4)

[16] Unfortunately, this include attempting to reconfigure the network after the failure of a device or the connections between devices.

```
zebra              2601/tcp                # zebra vty
ripd               2602/tcp                # ripd vty
ripngd             2603/tcp                # ripngd vty
ospfd              2604/tcp                # ospfd vty
bgpd               2605/tcp                # bgpd vty
ospf6d             2606/tcp                # ospf6d vty
ospfapi            2607/tcp                # OSPF-API
isisd              2608/tcp                # ISISd vty
pi@router2-2:~\$
```

11.6.2 Enabling Kernel Forwarding

Once Quagga is installed, the kernel must configured to forward packages that are IPv4 or IPv6. This is done by editing the file **/etc/sysctl.conf** to remove the comment mark # from the two statements about IP Forwarding as below:

```
# Uncomment the next line to enable packet forwarding for IPv4
net.ipv4.ip_forward=1

# Uncomment the next line to enable packet forwarding for IPv6
#   Enabling this option disables Stateless Address
      Autoconfiguration
#   based on Router Advertisements for this host
net.ipv6.conf.all.forwarding=1
```

11.6.3 Quagga Daemons

Configure the **/etc/daemons** file to automatically start **zebra** and **ripd**

```
pi@router2-2:~\$ sudo cat /etc/quagga/daemons
zebra=yes
bgpd=no
ospfd=no
ripd=yes
ripngd=no
babeld=yes
pi@router2-2:~\$
```

11.6.4 Quagga Configuration and Log Files

Quagga processes run under the user **quagga** which was created during the installation. We must insure that the configuration and log files exist and are owned by **quagga**. First the configuration files (do not be concerned if you have more or fewer configuration files at this point):

```
pi@router2-2:~\$ sudo chown quagga:quagga /etc/quagga/*.*
pi@router2-2:~\$ sudo chown quagga:quaggavty /etc/quagga/vtysh
   .*
pi@router2-2:~\$ ls -l /etc/quagga/
total 60
-rw-r--r-- 1 quagga quagga    126 Dec  6 15:35 babeld.conf
```

```
-rw-r--r--  1 quagga quagga     126 Dec   6 15:35 babeld.conf.
    sample
-rw-r-----  1 quagga quagga     126 Dec   7 18:08 bgpd.conf
-rw-r--r--  1 quagga quagga     126 Dec   6 15:34 bgpd.conf.
    sample
-rw-r--r--  1 quagga quagga      57 Dec  10 12:53 daemons
-rw-r--r--  1 quagga quagga      57 Dec   6 15:47 daemons.sample
-rw-r--r--  1 quagga quagga     125 Dec   6 15:32 isisd.conf.
    sample
-rw-r-----  1 quagga quagga       0 Dec   7 18:09 ospfd.conf
-rw-r--r--  1 quagga quagga     126 Dec   6 15:30 ospfv2.conf.
    sample
-rw-r-----  1 quagga quagga       0 Dec   7 18:09 pimd.conf
-rw-r--r--  1 quagga quagga     200 Dec   6 15:29 pimd.conf.
    sample
-rw-r-----  1 quagga quagga       0 Dec   7 18:08 ripd.conf
-rw-r--r--  1 quagga quagga     122 Dec   6 15:25 ripd.conf.
    sample
-rw-r-----  1 quagga quagga       0 Dec   7 18:09 ripngd.conf
-rw-r--r--  1 quagga quagga     127 Dec   6 15:28 ripngd.conf.
    sample
-rw-r-----  1 quagga quaggavty    59 Dec   6 15:17 vtysh.conf
-rw-r--r--  1 quagga quaggavty    59 Dec   6 15:17 vtysh.conf.
    sample
-rw-r-----  1 quagga quagga     126 Dec   6 15:19 zebra.conf
-rw-r--r--  1 quagga quagga     126 Dec   6 15:19 zebra.conf.
    sample
pi@router2-2:~\$
```

At this point, the critical files are **daemons**, **zebra.conf**, **ripd.conf**, and **vtysh.conf**. If the **daemons** file is missing, create it to agree with the one above. If any of the others are missing, create them by **sudo touch/etc/quagga/ripd. conf** for example. Then repeat the process of insuring that the files are owned by the **quagga** user.

Create the log files for all the processes that are part of Quagga. If the directories do not exist, create them.

```
pi@router2-2:/etc/quagga# cd /var/log/quagga
pi@router2-2:/var/log/quagga# sudo touch babeld.log
pi@router2-2:/var/log/quagga# sudo touch bgpd.log
pi@router2-2:/var/log/quagga# sudo touch isisd.log
pi@router2-2:/var/log/quagga# sudo touch ospfd.log
pi@router2-2:/var/log/quagga# sudo touch pimd.log
pi@router2-2:/var/log/quagga# sudo touch ripd.log
pi@router2-2:/var/log/quagga# sudo touch ripngd.log
pi@router2-2:/var/log/quagga# sudo touch zebra.log
pi@router2-2:/var/log/quagga# sudo chown quagga:quagga /var/
    log/quagga/*.*

pi@router2-2:/var/log/quagga# ls -l
total 0
-rw-r--r--  1 quagga quagga 0 Dec 10 13:32 babeld.log
-rw-r--r--  1 quagga quagga 0 Dec 10 13:32 bgpd.log
-rw-r--r--  1 quagga quagga 0 Dec 10 13:32 isisd.log
-rw-r--r--  1 quagga quagga 0 Dec 10 13:33 ospfd.log
-rw-r--r--  1 quagga quagga 0 Dec 10 13:33 pimd.log
-rw-r--r--  1 quagga quagga 0 Dec 10 13:33 ripd.log
-rw-r--r--  1 quagga quagga 0 Dec 10 13:33 ripngd.log
-rw-r--r--  1 quagga quagga 0 Dec 10 13:33 zebra.log
pi@router2-2:/var/log/quagga#
```

11.7 The Route Table

Table 11.1: A Sample Route Table

Source	Network	Next Hop	Interface	Cost[a]
C	127.0.0.1/8	127.0.0.1	lo	0
S	169.254.0.0/16	169.254.0.0	null0	0
C	172.30.0.0/16	172.30.0.0	eth0	0
R	172.31.0.0/16	172.30.0.20	eth0	1
C	192.168.1.0/24	192.168.1.0	wlan0	0

[a] How the cost is measured depends upon the source of the route. For example, RIP determines the cost by counting the number of router "hops" to a destination while OSPF uses the total costs of the links of a route as assigned by the network administrator.

The route table is used by the routing engine to determine where to send each packet. When a packet is received, the routing engine extracts the destination address from the packet and determines the network part of the destination. It is then a simple matter of searching down the route table for a matching network address. When a match is found, the routine engine has two choices. If the route in the route table is a directly connected network, the routing engine uses the ARP protocol to determine the MAC address that corresponds to the destination and sends the packet out the correct interface. If the route is not directly connected, the routing engine sends the packet to the interface on a different router that has a route in its route table that leads eventually to the destination network.

If the destination networkis *not* in the route table there is a problem. The router has no known path to the destination, so the packet is simply dropped and the next packet is processed. Routers do not attempt to find a path to the desired network and they are not required to notify any other device that the network is not reachable. These are Layer 4 responsibilities and routers do not function at Layer 4. [17]

How the route table is built and maintained will be discussed in Section 11.9.

11.8 The Optional Route Cache

Routers that maintain a dynamic route table usually maintain an optional cache of all routes to known networks even if these routes are not the best route to a known network. The actual meaning of the "best route" depends upon how the router learns the route but it is in some sense the cheapest route.

---

[17] The only time a router functions at the upper layers is for management purposes. Remember, the main function of a router is to route packets quickly.

11.9 Duties of a Router

As a quick summary, all routers must limit Layer 2 traffic to the correct network and route packets from one network to another in an attempt to send the packets to the proper destination network. With the exception of IP Forwarders, routers must also respond to changes in the network by maintaining a dynamic routing table.

11.9.1 Limiting Broadcast

Usually we do not think of a router as having any Layer 2 functions, but in fact the most important function of a router is to limit the scope of Layer 2 Broadcasts. Remember every NIC must process every Broadcast received. If the Internet was built at Layer 2, it would be one large LAN and every NIC would see, and possibly interfere with, every Layer 2 frame. A LAN of millions of NICs would have a throughput of zero frames due to collisions and be completely nonfunctional. Even if all the unicasts could be controlled, the Broadcasts would overwhelm the network. This is the reason for Layer 3 networks in the first place.

11.9.2 Routing Packets

Routers move packets toward known destination networks. If a router has an interface connected to the destination network, this is simple. If not, the router must place the packet into the payload of a frame and send the frame to a router that has announced a route to the destination. A "hop" is each time a packet goes from router to router. We will see that some routing protocols assume all links between routers are equal and measure the "best" route as the fewest hops.

11.9.3 Maintaining the Route Table

Routers use routing protocols, as explained in Chapter 12, to build and update both the Route Cache and Route Table. When a packet arrives at the router, the Routing Engine searches top–down the Route Table for a match between the destination networkand a known network. If a match is found, the router sends the packet to the proper location. If no match is found the packet is dropped.

In most cases it is desirable to have packets for unknown destinations sent somewhere for further processing. This is especially true if the router connects to an ISP as the ISP has good reasons to keep some of its routing private to hide the details of its proprietary network. It is therefore a good idea for every router to have a static route to some default gateway router that handles unknown destinations. This is discussed in detail in Chapter 12.

Projects

11.1 If a SOHO router is available, explore the management of the router and the following questions.

    a. Does the router run any routing protocols such as RIP or OSPF?

    b. Does the router run DHCP? If so, can the DHCP server be configured?

    c. Does the router have an embedded Layer 2 switch? How can you tell?

    d. Draw a best guess diagram of the router as it connects networks. Make an educated guess as to how many NICs the router has.

Exercises

11.1 What is the difference between a Route Table and a Route Cache?

11.2 Why might you want an assigned IP address to *not* survive a reboot?

11.3 Home routers has an RJ45 jack labeled "WAN" and jacks labeled "LAN". Why?

11.4 SOHO Routers do not have a static IP address for the interface that connect to the ISP. This address can change periodically. Why isn't this a problem?

11.5 A router is called a Layer 3 device, but it must also function as a Layer 2 and a Layer 1 device. Explain why.

11.6 Would you expect an IP Forwarder to maintain a route cache?

11.7 Would you expect an ISP router to maintain a route cache?

Chapter 12
Populating and Maintaining the Route Table

Overview

Populating and maintaining the Route Table (and Route Cache) is one of the most critical activities a network engineer performs. Static Routes are one way to populate the table, but Static Routes do not change as the network changes. At first this does not seem like a problem, but even medium–size networks can be too complex to maintain with Static Routes. Dynamic Routes allow the network as a whole to adapt easily to changes but require inter–router messaging that adds to the overhead of the network. Maintaining the Route Table comes down to a balancing act between static routing and the overhead of dynamic routing. Fortunately, the balance between the two is easy to find.

Entire courses are taught on maintaining the route table and routing protocols. Our goal is to understand these protocols and to be able to use them to build a complex network. For this a detailed knowledge of each of these protocols is not really necessary as those details can be found when needed. What is desired at this point is a firm basic knowledge and a grasp of the capabilities of the various routing protocols as these are the key building blocks for a network of routers.

12.1 Static Routing

Static Routes direct the Routing Engine to send all packets to a destination route to the same next hop and are determined by a network engineer. It is tempting to solve problems by creating a number of Static Routes to force packets to take a specific path, but there are major drawbacks to Static Routes that severely limit their efficacy. Static Routes are exactly that: static. They do not change and we

© Springer Nature Switzerland AG 2020
G. Howser, *Computer Networks and the Internet*,
https://doi.org/10.1007/978-3-030-34496-2_12

have already discussed the fact that the Internet is in a state of constant change. If a network is subject to changes, Static Routes are a problem[1].

The exact commands to enter static routes depends upon the whim of the person who designed the user interface for the router, but the information is usually the same. The network engineer must enter the destination networkand where packets for that network are to be sent by the router. Great care must be taken in the design of static routes to avoid some common problems.

1. Static routes can lead to packets being delivered to the wrong destination. If a guaranteed delivery such as TCP is being used over these routes, the resulting retries can cause wasted traffic. Fortunately, this usually results in an unreachable destination and the conversation dies.
2. Static routes can lead to a routing loop where the packets jump from router to router forever. IP is designed to deal with routing circles by attaching a TTL[2] to each packet. As a packet moves through a router the TTL is decremented. Packets with a TTL of zero are dropped.
3. Static routes may take a more expensive path by accident.
4. Static routes are "nailed down" and *do not change when the network changes.* Sooner or later, your network will change.

Static routes are to be discouraged except for two special static routes called the default route and routing networks to nowhere (a null interface). All of the private IPv4 networks should be "tanked" to avoid potential network routing issues and possible attack vectors.

12.1.1 Direct Routes and the Default Route

If static routing is such a bad idea, why does it exist? IP Forwarders use static routing exclusively and are some of the most common routers on the Internet. When a router becomes active (powered up), the Route Table is populated with static routes, referred to as "directly connected routes", to each of the interfaces configured on the router. These are the lowest possible cost routes to each of the directly connected networks because packets need only be placed inside a Layer 2 frame and sent using ARP. No other routing is needed.

IP Forwarders add another route from the local IP network to the port dedicated to the ISP connection. Any packets with destinations other than the local network are forwarded to the ISP making this the default route. For routers that can do both static and dynamic routing, this default route must be configured by the network engineer.

---

[1] There is a temptation to use complex Static Routes as a form of job security. Keep in mind that if part of the network fails, the Static Routes may cause network failures that the network engineer must fix. These failures seem to happen at inopportune times.
[2] Time To Live

In a display of the route table, the default route is listed with a destination of "0.0.0.0" for IPv4 and matches any destination network. For this reason, the default route is logically the last route in the route table. In my opinion, every router must have a default route[3].

12.1.2 Manual Entries

Warning: Manual entries into the Route Table are a security and functionality risk. It is possible to hijack an entire network by changing the configuration of a key router, either on purpose or by accident, which presents a security risk.

Static routes are also more likely to result in a functionality problem by causing packets to take a less than optimal route or by creating a "routing loop" where packets route in a cycle until they are dropped as having expired. One of the fields in a packet header is TTL which is designed to detect a packet that is looping or taking a route that is so many router hops that it cannot be delivered in a timely fashion. Each time a packet is forwarded by a router, the TTL is decremented. A packet with a zero value for TTL is automatically dropped by a router.

We will see later how to use static routes to increase network security by sending unwanted or dangerous packets to nowhere and cause the router to drop those packets.

12.2 Dynamic Routing and the Route Cache

Table 12.1: Example Route Sources

Type	
* or C	Directly connected
S	Static route
K	Kernal route
I	IS–IS route
R	RIP route
O	OSPF route
B	BGP route

Dynamic routing protocols allow routers to learn the network and to respond to network changes. Routers send update messages to each other and use them to

---

[3] At one time, this was also called "the network of last resort" which I thought was a brilliant name for the default route.

learn all the known destination networks and the best route to each one. As new routes are announced by other routers, a router adds those routes (with their cost!) to the route cache, see Algorithm 3. The exact details are up to the programmer, but this is what happens logically. If the route cache is updated, the routing engine processes the route cache to produce an updated route table using something similar to Algorithm 4.

The details of what is included in these route announcements and what triggers announcements depends upon the routing protocol in use. Most networks use only one routing protocol, but it is possible to use more than one when desired. For that reason, commands that display the route table or route cache usually include the source of the route. For example, one possibility is given in Table12.1.

Algorithm 3 Route Cache Update

1: **procedure** UPDATE ROUTE CACHE
2:     **while** There are unprocessed route announcements **do**
3:         **if** First announcement from this interface **then**
4:             Delete all routes learned from this interface
5:         **end if**
6:         Add this route to the route cache
7:     **end while**
8: **end procedure**

Algorithm 4 Cache Based Route Table Update

1: **procedure** BUILD ROUTE TABLE
2:     **while** There are unprocessed cache entries **do**
3:         Get an entry from the cache
4:         **if** Table contains an entry with the same destination network **then**
5:             **if** Cost is less than the cost of the existing route **then**
6:                 Replace the route with this one
7:             **else**
8:                 Do nothing
9:             **end if**
10:        **else**
11:            Add this route to the route table
12:        **end if**
13:    **end while**
14: **end procedure**

In the next chapter, we will introduce a number of open source and proprietary routing protocols that can be run on the Raspberry Pi. The protocols we will look at are all designed to meet the needs of a range of networks by balancing complexity and overhead.

Exercises

12.1 Would it be possible to write your own routing protocol to run on a private network?

12.2 Why would one expect most Intranets to run only one routing protocol?

Further Reading

The RFC below provide further information about static routes and default routes. This is *not* an exhaustive list and most RFC are typically dense and hard to read. Normally RFC are most useful when writing a process to implement a specific protocol.

RFCs Directly Related to This Chapter

default	Title
RFC 1397	Default Route Advertisement In BGP2 and BGP3 Version of The Border Gateway Protocol [99]
RFC 2461	Neighbor Discovery for IP Version 6 (IPv6) [172]
RFC 4191	Default Router Preferences and More-Specific Routes [232]

Other RFCs Related to Static Routes

For a list of other RFCs related to the static routes, but not closely referenced in this Chapter, please see Appendix B.

Part III
Dynamic Networks

Change is the handmaiden Nature requires to do her miracles with.
Roughing It

Mark Twain

Overview

The INTERNET is built by connecting large numbers of small networks together and correctly moving information (packets) from device to device. In order to do this effectively, these networks are connected using devices known as routers. Simply put, a router connects to two or more layer 3 networks and insures that packets move from network to network correctly. If networks were simple, or at least did not change, this would be a fairly easy thing to do.

The complexity of the INTERNET is beyond the ability of most to work with correctly and changes much too quickly for anyone to be able to keep up. Sites and machines go on the network and disappear from the network at an alarming speed. Entire networks connect to the INTERNET frequently and just as frequently change or completely disappear. The dynamic nature of the INTERNET insures that humans trying to configure devices to reflect the true status of the INTERNET are bound to fail. Fortunately, there are three main devices to handle these problems for us.

It is important to remember that while routers are the main building blocks of the INTERNET and to a large extent determine the topology of the INTERNET, the main function of a router is to move packets from one network to another. It is easy to get caught up in the details of how routers "learn" the network and forget that this is a secondary function. Routers move packets at high speed and limit the scope of Layer 2 broadcasts.

Classification of Protocols

There are two main classifications of routing protocols that can easily be implemented using routers: Distance–Vector protocols and Link–State protocols. Distance–Vector protocols measure the desirability of a route by advertised distance which is sometimes giving in "hops" which is the number of routers that must be traversed to reach the desired destination network. These protocols tend to store routes in the route cache as distance matched with the next router in the path. This cost and next router pairing is somewhat analogous to a vector, hence the name.

The other main classification we will use is Link–State protocol. Routers using Link–State protocols exchange information between router of links in the network and their associated cost. Each router constructs a weighted graph of the network from this information and then constructs a complete set of the shortest, i.e. least cost, to each known destination network. Link–State protocols only send messages when a link changes[4].

---

[4] Processes also exchange "hello" messages between each other to establish connections and to insure that the link and routing process on the other end are both still operational.

Chapter 13
Shortest Path Through the Network

Overview

Routing messages in a network can be thought of in terms of graphs[1] which allows routing protocols to use well known graph algorithms to determine the best way to reach a given network. There are many, many good books and on–line sources for graph theory [306, 308, 309] and graph algorithms [25, 308, 309] as they pertain to networking and routing.

Why do we care about graphs in networking? Networks lend themselves to being viewed as graphs and many graph problems are common networking issues [306]. For example, in routing the goal is to reach a distant network in the shortest time or the least cost. We will see how graph algorithms are able to quickly solve the problem of finding the lowest cost path, or shortest path, from the current router to any other router once the details of the network topology are known. These small time savings when routing packets add up quickly to significantly lower network costs.

Creating a spanning tree for a graph or finding the shortest path to all other nodes occurs frequently in routing and is usually explained by "running Dijkstra" or "running spanning tree" without really understanding what that means or why it is a good idea[2]. The goal of this chapter is to give you a feel for what happens when a router needs to run a shortest path first algorithm and how easily this can be done. It is hoped that you will also gain an appreciation as to why a link–state routing

---

[1] Technically a network with all symmetric, bidirectional connections with the same characteristics in both directions, can be thought of as a graph with edges. If even one connection is *not* the same both ways, the network is a Digraph (directional graph) with arcs instead of edges. We will not be picky about this distinction unless it is absolutely necessary. For our purposes it is best to think of all networks as directional graph (Digraph)s knowing that many of the connections between nodes are the same in both directions. A good example of a connection that is not symmetric is a home network with 100 megabits for downloading and 10 megabits for uploading.

[2] More than one expert has been unable to give me even a hint of what Dijkstra is or what is the point of a spanning tree when asked. I think it is best to have at least a high level understanding of what the algorithms do and why we need them.

© Springer Nature Switzerland AG 2020
G. Howser, *Computer Networks and the Internet*,
https://doi.org/10.1007/978-3-030-34496-2_13

protocol is desirable over a distance–vector one for complex networks. Networks become complex very quickly as they grow.

13.1 Graph Terms

Here we will make a few quick definitions. Please bear in mind these are "quick and dirty" definitions and may annoy some mathematicians, but our goal is to understand these as they pertain to a network.

Vertex   A vertex (also called node or point) is a part of a graph and *as such has no fixed location.* Nodes in a graph may be relabeled or moved at will without materially changing the graph. ,

Edge   An edge is a two–way connection between exactly two vertices. An edge has no fixed length, location, or direction[3]. An edge between nodes a and b is usually written as $e(a,b)$. In a graph $e(a,b) = e(b,a)$ for all edges. In networking, it is often best to use digraphs and change all edges to pairs of arcs. For example: replace $e(x,y)$ with two arcs $a(x,y)$ and $a(y,x)$.

Arc   An arc is a one–way connection between exactly two vertices. An arc has no fixed length or location. If you think of an arc as an "arrow" on a graph you are OK. An arc between two nodes x and y is written as $a(x,y)$ if it "points" from x to y. In a digraph $a(x,y) \neq a(y,x)$ and the fact that there is an arc $a(x,y)$ does *not* imply there is an arc $a(x,y)$. In networking terms think of a radio broadcast from the tower, t, to a radio, r. In this network there is an arc $a(t,r)$ but the radio cannot contact the tower because there is no connection $a(r,t)$.

Adjacent   Two nodes, x and y, are said to be adjacent in a graph if there exists an edge $e(x,y)$. If a digraph has an arc $a(a,b)$ but not an arc $a(b,a)$, then b is adjacent to a but a is *not* adjacent to b. In layman's terms, if there is a direct path between two nodes, they are adjacent. If there is an arc from b that takes leads directly to a, then a is adjacent to b and there must be another arc from a that points directly to b for b to be adjacent to a. The notation $adj(a)$ means the set of all vertices that can be reached *directly* from a, such as all the routers in a network with a link from a.

Graph   A graph, usually denoted by $G = (V, E)$ is a non–empty set of vertices (**V**) and a set of edges (**E**) which implies that a graph must have at least one vertex. By this definition, a single vertex is a graph. If there is a path from any vertex u in the graph to any vertex v in the graph, the graph is connected. While multiple connected graphs can be part of the same graph, we will consider all graphs as connected. It will be pointed out if a graph (or network) is not connected.

Empty Graph   The special graph with no edges or vertices[4].

---

[3] Think of edges as rubber bands.

[4] Exactly how we handle the empty graph definition, as well as the graph definition, can sometimes cause problems. We will ignore these as they will not cause us trouble when we talk about networks.

Directional Graph   A directional graph, or digraph, is a non–empty set of vertices (**V**) and a set of arcs (**A**). By this definition, a single vertex is also a digraph. We will only make a distinction between a digraph and a graph when it is important to networking. Bear in mind that most networks are digraphs even if all the links are bi–directional because a link *might* fail in one direction only. This type of failure can be hard to find.

Empty Digraph   The special graph with no arcs or vertices. Obviously the empty graph and empty digraph are interchangeable. If you are familiar with set theory, the empty graph is much like the empty set.

Weighted Graph   A graph, or digraph, where the edges or arc have an associated cost or weight which is designated $w(a,b)$. This could be distance, tolls, or relative desirability. Costs are of critical importance in networking.

Tree   A tree is a graph, or digraph, with no loops such that if any random edge or arc is removed the tree becomes broken into two or more trees. In networking, loops are great for redundancy but can cause *massive* problems with broadcasts, see Section 4.5.1 for an example.

Spanning Tree   A tree drawn on a graph (or network) that includes every node in the graph and the edges required to reach them. This is important in networking because a spanning tree reaches all nodes but does not lead to routing loops.

Minimum Spanning Tree   A minimum spanning tree (MST) is a spanning tree where the total of the costs from the root node to any other node is minimized. A graph may have *multiple* MSTs[5] but only one is required and no one cares which one because it is obvious that each MST in a graph has the same costs.

13.2 Shortest Path First

In some dynamic routing protocols such as OSPF and ISIS[6], each router develops a consistent graphical database of the network nodes and the costs associated with the connections (edges) between those nodes. From this data each router can easily determine the "best" path to each known subnetwork in the greater network. Typically this is done by building a MST with the local router as the root of the tree[8] as in Figure 13.1c.

A brief explanation is in order here. Figure 13.1a is a fairly simple weighted digraph with reasonable redundancy if it is a network. The leftmost node has been chosen as the root, or source, and the problem is to create a set of paths that touch each node with the least cost for each node. Figure 13.1b and Figure 13.1c both show solutions with the *exact* same costs to reach any node. It is not important which solution is chosen as both are equally correct. If the nodes are all routers,

---

[5] Minimum Spanning Trees

[6] ISIS[7]

[8] Technically any node in a tree may be chosen as root without changing the tree. I tell my students the only thing special about "root" is that "root" is the one we have chosen to call root.

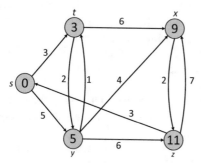

Fig. 13.1a: A weighted, directed graph with shortest-path weights from source *s*.

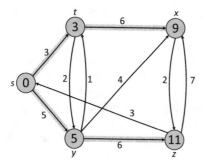

Fig. 13.1b: The shaded edges form a shortest-paths tree rooted at the source *s*.

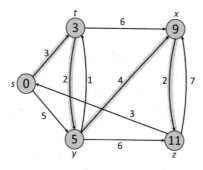

Fig. 13.1c: Another shortest-paths tree with the same root. [25]

router s could use either solution to minimize costs to send packets through the network.

13.3 Dijkstra's Algorithm

There are a number of good (fast) algorithms to find the shortest path to each node such as Dijkstra, Bellman–Ford, and other exotic ones. Because Dijkstra is very fast and commonly used in routing and switching, we will look at it in detail [30].

Constraint: For each $edge(u,v) \in E, weight(u,v) \geq 0$

Algorithm 5 Dijkstra Single Source Shortest Path

1: **procedure** DIJKSTRA$((G[],w[],s))$
2:     **for** each vertex $v \in G.V$ **do**
3:         $v.d = \infty$
4:         $v.\pi = $ null
5:     **end for**
6:     $s.d = 0$
7:     $S = \emptyset$
8:     $Q = G.v$
9:     **while** $Q \neq \emptyset$ **do**
10:         $u = $ EXTRACT-MIN(Q)
11:         $S = S \cup \{u\}$
12:         **for** each vertex $v \in G.Adj[u]$ **do**
13:             **if** $v.d > u.d + w(u,v)$ **then**
14:                 $v.d = u.d + w(u,v)$
15:                 $v.\pi = u$
16:             **end if**
17:         **end for**
18:     **end while**
19: **end procedure**

So how does this work? Briefly, the various parts of Dijkstra's algorithm[9] are fairly straight forward.

- The for–loop in lines 2–5 initialize the characteristics of each node in the graph by setting the distance to the source ($v.d$) to infinity and the predecessor ($v.\pi$) in the shortest path *back* to the source to a special value (null) to mark it as unknown.
- Line 6 sets the distance to the source from the source to zero ($s.d = 0$).

---

[9] For Dijkstra's algorithm the time complexity depends upon the number of arcs and the number of vertices: $O(V \log V + E)$

- Line 7 creates an empty set S to hold the vertices as they are processed.
- Line 8 creates a priority queue Q keyed upon the shortest path estimate for each vertex and populates it with all the vertices in the graph.
- While there are still vertices to be processed, the loop in lines 9–18 updates the estimated distance to each node from the source.

  - Lines 10 and 11 extract from the queue the vertex u with the shortest *estimated* path back to the source and then adds it to the set of processed vertices S.
  - The for–loop in lines 12–17 visits each vertex that can be reached directly from u and updates the estimated shortest distance back to s if needed. When all neighbors of u have been processed, control is returned to line 9 until the queue is empty.

- At this point the distance from each vertex back to the source s has been calculated and the exact path can be found by using $v.\pi$ to walk backwards to the source.

Dijkstra's algorithm will produce one possible tree of shortest paths but there may be many possible trees with the same shortest paths. In a network, or in a graph, this is not a problem as all the network really needs to do is to get the packets from the source to the destination as quickly (or cheaply) as possible. If two links in the network have the same cost it is assumed they are equally desirable, whatever that may mean. If this is not the case, they must have different links to reflect the difference. Where this is most often seen is where a more expensive leased link is held as a backup for a less expensive link with the same bandwidth.

If we run Dijkstra's Algorithm on the network in Figure 13.1a we find, after initialization, the queue is $Q = \{s.d = 0, t.d = \infty, x.d = \infty, y.d = \infty, z.d = \infty\}$ and $S = \{\}$ and the graph looks like Figure 13.2.

When the vertex with the smallest estimated distance back to s is extracted into vertex $u = s$, the queue is now $Q = \{t.d = \infty, x.d = \infty, y.d = \infty, z.d = \infty\}$ and after u is added to S we process all of the vertices adjacent, that is connected in with an arc from u, and update the estimated distances we arrive at Figure 13.3 and the queue is now $Q = \{t.d = 3, y.d = 5, x.d = \infty, z.d = \infty\}$.

As we continue to run the algorithm, we move through Figures 13.4, 13.5, 13.6, and Figure 13.7 after which the queue is empty and the shortest path from s to each vertex has been found.   If you would like another explanation of Dijkstra's Algorithm, see [41] or [25] Chapter 24.

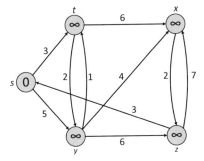

Fig. 13.2: Graph After Initialization

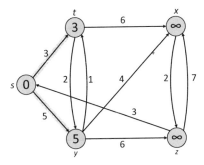

Fig. 13.3: Graph After Processing Root (s)

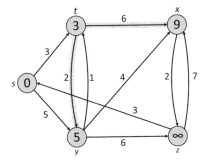

Fig. 13.4: Graph After Processing t

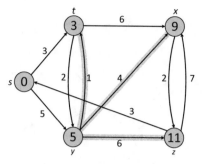

Fig. 13.5: Graph After Processing *y*

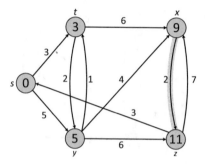

Fig. 13.6: Graph After Processing *x*

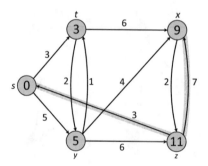

Fig. 13.7: Graph After Processing *z*

13.4 Bellman–Ford Algorithm

Another common algorithm encountered for quickly finding the best path through a network is the Bellman–Ford Algorithm presented below. We will not go over it in detail, but if you are interested see CLR[10] [25].

The Bellman-Ford algorithm[11] returns *true* if there are no negative weight cycles reachable from the root and returns the shortest path solution if there are no negative weight cycles.

Algorithm 6 Bellman-Ford Single Source Shortest Path

1: **procedure** BELLMAN-FORD((G[],w[],s))
2:     **for** each vertex $v \in G.V$ **do**
3:         $v.d = \infty$
4:         $v.\pi = $ null
5:     **end for**
6:     $s.d = 0$
7:     **for** $i = 1$ to $|G.V| - 1$ **do**
8:         **for** each $edge(u,v) \in G.E$ **do**
9:             **if** $v.d > u.d + w(u,v)$ **then**
10:                 $v.d = u.d + w(u,v)$
11:                 $v.\pi = u$
12:             **end if**
13:         **end for**
14:     **end for**
15:     **for** each $edge(u,v) \in G.E$ **do**
16:         **if** $v.d > u.d + w(u,v)$ **then**
17:             **return** FALSE
18:         **end if**
19:     **end for**
20:     **return** TRUE
21: **end procedure**

---

[10] A common way to refer to *Introduction to Algorithms* by Cormen, Leiserson, Rivest, and Stein. Poor Stein got left out of the acronym.

[11] Bellman-Ford runs in $O(V \times E)$.

Projects

13.1 (Optional) Write a program to read a graph from a file and output the results from Dijkstra.

13.2 (Optional) Write a program to read a graph from a file and output the results from Bellman–Ford.

13.3 Find another shortest path algorithm and compare it to Dijkstra's or Bellman–Ford.

Exercises

13.1 Why is there a constraint on Dijkstra's Algorithm? Why isn't this a problem in networking?

13.2 Using Figure 13.2 change the weight of $a(s,y) = 3$.

    a. Do you expect this to change the final distances back to s for any of the vertices?

    b. Run Dijkstra on the new graph showing your work.

    c. Run Bellman–Ford on the new graph showing your work. Would this lead to significant changes in routing on this network?

13.3 Reverse arc $a(y,t)$ in Figure 13.2 and run Dijkstra again.

13.4 If an edge is inserted $e(t,z) = 2$ into Figure 13.2, what changes must you make to run Dijkstra? Make the changes and run Dijkstra.

Further Reading

For another explanation of Shortest Path First algorithms, you might look at:

Dijkstra, E.W. (1959). A note on two problems in connection with graphs. Numerische mathematik 1(1), 269–271.

Cormen, T., C. Leiserson, R. L. Rivest, and C. Stein (2009). Introduction to Algorithms (Third ed.). MIT Press.[12]

Roberts, F. and B. Tesman (2009). Applied Combinatorics (Second ed.). Chapman and Hall/CRC[13].

Joshi, V. (2017, October). Finding the shortest path, with a little help from Dijkstra, (on line). Accessed: July 23, 2019.[15]

Sahni, S. (1998). Data structures, algorithms and applications in c++. Computer Science, Singapore: McGraw-Hill.

Sahni, S. (2005). Data structures, algorithms, and applications in Java. Universities Press.

---

[12] Chapter 24.3 in Edition 3

[13] Chapter 3 has a good introduction to the basics of Graph Theory while Chapters 11.1, 13.1, and 13.2 relate to the topics discussed in this chapter. These will provide more graph theory than you will need for understanding SPF[14]

[15] Ignore the child–like figures. This is a *good* resource. The author is correct and clear in all explanations. In my opinion, a job well done.

Chapter 14
Dynamic Host Configuration

Overview

Static assignment of names and addresses on a large network can consume a large amount of time and suffers from the same problems as any static configuration. Human errors by the network administrators can lead to duplicate IP addresses, IP addresses in the wrong ranges, and other errors. Some such errors can lead to the device not being able to function on the network and others can lead to the network crashing. To make matters worse, some of these human errors are difficult to find.

As devices are added to the network, or taken off of the network, static addressing cannot reflect those changes without human intervention. In the past this was not an issue for many networks, but with the explosive growth of mobile devices such as phones, tablets, and laptops, most modern networks change hourly or even faster. Static address assignment cannot keep up.

Fortunately it is relatively easy to assign addresses and other resources on an as needed–basis as well as provide the required IP configuration of address, network mask, and default gateway (router). To do this the DHCP was developed [113].

14.1 The Need for DHCP

In order to function on a network, a device needs a number of different resources, or key knowledge, about the network. The network needs to have knowledge of the device as well in order to determine what resources, *if any*, the device is allowed to use. DHCP, and the related protocol BOOTP[1], are used to facilitate a centralized, automatic means of controlling which devices are allowed and how those devices access the network.

---

[1] Bootstrap Protocol

© Springer Nature Switzerland AG 2020

G. Howser, *Computer Networks and the Internet*,

https://doi.org/10.1007/978-3-030-34496-2_14

14.2 BOOTP Protocol

While BOOTP [73] is quite different from DHCP the two protocols have much in common as both are intended to minimize the amount of configuration required for a client device and both rely on the client's MAC address to identify the client NIC. BOOTP provides a static IP address to a client so that it will have the exact same IP address each and every time it joins the local network, see also Section 14.3.3. BOOTP also provides the ability to force the client to load a missing configuration file or even an entire OS via TFTP[2]. At one time this was a great advantage for routers that had problems with corrupted configuration files[3] or OS.

BOOTP client	Message	BOOTP server
1. Unconfigured	BOOTP Request $\longrightarrow$	
2.		Searches files for matching MAC
3.	$\longleftarrow$ Response	Found match
4. Client retrieves:		
IP address		
Network address		
Default gateway		
TFTP Server address		
5.	TFTP Session request$\longrightarrow$	TFTP Server
6.		Checks tables
7.	$\longleftarrow$ TFTP Session start	Found match
8. Client configured,		
9. Possible file load		
10. Normal operations		Session ended

Fig. 14.1: The BOOTP Protocol

A device running BOOTP has no network configuration when the NIC initializes and must use the steps in Table 14.1. There are some things to consider at each step.

1. When a NIC becomes active the device can also initiate a BOOTP client. Whether or not a device uses BOOTP depends upon the hardware (NIC) and the software of the OS. As the NIC has no correct configuration[4], the BOOTP request must be sent as a Layer 2 Broadcast which contains the NIC MAC address. This is how the BOOTP server can identify the NIC and properly reply.
2. The server checks its BOOTP tables for an entry with a matching MAC address. If none is found and no default values are set, the server ignores the request. After enough BOOTP requests have been Broadcast (typically 5), the BOOTP

---

[2] Trivial File Transfer Protocol

[3] This was a common problem in the early 1990's with Bay Network's routers to the point that the configuration utility was sometimes called "Site Destroyer".

[4] If the NIC has a correct configuration, the device *might* still require BOOTP to download a configuration file or an OS. This is not common.

client simply gives up until the device or the NIC is rebooted. In this way, the BOOTP server can prevent an unknown NIC from being active on the network. If a match is found, the server sends a BOOTP response with the IP address, network mask, default gateway router address, and possibly other information. If the server file has a file name and TFTP server, these are sent as well in order that the client can initiate a download.

3. The client retrieves the desired information from the BOOTP response. If a TFTP server and file name are present, the client initiates a TFTP session with the TFTP server. If not, the protocol is terminated by the client as the client is now completely configured.

4. If the client needs to initiate a TFTP session, it sends a request with its address and the file it requires to the server address in the BOOTP response in Step 3.

5. The TFTP server checks its files for a matching client address and information to send to the client. If these are not found, the server may, or may not, send a negative reply[5].

6. If a match is found, the server initiates a session with the client and the required files[6] are transferred to the client device.

7. The files are transferred and the client device configured. The protocols is finished.

For a number of reasons BOOTP is *not* the first choice for dynamic configuration of client devices on a network. In a BOOTP environment each device must be individually configured by an administrator. If more is required than BOOTP can easily provide, PXE[7] (pronounced "pixie") is sometimes more appropriate than BOOTP.

If the device only needs to have an IP address, network mask, and internet gateway; DHCP is often the protocol of choice. The choice depends upon the situation and is usually quite obvious.

---

[5] Unfortunately, a negative response "leaks" some useful information to an adversary. It is best to not reply.

[6] These files can include configuration information or an entire operating system. This also serves as a backup of the files to be transferred. Most TFTP servers can be configured in such a way as to relieve the server of storing multiple copies of identical files. The configuration depends entirely upon the actual TFTP server used. Some are very touchy about the exact format of the files and commands for each device and can be a headache to set up and maintain.

[7] Preboot eXecution Environment

Table 14.1: Some Common DHCP Options

Option Name	Source	Description
1 Subnet Mask	RFC 2132	The subnet mask to apply to the address that is assigned to the client.
2 Time Zone Offset	RFC 2132	Informs the client about the time zone offset, in seconds. For example, Pacific Standard Time is GMT – 8 hours. This field would be filled with "- 28800". (Eight hours * 60 minutes/hour * 60 seconds/minute)
3 Gateway	RFC 2132	Tells the client which router is the default gateway router.
4 Time Server	RFC 2132	Tells the client the IP address of a time server that can determine the client's current time. This is related to the Time Zone Offset option.
6 Domain Name Server	RFC 2132	Carries the IP address(es) of the DNS servers that the client uses for name resolution.
7 Log Server	RFC 2132	Carries the IP address of the syslog server that receives the client's log messages.
12 Hostname	RFC 2132	Carries the hostname portion of a client's fully qualified domain name (FQDN). For example, the "www" part of "www.example.com".
15 Domain Name	RFC 2132	Carries the domain name portion of a client's fully qualified domain name (FQDN). For example, the "example.com" portion of "www.example.com".
51 Lease Time Option	RFC 2132	This defines the maximum amount of time that the client may use the IP address.
66 TFTP Server Name	RFC 2132	Carries the FQDN or IP address (or cluster identifier) that the device should use to download the file specified in option 67.[a]
67 Filename	RFC 2132	Carries the filename that is to be downloaded from the server specified in option 66.
82 Relay Agent	RFC 3046	This option carries many other sub-options that are added by relay agents and not the clients themselves.

[39]

[a] Note that often the data put into option 66 does not actually appear in the DHCP packet as option 66, but may have been moved into the "sname" field of the DHCP packet. Additionally, the FQDN may have been resolved to an IP address and also placed in the "siaddr" field of the DHCP packet.

14.3 DHCP Client and Server

Table 14.2: The DHCP Handshake

DHCP client	Message	DHCP server

DISCOVER$\longrightarrow$
The client sends our a DHCP discover
message to identify the server(broadcast)

$\longleftarrow$ **OFFER**
The DHCP server responds with an
available IP address and options(unicast)

REQUEST $\longrightarrow$
The client requests the IP address from
the server(unicast)

$\longleftarrow$ **ACKNOWLEDGE**
The server acknowledges the IP request
and completes the handsake(unicast)

The interactions between a DHCP client running on a NIC and a DHCP server running as a process on some connected device are very similar to those in the bootstrap protocol as shown in Table 14.2. One of the main differences between the two protocols is almost every IP network has a version of DHCP running to automatically manage IP addressing. Like BOOTP, DHCP can also provide other critical configuration information to the client as in Table 14.2.

This does still leave us with some issues. How do we decide what options to use? Obviously DHCP provides tools to handle many of the issues with providing services to anonymous clients that wish to join the network, but how do we control who and how many of these anonymous clients we allow on our network? These problems can be placed into two categories: Dynamic IP addressing and static IP addressing. DHCP provides tools to handle both of these possibilities as well as MAC level security (allow/disallow).

14.3.1 Duplicate DHCP Servers

The question of security and DHCP is very important. A rogue DHCP server can easily cripple a network and can be difficult to guard against and is a violation of many state and federal laws[8]. Equally dangerous is running multiple DHCP servers on the same subnetwork. Great care must be taken to insure that multiple DHCP servers give out the same options but do *not* give out any duplicate IP addresses. Should two DHCP servers give out different, and incorrect, options it is fairly easy to determine which server gave which information to a client. The issue of duplicate IP addresses is not as easy to track down and can create some rather novel problems. Even worse, the symptoms of duplicate IP addresses can come and go depending upon network traffic, activity by one of the duplicates, and other factors.

The safest way to deal with multiple servers is to not have them. Typically the load on a DHCP server is very light and can be handled by most devices. Should there be a legitimate reason for multiple DHCP servers it is best to have them administered by a single person and well documented as to what address pools each is using and the IP address (and location) of the other server(s). Again, avoid this situation when possible.

14.3.2 DHCP Dynamic IP Addressing

The most useful service provided by DHCP is dynamic addressing. No longer must an administrator intervene when a device wishes to join the network. Instead the administrator sets up an "address pool" of addresses to be shared by anonymous devices on a first–come–first–served basis. The administrator can control how these addresses are used without ever needing to touch the device.

When a DHCP client on a device requests a configuration from a DHCP server, the server provides the device with a "lease" on an IP address. The terms of this lease determine how long the IP address is reserved for a specific MAC. In this way resources are not tied up forever for a device that may, or may not, appear on the network in the future. Likewise a device will periodically request the lease to be renewed so that its IP address will be semi-predictable.

Another advantage of dynamic addressing occurs when the pool of addresses have all been assigned. The server keeps a database of all the assigned IP/MAC address pairs and when the lease expires. Should a new device request an address after the pool is full, the server deletes the oldest expired lease and assigns that address to the new device. Should there be no expired leases, the server can deny the request.

The only problem that remains is how to deal with devices that *must* always have a predictable address such as routers, servers, and printers. These devices require a static address.

---

[8] From my own experience this is not an easy problem to determine and fix.

14.3.3 DHCP Static IP Addressing

Obviously, any device that provides a service that is contacted by address must have a predictable address. If the address of a router interface changes each time the router boots, that router cannot be used by devices as a default gateway for Internet access. In both cases a number of addresses are reserved for specific devices and are not part of any address pool.

Method 1

The most reliable method is to reserve part of the available address pool and then manually configure important devices to use one of the reserved addresses. Typically router interfaces are given low host addresses with the best gateway to the Internet given a host address of decimal 1; however, this is only a custom and any address may be reserved for the router. It is wise to keep the address pool as one contiguous block of addresses even if there are multiple address pools. For example: pool1 10 → 35, pool2 36 → 50, pool3 50 → 150, and so on. It makes life easier for the network administrator when changes are required. To allow for future expansion, you could consider not allocating all the available addresses or placing them all in a single address pool that can be shrunk when creating new address pools.

Method 2

Most DHCP services can be configured with a list of IP/MAC addresses. This list is checked each time a new MAC sends a DHCP request. If a match is found, the request is answered with the reserved IP address and other paired information. In this way the device is guaranteed to always receive a predictable IP address. The major flaw with this method is left to the reader as an exercise, see Exercise 1.

14.4 Decentralized DHCP

It is common for a large network to have many smaller sub–networks that provide services to mobile devices. These devices must be able to obtain the information necessary to connect to many different networks. One solution to this problem is to have a decentralized schema to provide these services from a server local to that network. This requires the physical server to have a NIC for each sub–network for which it provides DHCP. For this reason some networks provide DHCP from routers because, in practice, every subnet includes at least one router.

14.4.1 Configuring a DHCP server on Raspbian

Table 14.3: DHCP Configuration Device Example

Interface	Address	Network	Host	DHCP server
eth0	10.2.2.2/16	10.2.0.0	0.2	yes
eth1	192.168.1.1/24	192.168.1.0	1	no
eth2	172.16.0.1/16	172.16.0.0	0.1	yes

Consider the Pi described in Table 14.3. This Pi has three interfaces in three very different networks and needs to supply DHCP on only two of its interfaces, 10.2.2.0, and 172.12.0.0. First let's do some quick calculations as to how many hosts are allowed on each network (see also Exercise 2.2 for a hint). First, the number of network bits can be found from the number of binary "1"s in the IPv4 subnet mask or the number after the "/" in CIDR notation. For example, 10.0.0.1/8 and 10.0.0.2, 255.0.0.0 both have 8 network bits and 24 host bits. The following pair of equations are used to find the number of allowed host addresses on a given network.

$$\text{Number of host bits } h = 32 - \textbf{network bits}. \tag{14.1}$$

$$\textbf{Number of hosts } n = 2^h - 2. \tag{14.2}$$

From this we can see that the network 10.2.2.2/16 has 65,534 possible host addresses and network 192.168.1.1/24 has only 254 allowed hosts. This gives us the maximum size of the address pool on each subnetwork. In most cases some of these addresses should be reserved for static IP addresses for routers and servers because client devices will need to know these addresses. For example, on network 192.168.1.0/24 it might be wise to use an address pool of 200 devices, say $26 \rightarrow 225$ to allow for 25 low addresses and 23 high addresses. A network administrator might choose a much smaller pool, or multiple pools, to allow for growth of the network.

14.4.2 DHCP on Raspbian

DHCP is provided by the **dhcpd** daemon which is configured by the **/etc/dhcp/dhcpd.conf**. Notice this server is being informed of a subnetwork that it should *not* supply DHCP. While this is not required, it is a good idea to document all the subnetworks for which this device has a NIC.

```
# No service will be given on this subnet, but declaring it
    helps the
# DHCP server to understand the network topology.
subnet 192.168.1.0 netmask 255.255.255.0 {
}

#subnet 10.152.187.0 netmask 255.255.255.0 {
```

```
#}
# This is a very basic subnet declaration.
#subnet 10.254.239.0 netmask 255.255.255.224 {
#   range 10.254.239.10 10.254.239.20;
#   option routers rtr-239-0-1.example.org, rtr-239-0-2.example
    .org;
#}
# A slightly different configuration for an internal subnet.
subnet 10.2.2.0 netmask 255.255.255.0 {
   range 10.2.2.10 10.2.2.200;
   option domain-name-server ns1.mydomain.com;
   option domain-name ``mydomain.com;
   option routers 10.2.2.1;
   option broadcast-address 10.2.2.255;
   default-lease-time 600;
   max-lease-time 7200;
}

subnet 172.16.0.0 netmask 255.255.0.0 {
range 172.16.0.50 172.16.10.200;
option domain-name-server ns1.anotherdomain.com;
option domain-name anotherdomain.com;
option routers 172.16.0.1;
option broadcast-address 172.16.255.255;
default-lease-time 600;
max-lease-time 7200;
}
# Hosts which require special configuration options can be
    listed in
# host statements.   If no address is specified, the address
    will be
# allocated dynamically (if possible), but the host-secific
    information
# will still come from the host declaration.

#host passacaglia {
#   hardware Ethernet 0:0:c0:5d:bd:95;
#   filename "vmunix.passacaglia";
#   server-name "toccata.example.com";
#}
```

14.5 Centralized DHCP

Many large networks with multiple sub–networks prefer to have only one DHCP server instead of multiple servers. This can be done by using a process called a DHCP helper. This process simply forwards all DHCP requests to a single central server. The **dhcp-helper** package is easy to configure by editing the file **/etc/dhcp-helper.conf** to inform the process of the IP address of the centralized DHCP server. All requests for DHCP are simply forwarded to the DHCP server and handled exactly like decentralized DHCP services[9].

---

[9] Personally, I do not see any advantages of centralized over decentralized except in the number of devices the administrators must touch. This is an opinion. When it is your network, you do what makes the most sense to you.

14.6 DHCP and Dynamic DNS

This section assumes that the hostname for each client is unique on the network. In Chapter 20 we discuss how to dynamically use the host name to construct the name by which other devices can find this device on the network by using name service.

Projects

14.1 Configure DHCP for each network owned by your group. Attach a laptop and check the setting via **sudo ifconfig -a** for Linux or **ipconfig /all** for windows. Take screenshots of correct results.

14.2 Research DHCP Helper and how it works.

14.3 (Advanced)Configure two DHCP servers for the *same* network:

   a. Each DHCP server has a range assigned to it so that no two devices can get that same address. Note the results. When could this be a good thing?

   b. Configure multiple DHCP servers to give out addresses in the exact same, or overlapping, ranges. NOTE: This is a bad thing. Note what problems, if any, occur.

   c. Change the problematic configurations so that DHCP works correctly.

Exercises

14.1 Using DHCP Server daemon configuration files to create a static IP address and MAC address pair has a major flaw. Briefly explain a situation where a router interface could legitimately obtain the wrong IP address from the daemon.

14.2 Why can only 254, not 256, hosts be assigned on a network 192.168.1.0?. What happened to the other 2 host addresses?

Further Reading

The RFC[10] below provide further information about BOOTP and DHCP. This is *not* an exhaustive list and most RFC are typically dense and hard to read. Normally RFC are most useful when writing a process to implement a specific protocol.

RFCs Directly Related to BOOTP and DHCP

BOOTP	Title
RFC 0951	BOOTSTRAP PROTOCOL (BOOTP) [73]
RFC 1395	BOOTP Vendor Information Extensions [98]
RFC 1497	BOOTP Vendor Information Extensions [106]
RFC 1532	Clarifications and Extensions for the Bootstrap Protocol [112]
RFC 1542	Clarifications and Extensions for the Bootstrap Protocol [114]
RFC 2132	DHCP Options and BOOTP Vendor Extensions [161]

DHCP	Title
RFC 0951	BOOTSTRAP PROTOCOL (BOOTP) [73]
RFC 1497	BOOTP Vendor Information Extensions [106]
RFC 1532	Clarifications and Extensions for the Bootstrap Protocol [112]
RFC 1541	Dynamic Host Configuration Protocol [113]
RFC 1542	Clarifications and Extensions for the Bootstrap Protocol [114]
RFC 2131	Dynamic Host Configuration Protocol [160]
RFC 2132	DHCP Options and BOOTP Vendor Extensions [161]
RFC 3046	DHCP Relay Agent Information Option [195]
RFC 3315	Dynamic Host Configuration Protocol for IPv6 (DHCPv6) [199]
RFC 3396	Encoding Long Options in the Dynamic Host Configuration Protocol (DHCPv4) [203]
RFC 3646	DNS Configuration options for Dynamic Host Configuration Protocol for IPv6 (DHCPv6) [220]
RFC 4361	Node-specific Client Identifiers for Dynamic Host Configuration Protocol Version Four (DHCPv4) [235]
RFC 4702	The Dynamic Host Configuration Protocol (DHCP) Client Fully Qualified Domain Name (FQDN) Option [245]
RFC 4703	Resolution of Fully Qualified Domain Name (FQDN) Conflicts among Dynamic Host Configuration Protocol (DHCP) Clients [246]
RFC 6842	Client Identifier Option in DHCP Server Replies [276]

---

[10] Request For Comments

DHCP	Title
RFC 7969	Customizing DHCP Configuration on the Basis of Network Topology [288]
RFC 8115	DHCPv6 Option for IPv4-Embedded Multicast and Unicast IPv6 Prefixes [290]
RFC 8353	Generic Security Service API Version 2: Java Bindings Update [293]
RFC 8415	Dynamic Host Configuration Protocol for IPv6 (DHCPv6) [296]
RFC 8539	Softwire Provisioning Using DHCPv4 over DHCPv6 [304]

Other RFCs Related to DHCP

For a list of other RFCs related to the OSPF protocol but not closely referenced in this chapter, please see Appendicies B–BOOTP and B–DHCP.

Chapter 15
Routing Protocols

Overview

Routing protocols are designed to automatically update routers when there are changes to the network. As a welcome side–effect, routing protocols are unable to create routing loops and always pick the "best" route depending upon how that protocol measures best. In this chapter, we will examine the open standard routing protocols which can run on a Raspberry Pi router. These protocols operate by updating the Route Cache and then running[1] Algorithm 4. The actual routing is done by routines in the Raspbian kernel. This is actually typical of routers as routing protocols do not directly deal with routing packets. Most of these protocols are able to route other Layer 3 packets, such as IPX, but any such support would be old and not updated on a regular basis. Fortunately any modern network should be able to restrict Layer 3 protocols to IPv4 and/or IPv6.

15.1 Proprietary Protocols

Some of the most effective routing protocols cannot be implemented on the Raspberry Pi or a Linux computer because they are proprietary to a specific company, usually Cisco Systems. Because they were developed to run on a specific set of hardware and a specific OS, these protocols are "tuned" to a given set of situations and gain some efficiencies because of this. Fortunately, these protocols must be able to interact with the open standards so they can be viewed as a cloud, or node, on the larger Internet. We will not discuss these in great detail. These are of some interest to us because the relative price of a dedicated router is part of the reason for the development of Zebra and Quagga [43].

---

[1] Running an algorithm or process is sometime referred to as "calling" the process. This comes from programming.

© Springer Nature Switzerland AG 2020
G. Howser, *Computer Networks and the Internet*,
https://doi.org/10.1007/978-3-030-34496-2_15

Proprietary routing protocols have a number of common characteristics. Most importantly, the details of how the protocols work are sometimes not well known. This is understandable, but it does present some problems at times as the user is dependent upon the vendor for support. Secondly, in order to make full use of a proprietary protocol, the network must be built upon one vendor's hardware and/or software which can present some embarrassing problems should the vendor drop support for a product or go out of business.

Please note that this book is not intended to cover *all* the protocols, open standard or proprietary, that can run on a router. Topics such as fail–over and load balancing are beyond the scope of this text.

15.1.1 IGRP

IGRP is a proprietary protocol for medium to large networks. The network routers are grouped into autonomous networks which are each assigned a network number by the network administrator[2] The routers establish a neighbor relationship and determine the best route using a balance between distance–vector (hops) and link bandwidth and availability. IGRP should not be used due to the fact that it does not support VLSM, much like RIPv1[3].

15.2 Open Standards Protocols

Open Standards Protocols provide a number of very good routing protocols such as RIP, OSPF, and BGP. Because these are open standards, anyone can write a process to run these standards. One of the better projects for these protocols is the Quagga project. Originally begun as a tool to help engineers study for Cisco Systems certification exams, Quagga[5] provides a Cisco Systems–like user interface to configure the router.

15.2.1 Enhanced Internal Gateway Routing Protocol (EIGRP)

At this time, EIGRP is being made an open standard protocol by Cisco Systems; however, there are still some parts of the protocol that have not been made public

---

[2] Smaller networks can have all the routers placed in one autonomous network. It seems to work just fine.

[3] While not exactly true, you should consider IGRP as deprecated or superseded by EIGRP[4].

[5] Quagga is a fork of the earlier Zebra project and runs very well on most Linux platforms such as Raspbian.

so EIGRP will be considered a proprietary standard for now. EIGRP is an enhanced version of IGRP and supports VLSM [287].

15.2.2 Route Interchange Protocol (RIP)

There are multiple versions of the RIP protocols: RIPv1, RIPv2[6], and RIPng[7] for IPv6. Fortunately, the three versions are very similar and RIPv1 has been superseded by RIPv2[8] so unless RIPv1 is noted specifically, we will use RIP to refer to RIPv2. We will explore RIP and implement it on a test network in Chapter 16.

15.2.3 Open Shortest Path First (OSPF)

OSPF [116] is a Link–State protocol and is more suited to larger and complex networks than those for which RIP is a better choice. Like RIP, OSPF is an open standard available for use on the Raspberry Pi; however, OSPF can be tuned to more closely model the actual network than RIP.

Unlike RIP, OSPF requires a detailed knowledge of the network design in order to realize the benefits of the power of the protocol. OSPF has significantly less overhead for networks of say 25 or more routers that can be grouped together into a bi–level network configuration. Also, OSPF allows the network administrator to mark each network link between routers with a relative cost in order that OSPF can determine a "best" network in terms of the actual total cost of the links in a path through the network. Like other routing protocols, OSPF quickly responds to changes in the network without intervention by the network administrator. We explore OSPF in depth in Chapter 17.

15.2.4 Itermediate System to Intermediate System

An interesting protocol used mainly by large ISPs is ISIS which uses NSAP[9] addresses, rather than IP, for messages between routers and can route many different protocols with one ISIS process. ISIS networks can be divided into two types: backbone and autonomous systems. ISIS builds the backbone using ISIS Level 2

---

[6] Route Interchange Protocol, Version 2

[7] Route Interchange Protocol for IPv6

[8] In fact, RIPv1 need only be discussed in terms of what it does *not* do. RIPv1 requires that all network masks be the natural network mask which break on octet boundaries while RIPv2 relaxes this requirement. Any situation that can be handled by RIPv1 on a modern network can easily be handled by RIPv2.

[9] Network Service Access Point

routers which find adjacent routers automatically. Likewise, autonomous systems are formed from routers with the same area in their addresses.

Because NSAP addressing is more Layer 2 than Layer 3, ISIS has complete transparency when routing packets. ISIS works, more or less, at Layer 2 but is able to connect various Layer 2 networks without actually bridging them into one network if different NSAP areas. For this reason, some networking people refer to NSAP as "Layer 2.5" which is irritating and obscures what is really happening.

In practice an ISIS network resembles an OSPF network, with some subtle differences that are beyond the scope of this text. You are not likely to ever encounter ISIS outside of a network run by a large ISP.

15.2.5 Border Gateway Protocol (BGP)

Unlike RIP or OSPF, BGP is designed to control the interconnections between large networks such as those developed by nationwide ISPs. BGP is used to control how routes are shared thus protecting the proprietary design of nationwide networks.

We will discuss BGP in more detail in Chapter 18 while we implement a BGP router between the groups of our test network. Keep in mind that BGP is designed to connect nation–wide ISPs and it is overkill to use it on our test network.

15.2.6 Babel

Babel is an emerging set of Layer 3 protocols that is reported to converge very quickly. Babel is especially useful when devices move from network to network or links change often such as WiFi[10]. Support for Babel appears somewhat limited and is still evolving.

15.3 Precedence of Routing Protocols

When a router "learns" routes from other routers in the network, it is highly likely that more than one protocol is advertising routes. The router places all of these into its Route Cache and determines the best route based upon a number of protocol specific characteristics. Each protocol has an administrative distance associated with routes it advertises. Routers "trust" the routes with the lowest administrative distance and prefer this route over one with a higher administrative distance. The Cisco Systems defaults, which are used by Quagga for inter–interoperability, are given in Table 15.3 [12].

---

[10] Wireless Network

Table 15.1: Administrative Distance of Common Routing Protocols

Routing protocol /Route source	Default Distance Values
Connected interface	0
Static route	1
EIGRP summary route	5
External Border Gateway Protocol (eBGP)	20
Internal EIGRP	90
IGRP	100
OSPF	110
Intermediate System-to-Intermediate System (IS-IS)	115
Routing Information Protocol (RIP)	120
Exterior Gateway Protocol (EGP)	140
On Demand Routing (ODR)	160
External EIGRP	170
Internal BGP	200
Unknown	255

15.4 Configuring Static Routes

Routes can be entered manually as static routes. Static routes are a potential hazard and should be avoided except in special circumstance because they are not learned via routing protocols and *they do not respond to changes in the network*. In other words, they are quite literally static. A good rule of thumb would be for all routers to have one static route unless the router needs to always dump certain traffic to block specific networks as we shall see below.

15.4.1 The Default Route

A common exception to the rule about avoiding static routes is the default route or "network of last resort[11]". This is signified by a destination networkof 0.0.0.0 for IPv4. Logically this network is at the tail end of the route table and matches any destination. Packets that are not sent anywhere else will match this entry and be passed along to the default route.

---

[11] I always like the older term, network of last resort, because it fit so well. The more current terms are "default gateway" or "default route" which serve the same purpose.

15.4.2 Blocking Private Networks

Another critical use of static routes is to drop all packets for the private IPv4 networks by sending them to a null interface or a black–hole[12]" If private networks are routed to the Internet bad things will happen, so it is common for routers that connect to the Internet to dump all traffic to the Class A, Class B, and Class C networks. This should always be done as such traffic could also be an attack. Likewise, you should consider blocking all traffic from outside your network that appears to match your interior networks. For example, if you were assigned the network 256.256.256.0/24 you should be suspicious of any traffic from the Internet with a source of 256.256.256.73[13].

15.5 Quagga Configuration, vtysh, and telnet

Quagga was designed to provide a low–cost alternative to dedicated routers such as those manufactured by Cisco Systems. Because it was to be used as a tool to help learn the "feel" of configuring a Cisco Systems router, Quagga is typically configured using a remote interface emulation provided by **vtysh** which acts like a ssh connection. This avoidance of configuration via editing a file is one of the strong points of the Cisco Systems philosophy. Changes to the configuration of a Quagga protocol happen immediately, just as with IOS[14].

Quagga has two command levels, Privileged and Unprivileged, which can be restricted based upon user. Typically configuration is limited to Network Administrators, but it is still useful to have a command level that can only report on the state of the router without the ability to make any changes. Accidental changes to a router configuration can cause catastrophic failures throughout the network. It can be especially embarrassing to change the network such that Quagga can no longer be contacted[15].

Before we can attempt to configure the Raspberry Pi for any routing protocol, it is best to examine how Quagga and vtysh[16] work together.

---

[12] Sometimes called the "bit bucket" after the trash bins that collected the chads punched out of punch cards.

[13] Yes, any IPv4 address starting with 256 is invalid anyway, but I did not want to use someones legitimate address space.

[14] Internet Operating System

[15] This is the reason Quagga and Cisco Systems routers are configured via commands that change the running configuration, not the configuration files. One simply calls a trusted person to power cycle the router, the router loads from the old configuration files, and all is well. That is, if you did not write the changes to the configuration files.

[16] Virtual Terminal Shell

15.5.1 Contacting Quagga

Quagga is designed to work with a terminal emulation program named **vtysh**. While the **zebra** daemon is running, it listens on a number of sockets, or ports, as defined in the **sysctl.config** file. In order to insure that the proper configuration files are updated, contact Quagga on port **2601**, which is the **zebra** socket or port. While it is possible to contact each routing daemon on a specific port, there is no guarantee the resulting configuration will run properly after a reboot of the router. Always contact the **zebra** daemon for reports or configuration changes by entering **sudo vtysh 2601**.

If you are not the network administrator, the proper way to contact Quagga is: **sudo telnet 127.0.0.1 2601**. This gives the user an experience close to that of using a Cisco Systems router via a remote connection[17]. The alternate method is **sudo vtysh 2601**, which gives the user complete access to all commands and bypasses the need to know the router passwords. This must be restricted to network administrators only.

> **Warning:** On a production system, **vtysh** can be a security violation and must be tightly controlled.

15.5.2 The Quagga Interface

By default **zebra** and the other daemons will only accept connections from users in the **quagga** group. This allows quite a bit of flexibility that is not needed for a test–bed network but can be used to allow inquiry–only access to some users and configuration access to a small group of network administrators as explained further in Subsection 15.5.10. In any case, members of the group allowed to use the **sudo** command can gain access to the privileged commands if they know the password.

All commands are first processed by the **zebra** daemon which either completes the command or passes it to the correct daemon for processing. If the **zebra** daemon is allowed to process all commands via port 2601, Quagga will produce the correct configuration files in the **/etc/quagga/** directory. While these configuration files can be edited directly, it is best to use the configuration interface.

Commands may be abbreviated as long as the abbreviation can be expanded to only one valid command. As an alternate option, the tab key can be used to complete any command. If there has not been enough entered to expand to only one command, Quagga will not complete the command. At any time a list of the currently valid commands, or options to complete the current command, can be listed by entering

---

[17] Do not make changes when you contact Quagga via telnet as they may not survive a reboot of the Pi. You were warned.

"?". For example, to find all the commands relevant to the **show** command you can enter **show ?**.

Once a command has been accepted by Quagga it cannot be edited[18]. The only solution is to delete the command and try again. Fortunately this is fairly easy to do as any command can be deleted by entering the keyword **no** followed by the command *exactly* as it was entered.

Commands that change how the router operates take effect immediately *but do not survive a reboot or a restart of the daemons*. This is in keeping with how IOS operates. In fact, this is one of the hidden strengths of Quagga. One can try out configuration changes and quickly back them out by restarting the daemons.

The current configuration can be listed on the screen with the command **write term** or other suitable abbreviation of **write terminal**. The current configuration can be written to the configuration files by the write command as well. This makes the current configuration permanent so that it survives a power cycle (on/off or reboot). Cisco Systems routers normally do not have a disk drive and keep their running configuration on flash memory; therefore, the command to save the configuration writes it to "memory" instead of "saving" it. The command is "write memory".

15.5.3 Unprivileged (Inquiry Only) Commands

If the prompt ends with ">", **zebra** is currently in the Unprivileged mode and only inquiry commands are active. It is still possible to obtain important information, but the running configuration is safe from any intentional or accidental changes. For example, it is possible to obtain information about the route table with the command **show ip route** as below:

```
router1-1> show ip route
Codes: K - kernel route, C - connected, S - static, R - RIP,
   O - OSPF, I - IS-IS, B - BGP, P - PIM, A - Babel,
   > - selected route, * - FIB route

K>* 0.0.0.0/0 via 10.3.0.1, eth1, src 10.3.0.1
C>* 10.1.0.0/24 is directly connected, eth0
R>* 10.2.0.0/24 [120/2] via 10.1.0.2, eth0, 00:19:46
R>* 10.2.2.0/24 [120/2] via 10.1.0.2, eth0, 00:19:46
C>* 10.3.0.0/24 is directly connected, eth1
C>* 127.0.0.0/8 is directly connected, lo
R>* 172.18.0.0/24 [120/3] via 10.1.0.2, eth0, 00:19:46
router1-1>
```

A full list of the Unprivileged Commands can be obtained by entering **?** at the prompt which will list all the commands that are valid at the current menu level. The **enable** command (with password) changes to the Privileged Command Menu.

---

[18] This can be a real problem for a poor typist or someone having a bad day.

15.5.4 Privileged Commands

The router configuration files are handled from the Privileged Command Menu and sub–menus. This menu is reached by the **enable** command and the prompt is changed to "#". Commands affect the running configuration immediately but do not survive a restart/reboot until the configuration is saved to "memory", or in other words the microSD card, with the **write memory** command. This menu allows one to configure the router interfaces, routing protocols, and configuration file security. No configuration commands can be entered until the command **configure terminal** is entered to inform **zebra** to accept configuration commands from the terminal.

Interface Configuration

The first step is to configure the router interfaces one at a time via the **interface** menu by entering : **interface <name of interface>** where name of interface is **eth0**, **eth1**, and so on. An important command for an interface is **shutdown** which shuts down the interface and **no shutdown** which brings the interface back up. It might be a good idea to shut down the **wlan0** interface to avoid creating a rouge wireless network[19].

This is the menu used to assign an IP address to the interface for the Route Engine to use for directly connected routes. **This *must* be the same IP address set in the dhcpcd.conf file or problems will occur.** It is also a good idea to set the **description** in order that the configuration is self–documenting. For example,

To leave this sub–menu, use the command **end**.

The other commands in this sub–menu are explained by entering **help**.

15.5.5 Sample Quagga Configuration Files

Quagga *should* install these files, but at least one version of the installation did not. Versions of these files can easily be found by searching the web, but these are included for your convenience.

---

[19] The other option is to talk with the local networking and security people about running a wireless network that goes nowhere. Wireless networks are a *huge* security risk. If you set up a network, set it up with an SSID that warns people not to connect and protect it. People will still try to connect because that's what people do.

15.5.6 Log Files

Quagga Log files must be created for each protocol in the **/var/log/quagga/** directory[20]. They can be created all at once or as needed with the **sudo touch** command as above. For example, **sudo touch /var/log/quagga/zebra.log**, will create the **zebra** log file.

Table 15.2: Quagga Log Files

Protocol	Daemon	Log File
babel	babeld	/var/log/quagga/babeld
BGP	bgpd	/var/log/quagga/bgpd
IS–IS	isisd	/var/log/quagga/isisfd
OSPF (IPv4)	ospfd	/var/log/quagga/ospfd
OSPF (IPv6)	ospf6d	/var/log/quagga/ospf6d
RIP (IPv4)	ripd	/var/log/quagga/ripd
RIP (IPv6)	ripngd	/var/log/quagga/ripngd
Zebra	zebra	/var/log/quagga/zebra

15.5.7 Files in /etc/quagga

On the Raspberry Pi these files should be in the directory **/etc/quagga**.

```
pi@CustomPi:~\$ cd /etc/quagga/
pi@CustomPi:/etc/quagga# ls -l
total 16
-rw-r--r-- 1 quagga quagga   0 Dec 22 15:58 babeld.conf.sample
-rw-r--r-- 1 quagga quagga   0 Dec 22 15:57 bgpd.conf.sample
-rw-r--r-- 1 quagga quagga  78 Dec 22 16:01 daemons
-rw-r--r-- 1 quagga quagga   0 Dec 22 15:58 isisd.conf.sample
-rw-r--r-- 1 quagga quagga   0 Dec 22 15:58 ospf6d.conf.sample
-rw-r----- 1 quagga quagga   0 Dec 22 15:57 ospfd.conf
-rw-r--r-- 1 quagga quagga   0 Dec 22 15:57 ospfd.conf.sample
-rw-r--r-- 1 quagga quagga   0 Dec 22 15:59 pimd.conf.sample
-rw-r----- 1 quagga quagga   0 Dec 22 15:56 ripd.conf
-rw-r--r-- 1 quagga quagga   0 Dec 22 15:57 ripd.conf.sample
-rw-r--r-- 1 quagga quagga   0 Dec 22 15:58 ripngd.conf.sample
-rw-r--r-- 1 quagga quagga 222 Dec 22 15:55 vtysh.conf
-rw-r--r-- 1 quagga quagga 222 Dec 22 15:55 vtysh.conf.sample
-rw-r--r-- 1 quagga quagga 508 Dec 22 15:50 zebra.conf.sample
pi@CustomPi:/etc/quagga#
```

daemons

```
pi@CustomPi:/etc/quagga# sudo cat daemons
zebra=yes
bgpd=no
ospfd=yes
```

---

[20] If the directory does not exist, create it with the command **sudo mkdir /var/log/quagga**

```
ospf6d=no
ripd=yes
ripngd=yes
isisd=no
babeld=yes
pi@CustomPi:/etc/quagga#
```

vtysh.conf.sample

```
1   pi@CustomPi:/etc/quagga# sudo cat vtysh.conf.sample
2   !
3   ! Sample vtysh.config file
4   !
5   ! Quagga.org recommends that the following line be commented
        -out
6   ! service integrated-vtysh-config !builds all configurations
        in Quagga.conf
7   hostname quagga-router
8   username root nopassword
9   !
10  pi@CustomPi:/etc/quagga#
```

zebra.conf

```
1    pi@CustomPi:/etc/quagga# sudo cat zebra.conf.sample
2    !
3    ! sample configuration file for zebra
4    !
5    ! !!Please report problems to: pi@gerryhowser.com
6    !
7    ! Version: 0.1
8    !
9    hostname Router
10   password zebra
11   enable password zebra!
12   ! sample configuration file for zebra
13   !
14   ! !!Please report problems to: pi@gerryhowser.com
15   !
16   ! Version: 0.1
17   !
18   hostname Router
19   password zebra
20   enable password zebra
21   !
22   !Interface's description
23   !
24   !interface lo
25   ! local host / hardware loopback
26   !
27   ! Static default route sample.
28   !
29   !ip route 0.0.0.0/0 10.0.0.1
30   !
31   !log file /var/logs/quagga/zebra.log
32   !
33   pi@CustomPi:/etc/quagga#
```

Blank Configuration Files

The following files simply need to exist and can be created by the command: **sudo touch** *filename* where *filename* is **ripd.conf**, **ospfd.conf**, **ospf6d.conf**, **ripng.conf**, **isisd.conf**, and **babel.conf**. These should be created as needed. After these files are created, the owner must be changed via the command `sudo chown quagga:quagga /etc/quagga/*` in order for the background processes to run properly under the username `quagga`.

Global Zebra Configuration

At this point you should enable logging for **zebra**. The log file must exist before **zebra** can write messages to the file.

```
sudo mkdir /var/log/quagga/
sudo touch /var/log/quagga/zebra.log
sudo chown quagga:quagga /var/log/quagga/*
```

Return to the router configuration and, after **config term**, enter `log file /var/log/quagga/zebra.log` to have all log messages sent to the same file.

15.5.8 Static Routes

If you need to set up static routes, especially the default route, enter the routes via the `ip route n.n.n.n/mb <destination>` where n.n.n.n is the IP network, mb is the mask, and ¡destination¿ is the destination for this network. The default route is set by `ip route 0.0.0.0/0 1.2.3.4` and packets can be "tanked" by sending them to a destination of "null0" or "blackhole". If the destination is "reject" the packets will be tanked and a message sent back to the source of the packets[21].

Routing Protocol Configuration

Each of the routing protocols has a separate sub–menu such as **router rip**, **router ospf**, and so on. The sub–menus have different options to fill the needs of each routing protocol and are explained in the sections of this book that reference configuring the various routing protocols.

To leave this sub–menu, use the command **end**

15.5.9 Saving the Configuration

Always remember to write the configuration to memory before you exit Quagga. It is a good idea to display the running configuration, **write term**, and verify it before it is saved with **write memory**. Quagga handles saving the running configuration

---

[21] On the Internet, packets to any of the private networks are typically routed to null0 to avoid a number of possible attacks.

as either one configuration file, **Quagga.conf**, or as a set of configuration files for zebra and one for each of the running routing protocols. Remember: if you don't save changes by **write terminal**, the changes are lost if the **ripd** is restarted or the Pi is rebooted.

15.5.10 Advanced Configuration Options

The default configuration options depend upon exactly how Quagga was installed, but for our purposes it does not really matter. If there are extra security concerns, the zebra configuration can be manually edited to force a password before a user can enter privileged mode. As a precaution **zebra** only connects from members of the **sudo** group on the Pi.

Projects

15.1 Use **sudo vtysh** without a port number and comment on the results.

Exercises

15.1 What are some of the advantages of using open standard routing protocols over using proprietary routing protocols?

15.2 In your own words, explain the rationale behind routing 10.0.0.0 to a null interface.

15.3 Where in the routing table would you expect to find the default route? Explain why it is in that location.

Chapter 16
Route Interchange Protocol

Overview

Route Interchange Protocol (RIP) is one of the simplest routing protocols and one of the first to be developed. For smaller networks it is the protocol of choice, even though RIP has many drawbacks. RIP is simple to implement, requires very little expertise, and is very common on the Internet. Routers running RIP quickly learn the reachable networks and the current best next step to reach those networks.

There are many different versions of RIP that are built to handle almost any routable Layer 3 protocol including IPX, AppleTalk, IPv4, and IPv6. Over the years, RIP has been updated to meet the needs of most small networks of about 25 or fewer routers. For larger networks, RIP route update traffic between routers becomes a significant overhead on the network and a burden on the routers involved.

In this chapter we will examine the advantages and disadvantages of RIPv1, RIPv2, and RIPng by building a RIP network of Raspberry Pi microcomputers.

16.1 The Route Table

Below is a typical Route Table containing routes learned by a Raspberry Pi router from a network of routers running RIP.

```
pi@router1-1:~\$ sudo vtysh 2601

Hello, this is Quagga (version 1.1.1).
Copyright 1996-2005 Kunihiro Ishiguro, et al.

router1-1# show ip route
Codes: K - kernel route, C - connected, S - static, R - RIP,
       O - OSPF, I - IS-IS, B - BGP, P - PIM, A - Babel,
       > - selected route, * - FIB route

K>* 0.0.0.0/0 via 10.3.0.1, eth1, src 10.3.0.1
C>* 10.1.0.0/24 is directly connected, eth0
R>* 10.2.0.0/24 [120/2] via 10.1.0.2, eth0, 00:00:49
R>* 10.2.2.0/24 [120/2] via 10.1.0.2, eth0, 00:00:49
C>* 10.3.0.0/24 is directly connected, eth1
```

© Springer Nature Switzerland AG 2020
G. Howser, *Computer Networks and the Internet*,
https://doi.org/10.1007/978-3-030-34496-2_16

```
C>* 127.0.0.0/8 is directly connected, lo
R>* 172.18.0.0/24 [120/3] via 10.1.0.2, eth0, 00:00:49
router1-1#
```

16.2 Overview of RIP

RIP has been referred to as "routing by rumor" or "routing by gossip" because it is a very simplistic and chatty protocol. However, RIP is a still a very useful protocol for small networks that are not highly complex. A general rule of thumb is that RIP works well for small networks of 25 or fewer routers. What makes RIP so useful is that it is straight–forward and easy to implement.

16.3 Best Route

If a router does not have an interface on the destination networkof a packet, the packet must be passed along to another router that has announced a known route to the destination network. Should the router *not* have a route to the destination network, the packet is either sent to the default route or dropped. The router does not have to notify any device of the problem, although it has the option to do so. Dropped packets are noted by the Transport Layer, Layer 4, and handled there.

Each time a packet must be passed along to a different router counts as one "hop". Hops are critical in RIP as the number of hops determines the best route to a destination network. This can lead to inefficient routes in some cases, such as two high–speed links being faster than one low–speed link, but this is not a concern in RIP networks.

As an analogy, a direct flight (one hop) from Pittsburgh to Atlanta may arrive at 2:00 a.m. while a flight from Pittsburgh to Columbus to Atlanta (two hops) might arrive at 7:30 p.m. RIP would always pick the direct flight while a person might be willing to make the longer journey to arrive earlier in the evening.

The number of hops, or hop count, is also important for packets that take a complex route of many hops. RIP networks have a fixed diameter of 30 hops and any route of 16 or more hops is considered unreachable. The network administrator can set the diameter, or horizon, to a larger value if the network links are high–speed and a packet can therefore travel more hops.

16.4 Routing by Rumor

RIP is a routing protocol based on the Bellman–Ford (or distance vector) algorithm [25]. This algorithm has been used for routing computations in computer networks since the early days of the ARPANET[12]. The particular packet formats and protocol described in RFC 2453 are based on the program **routed**, which is included all distributions of UNIX Operating System [171]. RIP has been called routing by rumor because the protocol propagates known routes by the routers announcing all of their known routes and assuming any announcement received is completely valid.

16.4.1 RIP Route Announcements

By default, RIP sends out route announcements every 30 seconds, but this value can be changed by the Network Administrator. This value could be increased on a stable network or decreased on a network that experiences a large number of connection changes. The trade–off is that larger values lead to longer network convergence times and makes the network less responsive to changes, or failures, which leads to more lost messages in the case of a link failure. If the time is reduced, the network is more responsive to changes but more of the traffic on the network is RIP route announcements. As most networks running RIP are relatively stable and only moderately complex, the default value seems reasonable and should not be changed.

Likewise, the default diameter of a RIP network can be changed by the Network Administrator by adjusting the maximum number of hops allowed before a destination networkis considered unreachable. The default value of 15 hops is reasonable but was determined at a time when network links between routers typically had speeds of 10 mBits or less. The maximum number of hops could be increased but normally if the RIP network has routes of more than 15 hops a different routing protocol should be considered.

RIP announcements contain basic information about every route in the router's Route Table such as the destination networkaddress, the IP address of the router interface, and the number of hops to the destination. Note: These routes are in the Route Table, not the Route Cache, and represent the best known next hop to reach the destination. Every 30 seconds each router sends out a route announcement for *every* route in its route table[3] using Algorithm 7.

---

[1] Advanced Research Projects Agency Network

[2] ARPANET was the original project, the Advanced Research Projects Agency Network, that grew into the Internet as we know it. This would not have happened, or at least not as quickly, without then Senator Al Gore. In many ways, Al Gore is the father of the Internet.

[3] A router can be configured to announce, or not announce, its default route by the Network Administrator on a router by router basis.

Algorithm 7 Route Announcements

```
 1: procedure RIP ANNOUNCE
 2:     while Running RIP do
 3:         Wait default time
 4:         while Routes in Route Table do
 5:             while Interfaces to process do
 6:                 if Best route does not use this interface then
 7:                     Send: I have a route to network x with n hops
 8:                 else
 9:                     Skip this route for this interface
10:                 end if
11:                 next interface
12:             end while
13:             next route in Route Table
14:         end while
15:     end while
16: end procedure
```

16.5 Processing RIP Announcements

When a router receives an announcement from a neighbor router, i.e. one to which it has a direct link, the router uses the announcement to update its local Router Cache. No efforts are made to determine if the route announcement is correct or believable. All announcements are assumed to be correct and valid[4]. When an announcement is received, the router uses Algorithm 8. The RIP process then triggers a Route Table update using an Algorithm such as Algorithm 4 in Chapter 12.

16.6 Convergence of a RIP Network

The two algorithms used to exchange updates between routers work very well with a few exceptions. Because updates only happen every 30 seconds (default) and it can take a number of update cycles to propagate a new route through a network, RIP networks are notorious for the length of time it takes for all the routers to acquire critical information about the network. During this time packets can be routed over inefficient paths through the network or even dropped incorrectly. The goal of RIP is to dynamically adjust all of the route tables to facilitate the routing of packets in the most effective manner possible. What is even worse, it can take much more time for information about routes that no longer exist to get to all routers. Should a destination network disappear, it can be some time before that information reaches

---

[4] The Network Administrator *can* configure a router to accept, or ignore, announcements from a specific router or received on a specific local interface. For the Raspberry Pi, this is done by configuring the **neighbor** routers.

Algorithm 8 Route Cache Updates

```
 1: procedure RIP UPDATE
 2:     Wait until an announcement is received
 3:     if Trusted⁵ then
 4:         while new Route Announcement do
 5:             if Next–hop/Destination pair not in cache then
 6:                 Add 1 to hop count for this route
 7:                 Mark this route as "missed" zero times
 8:                 Add route entry to Route Cache
 9:             else
10:                 Skip this route announcement
11:             end if
12:         end while
13:         Start scan of Route Cache
14:         while Route Cache entry do
15:             if Next–hop matches source of updates then
16:                 if Route not in this update then
17:                     increase "missed" by one
18:                     if "missed" greater than 3 then
19:                         Delete this route from the cache
20:                     end if
21:                 end if
22:             end if
23:         end while
24:         Trigger Route Table update (Algorithm 4)
25:     end if
26: end procedure
```

---

all the routers and during this time packets might still be routed towards the missing networks.

When all the routers have learned the best routes to all known destination networks, the network has converged. At this point packets are being routed correctly and quickly. RIP routers continue to send route announcements every 30 seconds, but route caches, and route tables, are stable. This can take seconds on small networks or hours on large, complex networks with many routers and many network connections. Because of this, RIP is known to not scale well to large networks.

Some networks can never converge. Consider a network with a damaged, but working, link that appears and disappears from a router. This can happen because of hardware failure, convergence issues on another network, or something interrupting the signal[6]. When the link is up, the destination network "appears" and when the link goes down the link "disappears". If this happens often enough, routes to that destination change regularly and route tables never stabilize. This is called "route flapping" and can be very difficult to find because the problem changes instead of being fully broken. Route flapping can also happen due to enemy action.

---

[6] A common problem is a loose cable or a satellite link with a tree in the way.

16.7 Advantages of a RIP Network

RIP has many advantages. It is a simple, easy to understand protocol. This means that it is quite easy to set up a RIP network and there are few places a Network Administrator can make a mistake. This is more important than it might seem as human mistakes are difficult to track down and fix, especially if you are the person that made them[7].

Like all routing protocols, RIP avoids routing packets in circles, a routing loop, or to incorrect destinations. It is nearly impossible for RIP to lead to an inefficient network within its normal capacities and it reacts well to dynamic changes in the network. If the network itself is stable, RIP converges and then is hopefully a minor overhead for each router.

While there is no way to really know, RIP is most likely one of the most common routing protocols in use today.

16.8 Disadvantages of a RIP Network

Like anything else, RIP is far from perfect. There are many common network situations that RIP does not handle well.

1. Each RIP router generates updates every 30 seconds, even if the Route Table has not changed. Indeed, the protocol is not aware of changes to the Route Table at all.
2. RIP route announcements can be a large overhead on the network. It is not uncommon for route announcements to exceed 10% of the network traffic.
3. RIP is slow to converge for larger networks. Net networks are not announced when they are discovered nor are they announced when they disappear. Changes are announced during the normal announcements.
4. RIP networks are extremely slow to drop broken routes. A route must be missing from three consecutive announcements before it is deleted from the Route Cache and possibly the Route Table.
5. RIP cannot take into account the speed of a network link when deciding the best route. Only router hops are considered.
6. RIP networks have a maximum hop count, the default is 15, past which no network is reachable. That means a packet that needs to pass through 16 routers to reach a destination will not be sent. These packets are dropped.
7. Like most protocols, RIP assumes all routers can be trusted unless configured to ignore certain routers.

---

[7] Network Administrators, like everyone else, tend to see what they intended to do rather than what they *actually* did.

16.9 RIP Versions

There are three different versions of the RIP protocol available on most routers to-day. RIPv1 is an older version of IPv4 routing that pre–dated VLSM and did not allow networks to use any mask other than the natural subnet mask. This restriction was relaxed in RIPv2 and we should consider RIPv1 as superseded by version 2; however, routers *must* be informed as to which version of the RIP route announcements to expect on each interface. The current version of the **ripd** provided as part of the Quagga package requires all RIP IPv4 announcements to be the same version, which is not a problem as there is no good reason to run version 1. Version 1 should not be used unless there is a compatibility issue with an existing network, and then should only be run until the entire network can be updated to version 2.

The third version of the RIP protocol, RIPng, works essentially the same as version 2 but is for IPv6 addressing only. Like the IPv6 version, RIPng is extremely easy to configure and places no constraints on the network. As the network becomes more and more complex, RIPng leads to issues with network overhead.

16.10 RIP on the Pi

All three versions of the RIP protocol are services installed as part of the Quagga package. **ripd** handles both RIPv1 and RIPv2 while **ripng** handles RIPng for IPv6 networks. Like most routing protocols, RIP is dependent upon connectivity between routers more than it is dependent upon the physical or logical topology of the underlying network. Therefore, either the ring or star version of the laboratory network may be used to explore the details of RIP on the Raspberry Pi.

Like all the routing services installed with Quagga, the RIP services only handle route announcements while the **zebra** service updates the Route Table (and Route Cache) by passing routes to the Linux kernel where IP Tables and IP Forwarding do the actual routing. **NOTE:** Double check that the IP addresses for each interface is set correctly as in 11.3.2 to avoid a potential issue with network **169.254.x.x**. It appears as if you can correct these addresses with the configuration of the router, but **you cannot.**

16.10.1 IP Forwarding in the Kernel

The state of IP Forwarding is controlled by the **systemctl** configuration file found at **/etc/sysctl.conf**. Edit the file to remove the **#** comment character from the front of the following lines to enable IPv4 and IPv6[8] routing:

---

[8] You need not enable IPv6 routing unless you wish, but turning this on in the kernel has no significant effect if IPv6 is not present.

```
##############################################################3
# Functions previously found in netbase
#

# Uncomment the next two lines to enable Spoof protection (
    reverse-path filter)
# Turn on Source Address Verification in all interfaces to
# prevent some spoofing attacks
#net.ipv4.conf.default.rp_filter=1
#net.ipv4.conf.all.rp_filter=1

# Uncomment the next line to enable TCP/IP SYN cookies
# See http://lwn.net/Articles/277146/
# Note: This may impact IPv6 TCP sessions too
#net.ipv4.tcp_syncookies=1

# Uncomment the next line to enable packet forwarding for IPv4
net.ipv4.ip_forward=1

# Uncomment the next line to enable packet forwarding for IPv6
#  Enabling this option disables Stateless Address
    Autoconfiguration
#  based on Router Advertisements for this host
net.ipv6.conf.all.forwarding=1
```

Reboot the Pi and check the IP addresses of all interfaces. Interfaces will not have complete information if the physical NIC does not have an active link to another device.

16.10.2 Contact and Configure the Router

Before the Pi router can be configured, the **daemons**, configuration, and log files must be created and their ownership set properly. A file "myfile.txt" is created by the command **sudo touch myfile.txt** or **sudo vi myfile.txt**, either is acceptable. "Touch" creates an empty file of size zero which can then be written or edited.

If the following files do not exist, create them with the **touch** command or by copying the sample files to the correct name[9].

```
    pi@router1-1:/etc/quagga$
pi@router1-1:/etc/quagga$ sudo touch /etc/quagga/daemons
pi@router1-1:/etc/quagga$ sudo touch /etc/quagga/zebra.conf
pi@router1-1:/etc/quagga$ sudo touch /etc/quagga/vtysh.conf
pi@router1-1:/etc/quagga$ sudo touch /etc/quagga/ripd.conf
pi@router1-1:/etc/quagga$ sudo touch /etc/quagga/ripngd.conf
pi@router1-1:/etc/quagga$ sudo touch /etc/quagga/bgpd.conf
pi@router1-1:/etc/quagga$ sudo touch /etc/quagga/isisd.conf
pi@router1-1:/etc/quagga$ sudo touch /etc/quagga/ospfd.conf
pi@router1-1:/etc/quagga$ sudo touch /etc/quagga/ospf6d.conf
pi@router1-1:/etc/quagga$ sudo chown quagga:quagga /etc/quagga
    /*
pi@router1-1:/etc/quagga$ sudo chown quagga:quaggavty /etc/
    quagga/vtysh.conf
pi@router1-1:/etc/quagga$ ls -l /etc/quagga/
total 0
-rw-r----- 1 quagga quagga        0 Dec 17 12:56 bgpd.conf
-rw-r--r-- 1 root   root          0 Dec 17 12:26 daemons
```

---

[9] If the sample files do not exist, use the touch method.

```
-rw-r----- 1 quagga quagga      0 Dec 17 12:56 isisd.conf
-rw-r----- 1 quagga quagga      0 Dec 17 12:56 ospfd.conf
-rw-r----- 1 quagga quagga      0 Dec 17 12:56 ospf6d.conf
-rw-r----- 1 quagga quagga      0 Dec 17 12:56 ripd.conf
-rw-r----- 1 quagga quagga      0 Dec 17 12:56 ripngd.conf
-rw-r----- 1 quagga quaggavty   0 Dec 17 12:27 vtysh.conf
-rw-r----- 1 quagga quagga      0 Dec 17 12:56 zebra.conf
pi@router1-1:/etc/quagga$
```

To maximize the similarity with a Cisco Systems router, the Pi router is configured "remotely" by using **sudo vtysh 2601** to emulate configuring a router over a network.

16.11 Pi RIP Configuration

Verify that the interfaces that will run RIP are configured with the correct IP addresses via **write terminal**. While it is possible to add static IPv4 addresses during RIP configuration, this is not a good idea. These interface addresses should have already been configured during the first phase of router configuration in Subsection 11.3.2.

16.11.1 Quagga RIP Commands

From an **enable** prompt, start the RIP process by entering **router rip** which also opens the RIP sub–menu and gives you access to the following set of commands:

allow–ecmp	Allow Equal Cost MultiPath
default–information	Control distribution of default route
default–metric	Set a metric of redistribute routes
distance	Administrative distance
end	End current mode and change to enable mode
exit	Exit current mode and down to previous mode
list	Print command list
neighbor	Specify a neighbor router
network	Enable routing on an IP network
no	Negate a command or set its defaults
offset–list	Modify RIP metric
passive–interface	Suppress routing updates on an interface
quit	Exit current mode and down to previous mode
redistribute	Redistribute information from another routing protocol
route	RIP static route configuration
route–map	Route map set
timers	Adjust routing timers
version	Set routing protocol version

Most of these commands are beyond the scope of this book, but some are of interest to us even if we never plan to run RIP in the real world. The order in which the commands are issued does not really matter as Quagga will write out the configuration from the running configuration in a specific order and the commands control settings independently of each other. We will skip the commands that are more advanced than we need for our test–bed network.

default–information

This command controls whether the Pi will include its default route as part of all route announcements. The default setting is to *not* announce the default route. This can be set to **default--information originate** to cause this router to announce the default route. It is up to the group to decide whether or not to announce this router's default route. In reality, this setting rarely makes any difference in a working network.

default-metric

This command can be used to change the number of hops added to the hop–count of learned routes. Leave this as the default of 1.

distance

This command does not have a direct bearing on a network of purely RIP routers, but does effect the precedence of routes learned via RIP if a router can learn routes from more than one protocol. If a route cache includes multiple routes to the same network, the Route Table is populated with the route that has the lowest administrative distance. By default, RIP routes have an administrative distance of 120. Other protocols have different values for administrative distance. There should be no need to change this value.

end and quit

When a sub–menu or command is no longer needed, the **end** and **quit** commands close the current command and returns to the enabled mode. It is tricky to know when to use one over the other. If one does not work, try the other.

exit

This command returns the user to the previous menu level. If the user is returned to the top level of the configuration menu, the mode is set to Unprivileged.

list

This command lists *all* available commands in a less than useful manner. Use "**?**" instead to list all the commands that are available at the current menu state as this list is much more useful.

neighbor

In order to insure the router exchanges route announcements with a directly connected router, each neighbor router should be noted with this command. For example, if this router is directly connected with a router whose IP address is 10.1.1.1 it should be noted with the command **neighbor 10.1.1.1/24** where 24 is the length of the subnet mask. Obviously this router must have an interface in the 10.1.1.0 network in order to have such a neighbor. When you configure your router list each neighbor one at a time using this command.

network

The **ripd** process must be told which interfaces will be allowed to receive and send routing announcements by the **network <interface>** command. The actual value for **<interface>** may be specified by either name or IP address. For example, the interface **eth0** with an address of 192.16.1.5 could be specified either as **interface eth0** or **interface 192.168.1.5** as both refer to the same interface and both are self–documenting. However, all interfaces should be entered either by name or address. Do not mix the two forms[10].

passive–interface

This command is used to inform **ripd** not to send routing announcements out a specific interface. You could, for example, set interface **lo** to passive–interface.

---

[10] It is your router, but it seems easier to read if the two formats are not mixed.

redistribute

This command is used to force **ripd** to add information learned from another protocol, say OSPF, to outgoing route announcements. This allows a path through the network to utilize any open links regardless of what protocol announced that next link. In general, if this router is running multiple routing protocols you will want to redistribute all known routes.

version (1 or 2)

As has been noted before, Quagga and **ripd** can handle either RIPv1 or RIPv2 and can even handle a mix of the two on the same router. In my opinion, a network administrator should attempt to eradicate RIPv1 from the network. Set this to **version 2** as the first step of the configuration.

16.12 Exploring RIP and RIP Convergence

In this section we will re–configure the test–bed network and explore how that changes RIP convergence. RIP works well for small networks which makes it difficult to actually see RIP converge, but a more complex network of routers *might* lead to enough complexity to see RIP converge. This section can be treated as a class lab project or in–class exercise. Below is a sample session to set up **ripd** for **router1-1** for the ring version of the laboratory network *after* **zebra** then **ripd** have both been restarted and are correctly running.

16.12.1 Example Configuration Steps

One possible way to configure the router for RIP would be to set the version, configure all the networks, configure the neighbors, and then set up redistribute messages. Suppose we have a router with two interfaces (**eth0 192.168.1.1** and **eth1 172.16.0.7**), three neighbor routers (192.168.1.2, 172.16.0.1, and 172.16.0.27), and is also running **ospfd**. The network administrator would enter the following commands:

```
router rip
version 2
network eth0
network eth1
neighbor 192.168.1.2
neighbor 172.16.0.1
neighbor 172.16.0.27
redistribute ospf
end
write terminal
write memory
```

The script for router 1-1 (Group 1, Pi #1) using the ring lab network might look like the following[11]:

16.12.2 Set Logging for Zebra

```
 1   pi@router1-1:~$ sudo vtysh 2601
 2
 3   Hello, this is Quagga (version 1.2.4).
 4   Copyright 1996-2005 Kunihiro Ishiguro, et al.
 5
 6   router1-1# write term
 7   Building configuration...
 8
 9   Current configuration:
10   !
11   !
12   interface eth0
13   !
14   interface eth1
15   !
16   interface eth2
17   !
18   interface lo
19   !
20   interface wlan0
21   !
22   line vty
23   !
24   end
25   router1-1# config term
26   router1-1(config)# log file /var/log/quagga/zebra.log
27   router1-1(config)# exit
28   router1-1# write term
29   Building configuration...
30
31   Current configuration:
32   !
33   log file /var/log/quagga/zebra.log
34   !
35   interface eth0
36   !
37   interface eth1
38   !
39   interface eth2
40   !
41   interface lo
42   !
43   interface wlan0
44   !
45   line vty
46   !
47   end
48   router1-1# write mem
49   Building Configuration...
50   Configuration saved to /etc/quagga/zebra.conf
51   Configuration saved to /etc/quagga/ripd.conf
52   [OK]
53   router1-1#exit
54   pi@router1-1:~$
```

---

[11] In reality, the order of the RIP configuration steps is immaterial.

16.12.3 Set the IPv4 Addresses for all Interfaces

```
pi@router1-1:~$ sudo vtysh 2601

Hello, this is Quagga (version 1.2.4).
Copyright 1996-2005 Kunihiro Ishiguro, et al.

router1-1# config term
router1-1(config)# interface eth0
router1-1(config-if)# ip address 192.168.1.1/24
router1-1(config-if)# quit
router1-1(config)# interface eth2
router1-1(config-if)# ip address 192.168.6.1/24
router1-1(config-if)# quit
router1-1(config)# interface eth1
router1-1(config-if)# ip address 10.1.0.1/24
router1-1(config-if)# quit
router1-1(config)# exit
router1-1# write term
Building configuration...

Current configuration:
!log file /var/log/quagga/zebra.log

!
interface eth0
ip address 192.168.1.1/24
!
interface eth1
ip address 10.1.0.1/24
!
interface eth2
ip address 192.168.6.1/24
!
interface lo
!
interface wlan0
!
line vty
!
end
router1-1# write mem
Building Configuration...
Configuration saved to /etc/quagga/zebra.conf
Configuration saved to /etc/quagga/ripd.conf
[OK]
router1-1#exit
pi@router1-1:~$
```

16.12.4 Configure the RIP Daemon

```
router1-1# config term
router1-1(config)# router rip
router1-1(config-router)# version 2
router1-1(config-router)# network eth0
router1-1(config-router)# network eth1
router1-1(config-router)# network eth2
router1-1(config-router)# neighbor 192.168.1.2
router1-1(config-router)# neighbor 192.168.6.6
router1-1(config-router)# neighbor 10.1.0.2
router1-1(config-router)# quit
router1-1(config)# exit
router1-1# write term
Building configuration...

Current configuration:
!
```

```
17   log file /var/log/quagga/zebra.log
18   !
19   interface eth0
20   ip address 192.168.1.1/24
21   !
22   interface eth1
23   ip address 10.1.0.1/24
24   !
25   interface eth2
26   ip address 192.168.6.1/24
27   !
28   interface lo
29   !
30   interface wlan0
31   !
32   router rip
33   version 2
34   network eth0
35   network eth1
36   network eth2
37   neighbor 10.1.0.2
38   neighbor 192.168.1.2
39   neighbor 192.168.6.6
40   !
41   line vty
42   !
43   end
44   router1-1# write mem
45   Building Configuration...
46   Configuration saved to /etc/quagga/zebra.conf
47   Configuration saved to /etc/quagga/ripd.conf
48   [OK]
```

It is a good idea to write the configuration to the screen for a visual check before it is written to memory to become permanent. After the configuration has been saved, **zebra** and **ripd** should be restarted. It is not necessary to reboot the Pi, instead the following commands will restart the router with the new configurations as a final check for errors.

16.12.5 Restarting RIP

While the changes *do* take effect immediately, it is still a really good idea to restart both **zebra** and all other routing protocols after configuring the router. An alternate plan would be to reboot the Pi.

Important: always start or restart zebra before any other routing protocols. Failure to restart zebra first may lead to instability and failure of the router.

```
pi@router1-1:~\$ sudo systemctl restart zebra
pi@router1-1:~\$ sudo systemctl restart ripd
pi@router1-1:~\$ sudo systemctl status zebra
zebra.service - GNU Zebra routing manager
Loaded: loaded (/lib/systemd/system/zebra.service; enabled;
    vendor preset: en
Active: active (running) since Wed 2018-12-19 17:47:09 EST;
    13s ago
Docs: man:zebra
Process: 1001 ExecStart=/usr/sbin/zebra -d -A 127.0.0.1 -f /
    etc/quagga/zebra.c
Process: 998 ExecStartPre=/bin/chown -f quagga:quaggavty /
    etc/quagga/vtysh.con
Process: 996 ExecStartPre=/bin/chown -f quagga:quagga /run/
    quagga /etc/quagga/
```

```
Process: 993 ExecStartPre=/bin/chmod -f 640 /etc/quagga/
    vtysh.conf /etc/quagga
Process: 990 ExecStartPre=/sbin/ip route flush proto zebra (
    code=exited, statu
Main PID: 1003 (zebra)
CGroup: /system.slice/zebra.service
1003 /usr/sbin/zebra -d -A 127.0.0.1 -f /etc/quagga/zebra.
    conf

Dec 19 17:47:09 router1-1 systemd[1]: Starting GNU Zebra
    routing manager...
Dec 19 17:47:09 router1-1 systemd[1]: Started GNU Zebra
    routing manager.
pi@router1-1:~\$ sudo systemctl status ripd
ripd.service - RIP routing daemon
Loaded: loaded (/lib/systemd/system/ripd.service; enabled;
    vendor preset: ena
Active: active (running) since Wed 2018-12-19 17:47:09 EST;
    20min ago
Docs: man:ripd
Process: 1017 ExecStart=/usr/sbin/ripd -d -A 127.0.0.1 -f /
    etc/quagga/ripd.con
Process: 1009 ExecStartPre=/bin/chown -f quagga:quagga /etc/
    quagga/ripd.conf (
Process: 1005 ExecStartPre=/bin/chmod -f 640 /etc/quagga/
    ripd.conf (code=exite
Main PID: 1021 (ripd)
CGroup: /system.slice/ripd.service
1021 /usr/sbin/ripd -d -A 127.0.0.1 -f /etc/quagga/ripd.conf

Dec 19 17:47:09 router1-1 systemd[1]: Starting RIP routing
    daemon...
Dec 19 17:47:09 router1-1 systemd[1]: Started RIP routing
    daemon.
pi@router1-1:~\$
```

Remember: The order of the commands does not matter, but should be done in some reasonable order to help the network administrator to enter all of the required information. Develop your own order that makes sense to you as Quagga does not require any particular order.

Configure the other routers in the Group following the same general procedure. When this is completed, check the route tables with the **show ip route** command. You should also be able to ping all the other Group's Pi#1s.

16.13 RIPng on the Pi

Either the Ring or Star Laboratory Network is a good test network to explore RIPng and IPv6. Like RIP, RIPng uses **zebra** to insert routes into the Linux kernel and Quagga to provide configuration support via **vtysh**. Regardless of which Laboratory Network is used, the group addressing will use the values from Table 16.1 for subnetting the group networks. Each group can assign new subnets as needed as long as they start with the group number for the first byte[12]. Remember to use the correct global ID instead of 869b29e5e1which is used only as an example.

---

[12] For more than nine groups the group number is *two* digits, not one.

Table 16.1: Lab Network IPv6 Addresses

	Network Part
backbone	fd86:9b29:e5e1:ff00::/64
Group 1	fd86:9b29:e5e1:1000::/64
Group 2	fd86:9b29:e5e1:2000::/64
Group 3	fd86:9b29:e5e1:3000::/64
Group 4	fd86:9b29:e5e1:4000::/64
Group 5	fd86:9b29:e5e1:5000::/64
Group 6	fd86:9b29:e5e1:6000::/64

16.13.1 Quagga RIPng Interface Commands

In order to communicate between routers, each interface that is to participate in
RIPng must have a valid IPv6 address. To find out all the commands that can be
used to configure an interface for RIPng, try the following commands:

```
pi@FakeISP:~$ sudo vtysh

Hello, this is Quagga (version 1.2.4).
Copyright 1996-2005 Kunihiro Ishiguro, et al.

FakeISP# configure terminal
FakeISP(config)# interface eth0
FakeISP(config-if)# ipv6
address   Set the IP address of an interface
nd        Neighbor discovery
nhrp      Next Hop Resolution Protocol functions
ospf6     Open Shortest Path First (OSPF) for IPv6
ripng     Routing Information Protocol
router    IS-IS Routing for IP
FakeISP(config-if)#
pi@FakeISP:~$
```

Most of these commands refer to configuration options for other routing protocols. The only commands that pertain to RIPng are:

Command	Function
address	Set the IP address of an interface
ripng	Routing Information Protocol (fine tuning)

Because the network, either star or ring, is not very complex, we should not need to
use the **ripng** sub-menu to fine tune route announcements and neighbor communications. This leaves only the address command to set the IPv6 address for each
interface. Unlike IPv4, IPv6 can require multiple addresses for an interface to function correctly on a link. These are generated "automagically" and are not a concern.
The only address that needs to be configured is one to give the Group's prefix and
host ID as in Table 16.2 for the Ring Laboratory Network or Table 16.3.

At this point, the Laboratory Network should be up and running IPv4. Depending upon which version of the Laboratory Network is configured, follow either Subsection 16.13.2 for the Ring network or Subsection 16.13.3.

16.13.2 RIPng Ring Laboratory Network

Table 16.2: Ring backbone IPv6 Network Prefix

Pi #1	Interface	IPv6 Address
Group 1, Pi #1	eth0	fd86:9b29:e5e1:ff01::110/64
	eth1	fd86:9b29:e5e1:1xxx::111/64
	eth2	fd86:9b29:e5e1:ff06::112/64
Group 2, Pi #1	eth0	fd86:9b29:e5e1:ff02::210/64
	eth1	fd86:9b29:e5e1:2xxx::211/64
	eth2	fd86:9b29:e5e1:ff01::212/64
Group 3, Pi #1	eth0	fd86:9b29:e5e1:ff03::310/64
	eth1	fd86:9b29:e5e1:3xxx::311/64
	eth2	fd86:9b29:e5e1:ff02::312/64
Group 4, Pi #1	eth0	fd86:9b29:e5e1:ff04::410/64
	eth1	fd86:9b29:e5e1:4xxx::411/64
	eth2	fd86:9b29:e5e1:ff03::412/64
Group 5, Pi #1	eth0	fd86:9b29:e5e1:ff05::510/64
	eth1	fd86:9b29:e5e1:5xxx::511/64
	eth2	fd86:9b29:e5e1:ff04::512/64
Group 6, Pi #1	eth0	fd86:9b29:e5e1:ff06::610/64
	eth1	fd86:9b29:e5e1:6xxx::611/64
	eth2	fd86:9b29:e5e1:ff05::612/64

For this example, we will look at configuring Group 3 to run RIPng along with a running RIP network. It is a good idea to check the configuration of each interface from the command line:

```
pi@router3-1:~$ sudo ifconfig -a
eth0: flags=4163<UP,BROADCAST,RUNNING,MULTICAST>  mtu 1500
inet 192.168.3.3  netmask 255.255.255.0  broadcast
    192.168.3.255
inet6 fe80::c607:47e4:a82d:eb28  prefixlen 64  scopeid 0x20<
    link>
inet6 fd86:9b29:e5e1:ff03::310  prefixlen 64  scopeid 0x0<
    global>
ether b8:27:eb:15:f1:54  txqueuelen 1000  (Ethernet)
RX packets 23  bytes 2814 (2.7 KiB)
RX errors 0  dropped 0  overruns 0  frame 0
```

```
TX packets 66  bytes 5876 (5.7 KiB)
TX errors 0  dropped 0 overruns 0  carrier 0  collisions 0

eth1: flags=4163<UP,BROADCAST,RUNNING,MULTICAST>  mtu 1500
inet 10.3.3.1  netmask 255.255.255.0  broadcast 10.3.3.255
inet6 fe80::6512:b614:b136:ba35  prefixlen 64  scopeid 0x20<
    link>
inet6 fd86:9b29:e5e1:3310::311  prefixlen 64  scopeid 0x0<
    global>
ether 00:00:00:06:be:4f  txqueuelen 1000  (Ethernet)
RX packets 37  bytes 1954 (1.9 KiB)
RX errors 0  dropped 0  overruns 0  frame 0
TX packets 31  bytes 3846 (3.7 KiB)
TX errors 0  dropped 0 overruns 0  carrier 0  collisions 0

eth2: flags=4163<UP,BROADCAST,RUNNING,MULTICAST>  mtu 1500
inet 192.168.2.3  netmask 255.255.255.0  broadcast
    192.168.2.255
inet6 fe80::a8b0:1873:81fe:b824  prefixlen 64  scopeid 0x20<
    link>
inet6 fd86:9b29:e5e1:ff02::312  prefixlen 64  scopeid 0x0<
    global>
inet6 fe80::205:1bff:fe24:6f3e  prefixlen 64  scopeid 0x20<
    link>
ether 00:05:1b:24:6f:3e  txqueuelen 1000  (Ethernet)
RX packets 80  bytes 7709 (7.5 KiB)
RX errors 0  dropped 0  overruns 0  frame 0
TX packets 101  bytes 13817 (13.4 KiB)
TX errors 0  dropped 0 overruns 0  carrier 0  collisions 0

lo: flags=73<UP,LOOPBACK,RUNNING>  mtu 65536
inet 127.0.0.1  netmask 255.0.0.0
inet6 ::1  prefixlen 128  scopeid 0x10<host>
loop  txqueuelen 1000  (Local Loopback)
RX packets 20  bytes 1900 (1.8 KiB)
RX errors 0  dropped 0  overruns 0  frame 0
TX packets 20  bytes 1900 (1.8 KiB)
TX errors 0  dropped 0 overruns 0  carrier 0  collisions 0

pi@router3-1:~$
```

Everything appears to be in order. Notice the addresses beginning with *fe*80 are link addresses that are generated automatically for each interface to allow for IPv6 on a link and are *not* routed by any router. These are simply housekeeping for the link and are of no concern to us. Before any configuration is done via Quagga, you should check the existence of **/etc/quagga/ripngd.conf** and that the **ripngd** daemon is running; otherwise Quagga will not save any changes made to the RIPng configuration. The following steps will configure the router for Group 3 on the IPv6 Ring Laboratory Network.

```
pi@router3-1:~$ sudo vtysh

Hello, this is Quagga (version 1.2.4).
Copyright 1996-2005 Kunihiro Ishiguro, et al.

router3-1# configure terminal
router3-1(config)# interface eth0
router3-1(config-if)# ipv6 address fd86:9b29:e5e1:ff03::310/64
router3-1(config-if)# quit
router3-1(config)# interface eth1
router3-1(config-if)# ipv6 address fd86:9b29:e5e1:3310::311/64
router3-1(config-if)# quit
router3-1(config)# interface eth2
router3-1(config-if)# ipv6 address fd86:9b29:e5e1:ff02::312/64
router3-1(config-if)# quit
router3-1(config)# router ripng
router3-1(config-router)# network eth0
```

```
router3-1(config-router)# network eth1
router3-1(config-router)# network eth2
router3-1(config-router)# quit
router3-1(config)# exit

router3-1# write term
Building configuration...

Current configuration:
!
log file /var/log/quagga/zebra.log
!
debug ospf6 lsa unknown
!
interface eth0
ip address 192.168.3.3/24
ipv6 address fd86:9b29:e5e1:ff03::310/64
!
interface eth1
ip address 10.3.3.1/24
ipv6 address fd86:9b29:e5e1:3310::311/64
!
interface eth2
ip address 192.168.2.3/24
ipv6 address fd86:9b29:e5e1:ff02::312/64
!
interface lo
!
router rip
version 2
network eth0
network eth1
network wth2
neighbor 192.168.1.1
neighbor 192.168.2.2
neighbor 192.168.5.5
!
router ripng
network eth0
network eth1
network eth2
!
line vty
!
end
router3-1# write mem
Building Configuration...
Configuration saved to /etc/quagga/zebra.conf
Configuration saved to /etc/quagga/ripd.conf
Configuration saved to /etc/quagga/ripngd.conf
[OK]
router3-1#
```

The router should now be routing both RIP and RIPng which can easily be verified by Quagga[13].

```
router3-1# show ipv6 route
Codes: K - kernel route, C - connected, S - static, R - RIPng,
O - OSPFv6, I - IS-IS, B - BGP, A - Babel, N - NHRP,
> - selected route, * - FIB route

C>* ::1/128 is directly connected, lo
R>* fd86:9b29:e5e1:1000::/64 [120/3] via fe80::60cf:f079:c841:
    ae20, eth0, 00:06:27
R>* fd86:9b29:e5e1:2000::/64 [120/2] via fe80::2c18:2873:9c0c
    :6792, eth2, 00:06:17
```

---

[13] The "missing" routes are because this network is not a complete ring of 5 groups. Group 4 is
missing because I ran out of Raspberry Pis.

```
K * fd86:9b29:e5e1:3310::/64 is directly connected, eth1
C>* fd86:9b29:e5e1:3310::/64 is directly connected, eth1
R>* fd86:9b29:e5e1:ff01::/64 [120/2] via fe80::2c18:2873:9c0c
    :6792, eth2, 00:06:17
K * fd86:9b29:e5e1:ff02::/64 is directly connected, eth2
C>* fd86:9b29:e5e1:ff02::/64 is directly connected, eth2
K * fd86:9b29:e5e1:ff03::/64 is directly connected, eth0
C>* fd86:9b29:e5e1:ff03::/64 is directly connected, eth0
R>* fd86:9b29:e5e1:ff05::/64 [120/2] via fe80::60cf:f079:c841:
    ae20, eth0, 00:06:27
C * fe80::/64 is directly connected, eth1
C * fe80::/64 is directly connected, eth2
C>* fe80::/64 is directly connected, eth0
router3-1#
```

Now configure the rest of the Pis in the group to the correct IPv6 addresses. If a Pi is a router, configure it to run RIPng as well. Put some thought and group discussion into choosing the proper subnet remembering that the first hex digit is your group number.

16.13.3 RIPng Star Laboratory Network

Table 16.3: Star backbone IPv6 Network Prefix

Pi #1	Interface	IPv6 Address
Group 1, Pi #1	eth0	fd86:9b29:e5e1:ff00::110/64
	eth1	fd86:9b29:e5e1:1xxx::111/64
Group 2, Pi #1	eth0	fd86:9b29:e5e1:ff00::210/64
	eth1	fd86:9b29:e5e1:2xxx::211/64
Group 3, Pi #1	eth0	fd86:9b29:e5e1:ff00::310/64
	eth1	fd86:9b29:e5e1:3xxx::311/64
Group 4, Pi #1	eth0	fd86:9b29:e5e1:ff00::410/64
	eth1	fd86:9b29:e5e1:4xxx::411/64
Group 5, Pi #1	eth0	fd86:9b29:e5e1:ff00::510/64
	eth1	fd86:9b29:e5e1:5xxx::511/64
Group 6, Pi #1	eth0	fd86:9b29:e5e1:ff00::610/64
	eth1	fd86:9b29:e5e1:6xxx::611/64

For this example, we will look at configuring Group 3 to run RIPng along with a running RIP network. It is a good idea to check the configuration of each interface from the command line:

```
pi@router3-1:~$ sudo ifconfig -a
eth0: flags=4163<UP,BROADCAST,RUNNING,MULTICAST>  mtu 1500
```

```
inet 192.168.3.3  netmask 255.255.255.0  broadcast
    192.168.3.255
inet6 fe80::c607:47e4:a82d:eb28  prefixlen 64  scopeid 0x20<
    link>
inet6 fd86:9b29:e5e1:ff00::310  prefixlen 64  scopeid 0x0<
    global>
ether b8:27:eb:15:f1:54  txqueuelen 1000  (Ethernet)
RX packets 23  bytes 2814 (2.7 KiB)
RX errors 0  dropped 0  overruns 0  frame 0
TX packets 66  bytes 5876 (5.7 KiB)
TX errors 0  dropped 0 overruns 0  carrier 0  collisions 0

eth1: flags=4163<UP,BROADCAST,RUNNING,MULTICAST>  mtu 1500
inet 10.3.3.1  netmask 255.255.255.0  broadcast 10.3.3.255
inet6 fe80::6512:b614:b136:ba35  prefixlen 64  scopeid 0x20<
    link>
inet6 fd86:9b29:e5e1:3310::311  prefixlen 64  scopeid 0x0<
    global>
ether 00:00:00:06:be:4f  txqueuelen 1000  (Ethernet)
RX packets 37  bytes 1954 (1.9 KiB)
RX errors 0  dropped 0  overruns 0  frame 0
TX packets 31  bytes 3846 (3.7 KiB)
TX errors 0  dropped 0 overruns 0  carrier 0  collisions 0

lo: flags=73<UP,LOOPBACK,RUNNING>  mtu 65536
inet 127.0.0.1  netmask 255.0.0.0
inet6 ::1  prefixlen 128  scopeid 0x10<host>
loop  txqueuelen 1000  (Local Loopback)
RX packets 20  bytes 1900 (1.8 KiB)
RX errors 0  dropped 0  overruns 0  frame 0
TX packets 20  bytes 1900 (1.8 KiB)
TX errors 0  dropped 0 overruns 0  carrier 0  collisions 0

pi@router3-1:~$
```

Everything appears to be in order. Notice the addresses beginning with $fe80$ are link addresses that are generated automatically for each interface to allow for IPv6 on a link and are *not* routed by any router. These are simply housekeeping for the link and are of no concern to us. Before any configuration is done via Quagga, you should check the existence of **/etc/quagga/ripngd.conf** and that the **ripngd** daemon is running; otherwise Quagga will not save any changes made to the RIPng configuration. The following steps will configure the router for Group 3 on the IPv6 Star Laboratory Network.

```
pi@router3-1:~$ sudo vtysh

Hello, this is Quagga (version 1.2.4).
Copyright 1996-2005 Kunihiro Ishiguro, et al.

router3-1# configure terminal
router3-1(config)# interface eth0
router3-1(config-if)# ipv6 address fd86:9b29:e5e1:ff00::310/64
router3-1(config-if)# quit
router3-1(config)# interface eth1
router3-1(config-if)# ipv6 address fd86:9b29:e5e1:3310::311/64
router3-1(config-if)# quit
router3-1(config)# router ripng
router3-1(config-router)# network eth0
router3-1(config-router)# network eth1
router3-1(config-router)# quit
router3-1(config)# exit

router3-1# write term
Building configuration...

Current configuration:
!
```

```
log file /var/log/quagga/zebra.log
!
interface eth0
ip address 192.168.3.3/24
ipv6 address fd86:9b29:e5e1:f000::310/64
!
interface eth1
ip address 10.3.3.1/24
ipv6 address fd86:9b29:e5e1:3310::311/64
!
interface lo
!
router rip
version 2
network eth0
network eth1
neighbor 192.168.1.1
neighbor 192.168.2.2
!
router ripng
network eth0
network eth1
!
line vty
!
end
router3-1# write mem
Building Configuration...
Configuration saved to /etc/quagga/zebra.conf
Configuration saved to /etc/quagga/ripd.conf
Configuration saved to /etc/quagga/ripngd.conf
[OK]
router3-1#exit
pi@router3-1:~$
```

Now configure the rest of the Pis in the group to the correct IPv6 addresses. If a Pi is a router, configure it to run RIPng as well. Put some thought and group discussion into choosing the proper subnet remembering that the first hex digit is your group number.

Projects

Project 16.1 (Ring backbone) Disconnect cables from odd numbered group's **eth1**
on Pi #1 and from even numbered group's **eth0** so that groups 1 and
2, 3 and 4, 5 and 6, and so on are in still connected to each other but to
no other networks. If this leaves one group completely disconnected,
it should be left as a stand–alone group. Reboot all, connect to your
routers via **vtysh 2601**, and check the route tables. How have they
changed? Attempt to ping some address in each group. Did you get the
results you expected?

Project 16.2 (Star backbone) Disconnect cables from odd numbered group's **eth0**
on Pi #1 so that the even numbered groups are still connected to the
backbone. Reboot all, connect to your routers via **vtysh 2601**, and
check the route tables. How have they changed? Attempt to ping some
address in each group. Did you get the results you expected? Do your
results depend upon the other Pi routers?

Project 16.3 If some of the other Group's routers are reachable, attempt a trace route
(**tracert** for Windows or **traceroute** for Linux) to one of them
and record the results. Is this what you would have expected?

Exercises

16.1 Start a ping of an address starting with 172 that is not owned by your group.
Once everyone has a ping going, randomly disconnect an Ethernet cable between two groups and document the results. The results will vary from group
to group.

16.2 When would you set RIP to version 1?

16.3 When would you use RIPng?

16.4 It is entirely possible for RIP to pick a less than optimal path for a packet.
Briefly describe a possible network where this would happen.

Further Reading

The RFCs below provide further information about RIP. This is *not* an exhaustive list and most RFCs are typically dense and hard to read. Normally RFCs are most useful when writing a process to implement a specific protocol.

RFCs Directly Related to RIPv2 in This Chapter

RIP	Title
RFC 1058	Routing Information Protocol [79]
RFC 1387	RIP Version 2 Protocol Analysis [96]
RFC 1388	RIP Version 2 Carrying Additional Information [97]
RFC 1723	RIP Version 2 Carrying Additional Information [127]
RFC 2453	RIP Version 2 [171]

RFCs Directly Related to RIPng in This Chapter

RIPng	Title
RFC 2080	RIPng for IPv6 [158]
RFC 2081	RIPng Protocol Applicability Statement [159]

Other RFCs Related to RIP and RIPng

For a list of other RFCs related to the RIP protocols but not closely referenced in this Chapter, please see Appendicies B–RIP, B–RIPv2, and B–RIPng.

Chapter 17
Open Shortest Path First

Overview

Enterprise networks are large, complex, and typically organized along enterprise lines. While RIP is fine for smaller networks, at the enterprise level RIP becomes very problematic. In some cases, RIP updates can comprise more than 25% of the network traffic. Even worse, enterprise level networks often contain redundant routes for fail–safe operations. RIP will choose the best route based solely upon the number of router hops which quite often leads to inefficient choices.

To address the problems created when RIP is applied to large networks and to better reflect organizational structure, Open Shortest Path First (OSPF) routing protocol was created. OSPF is a link–state protocol [116] [89] [91] and as such only announces updates to link states (up or down) which drastically reduces the amount of overhead needed to keep the network up to date. Furthermore, with OSPF each router has the same view of the local area of routers and links and can choose the best route based upon more characteristics than a simplistic count of router hops.

OSPF is designed to reflect the organization chart of a large enterprise, if desired. OSPF is designed to make use of the Principle of Locality [51]; that is, most traffic on the network is expected to be local traffic between routers in the same general "location[1]." The network is broken up into "areas[2]. which concentrate traffic and updates within a smaller group of routers and networks dedicated to a portion of the overall enterprise. Traffic from one part of the enterprise to another must cross through the highest level area, area 0, which shares an interface with each local area. This more closely resembles the actual organization of the enterprise than does a typical RIP network.

In order to handle larger, more complex networks, OSPF is a fairly complex protocol and requires more planning to run effectively and therefore is not a good choice for a network of fewer than about 25 routers. For large networks with mul-

---

[1] This is also known as the 10/90% rule or the Switchboard Model.

[2] When speaking of OSPF, the terms "area", "Autonomous System", and "AS" are often used interchangeably.

© Springer Nature Switzerland AG 2020
G. Howser, *Computer Networks and the Internet*,
https://doi.org/10.1007/978-3-030-34496-2_17

tiple wide–area network links, multiple different speed links, or leased facilities, OSPF is frequently the best choice in routing protocols.

17.1 Overview of OSPF

OSPF is a link-state protocol. We could think of a link as being an interface on the router. The state of the link is a description of that interface and of its relationship to its neighboring routers. A description of the interface would include, for example, the IP address of the interface, the mask, the type of network it is connected to, the routers connected to that network and so on. The collection of all these link-states would form a link-state database [14].

OSPF has a very ridged structure which at first glance appears to be a distinct disadvantage, but it is in fact one of the advantages of OSPF as the structure of an OSPF network mimics the structure of most large networks. The network is divided into a number of areas in a two–level hierarchy with the single top layer being designated as "area 0". All other areas must share at least one router with area 0 and all connections between areas are made via area 0[3].

All routers need only learn the network in their area and a router, called a "border router", that connects to area 0. This greatly limits the number of routes any router must learn. In any area, all the routers learn the same exact view of the local area and all the links between routers. When, and only when, a link between two routers in the area changes, this change is announced to all the routers in the local area and not to the entire network.

Announcements are made across area 0 only if the changed link is in area 0 or a new network is learned in one of the areas. Area border routers announce new networks into their area as a new link which all the routers in that area add to their view of the network.

---

[3] Some extensions to OSPF allow connections between areas that do not go through area 0 such as the Cisco Systems implementation.

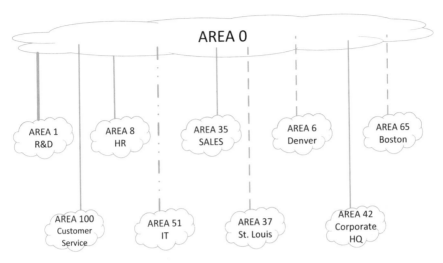

Fig. 17.1: A Typical OSPF Network

17.2 OSPF Areas

Each area of an OSPF network acts as a semi-autonomous network. The members of an area should share either geographical locality, a common purpose in the organization, or traffic locality[4]. Great care should be taken in the initial design of the network to avoid excessive traffic crossing area 0 to keep routes low cost.[5]

17.3 Area Border Routers

Each area, except area 0, must have at least one router that has an additional interface into area 0. This router is referred to as the ABR[6] and typically runs one copy of

---

[4] That is, the majority of the traffic in the area should stay in the area. If this is not the case, too much traffic will need to cross area 0 which cuts down on efficiency.

[5] Some implementations of OSPF allow for areas that do *not* connect directly to area 0 but rather connect via a special area called a transit area. This is beyond the scope of this book. The author, in general, prefers to limit himself to open standards as he has no way of knowing all the vendors who supply your network.

[6] Area Border Router

ospfd for area 0 and one for the local area. There is more overhead on this router as it must handle additional LSAs[7] and two Link State databases, but this overhead is not very great. In a production network, it is common to have multiple ABRs for load–balancing and fault tolerance.

17.4 Best Route

Because of the possibility of WAN links in large networks, each link is given a "cost" by the network administrator that reflects the desirability of using that link[8]. Each router in the area learns the cost of the links via link announcements and uses these costs to determine all the best routes in the network by lowest total cost. After the network converges, each router has the lowest cost route to all known networks in the route table and can easily forward a packet to the next router on the lowest total cost route.

The network administrator need only configure each router with the appropriate area information and each router–to–router link with the proper costs. This requires more initial effort than a RIP network, for example, but leads to less overhead in terms of announcements. For large networks, this is a significant improvement. If no links change, or there are a reasonable amount of link changes, it is possible to build networks that are literally too large or complex to run some other routing protocols such as RIP.

17.5 OSPF Adjacency Relationship

Any two routers that share an active link attempt to form an OSPF Neighbor relationship by sending an "hello" message. The link is only considered to be "up" if an active neighbor relationship exists, so it is critical that both routers know that the link is active between two neighbors. An unresponsive router can be detected when the adjacency relation fails and that router is not sent any packets.

17.5.1 Forming the Adjacency Relationship

When a link is sensed as being "up", the OSPF process, typically ospfd on a Raspbian router, performs Algorithm 9 while it is running. It is important for the process to constantly identify the relationship with any adjacent routers as a link

---

[7] Link State Announcements

[8] Quagga uses a current default cost based upon a cost of 1 for 100mps. Any link faster than 100mps is given a cost of 0. These default values can be changed but they *must* be the same for all routers in the network or the routers will not form any adjacency relationships.

is only usable if an adjacency, or two–way neighbor, relationship can be formed. When this relationship is broken, the link is marked as "down" and an LSA is sent to the DR[9] for flooding into the area regardless of the actual state of the link.

Algorithm 9 Neighbor Relationship

```
1:  procedure OSPF NEIGHBOR
2:      while Link sensed do
3:          Wait(hello  timeout)
4:          if Two–way Neighbor then
5:              Send two–way neighbor hello
6:          else
7:              Send one–way neighbor hello message
8:              if hello received then
9:                  if Message confirms relationship then
10:                     Two–way Neighbor
11:                 else
12:                     One–way Neighbor
13:                     Send Two–way neighbor message
14:                     Two–way Neighbor
15:                 end if
16:             else
17:                 if Missed too many responses then
18:                     Mark link down until an hello message
19:                 end if
20:             end if
21:         end if
22:     end while
23: end procedure
```

17.5.2 Exchanging Route Information with a Neighbor

Once a two–way neighbor relationship has been established, the two routers exchange database timestamps. If one router has a Link State database that is out of date, the two routers exchange LSAs until both routers have the same view of the network, or in other words the same entries in their LSDs[10], see Figure 17.2. This exchange is initiated when one of the routers sense a link is active, much like the LED turning on when you plug in a cable.

The steps are as follows:

1. Both routers have the link in the "down" state because no media is detected.

---

[9] Designated Router

[10] Link State Databases

2. Router 1 senses the media is present and changes the state of the link to "initializing" then sends a message with its **router-id** across the link. Router 2 senses the message and changes its link state to "initializing" as well.

3. When Router 2 receives the message it makes note of the **router-id** and changes its link state to "1–WAY" adjacency.

4. Router 1 changes its link state to "1–WAY" adjacency when it receives a message from Router 2.

5. If the message contains the correct **router-id** for Router 1, Router 1 changes its link state to "2–WAY" adjacency and sends a message "Hello, Router 2".

6. If the message contains the correct **router-id** for Router 2, Router 2 changes its link state to "2–Way" adjacency and replies with a message containing the correct **router-id** for both routers.

7. The adjacency has been correctly formed for this link. Router 1 initiates the synchronization of the two LSDs by changing the link state to "Exchange start". Router 1 then sends a message with its LSD highest sequence number and requesting to be the master for the synchronization.

8. Router 2 acknowledges the request by replying with its LSD and also requesting to be the master.

9. Router 1 then sets the link state to "EXCHANGE" and compares the two LSD sequence numbers.

   a. If Router 1 has the lower of the two LSD sequence numbers, as in the example given in Figure 17.2, it sends a message with Router 2's LSD sequence number.This also acknowledges that Router 2 has the older LSD and will be the secondary, or slave, in this process.

   b. If Router 2 has the higher LSD sequence number, then it will send a message to Router 2 with its own LSD sequence number and newer database. Router 2 also declares it will control the exchange as master[11].

10. Router 2 sets its link state to "EXCHANGE" and confirms by sending Router 1 a new LSD sequence number, typically $y + 1$.

11. Router 1 confirms the new sequence number and that it will act as the secondary, or slave, process.

12. Router 1 changes its link state to "LOADING" and begins requesting link state updates.

13. When Router 2 receives the first link state request from Router 1, it changes the link state to "LOADING" and replies with the proper link state update.

14. The request/update process continues until Router 1 has all the required updates.

15. The LSDs are now synchronized. Router 1 sets its link state to "FULL" to signify a full adjacency or "two–way" neighbor relationship. Router 1 then sends a message, LS ACK, to Router 2.

16. Router 2 sets its link state to "FULL" and replies with the same message.

17. The procedure is now complete.

---

[11] This is a classic example of a bully algorithm with only two processes competing.

Router 1		Router 2
Link state		Link state
1. Down		Down
2. Init	HELLO $\longrightarrow$	Init
3.	$\longleftarrow$ HELLO Nbr=R1id	1–WAY
4. 1-WAY		
5. 2–WAY	HELLO Nbr=R2id $\longrightarrow$	
6.	$\longleftarrow$ HELLO Nbr=R1id	2–WAY
7. EXSTART	LS Db Seq=x msg, Master $\longrightarrow$	
8.	$\longleftarrow$ LS Db Seq=y msg, Master	EXSTART
9. EXCHANGE	LS Db Seq=y msg, Slave $\longrightarrow$	
10.	$\longleftarrow$ LS Db Seq=y+1 msg, Master	EXCHANGE
11.	LS Db Seq=y+1 msg, Slave $\longrightarrow$	
12. LOADING	LS Request $\longrightarrow$	LOADING
13.	$\longleftarrow$ LS Update	
14.	LS Request $\longrightarrow$	
15.	$\longleftarrow$ LS Update	
16. FULL	LS Ack $\longrightarrow$	FULL
17.	$\longleftarrow$ LS Ack	

Fig. 17.2: Router 1 Initating an Adjacency with Router 2

17.5.3 Keeping the Adjacency Relationship Active

Each router must know if a link is up and the neighbor relationship is active. If one of the routers is "hung" and not processing, the link may still be sensed by the other router and packets sent to the "hung" router will be lost. To eliminate this possibility, each router sends an "hello" message on the link every time a counter expires, typically every 30 seconds, as in line 5 of Algorithm 9. If there is no response to a given number of "hello" messages, the router marks the link as down and sends out LSAs to its other neighbors. The neighbor relationship is then broken until an "hello" message is received from that router.

17.5.4 Designated Router

Once adjacencies have been formed, the routers in the local area have an election to determine the DR and BDR[12] for the area. This is usually done via a simple bully algorithm. The router with the highest **router-id** is elected DR and the second

---

[12] Backup Designated Router

highest is elected BDR. Should the DR be unable to fulfill its duties (flooding certain updates to the local area, see Section 17.5.5), the BDR assumes the role of the DR and a new election is held. Many implementations of OSPF hold a new election each time a new router is sensed in the local area.

17.5.5 Link State Announcements

When a router senses a change in state on a link, it checks its internal configuration for information about that link such as IP network, cost, and so on. It then forms a LSA and sends it to the DR which then floods the announcement via an IP multicast to all of the routers in the area. IP Multicasts are used so that devices in the area may easily ignore the LSA if they are not running OSPF. If needed, the ABR floods the update to all the ABRs in area 0.

There are five types of LSAs to exchange information about link changes.

Type 1 Router Links Advertisement   These are flooded within the local area and contain neighbor routers' link status and cost.

Type 2 Network Links Advertisement   These are flooded within the local area by the DR (via multicast). These are *only* generated when router notifies the DR that it has sensed a change in a link state.

Type 3 Summary Links Advertisement   These are flooded into the local area by the ABR when a new network is reachable from outside the local area.

Type 4 Area Boundary Router Summary Link Advertisement   These are flooded into the local area by the ABR with the cost from this router to another ABR.

Type 5 Area External Link Advertisement   These are flooded to all areas and describe an external network reachable via the ABR that generated it.

17.6 OSPF Link State Database

In reality, the Link State Database is a directed, weighted digraph of the network with the link costs being the weight on each arc. It is very easy for the routers to run Dijkstra [30] on this digraph (using the current router as the source of the digraph) to obtain the single source, directed minimum spanning tree with the current router as the source and the adjacent nodes as the next step to the known networks. Convergence is reached when each router constructs the same digraph for the area.

Border Routers maintain two Link State Databases, one for the local area and one for area 0. Typically, Border Routers run two copies of **ospfd** but this is up to the programmer who implemented OSPF on the router.

17.7 Convergence of an OSPF Network

As always, the network is said to have converged when the Route Tables are stable. When an OSPF network has converged, all of the Link State databases kept by the routers reflect the exact same information for each area. This implies that each area converges separately from the state of other networks with the exception of Area Border Routers.

Once the network has converged, the only overhead generated by OSPF consists of **hello** messages between neighbors. This overhead is worth the expense because of the ability of OSPF to detect a "hung" routing process. If the network is extremely stable and reliable, the **hello** timeout can be lengthened to minimize the overhead. By the same token, if the network has unreliable links or links that change often, the **hello** timeout can be shortened to minimize the time required to detect a "hung" routing process. Care must be taken when adjusting this parameter[13].

17.8 Advantages of a OSPF Network

OSPF has many advantages over a distance–vector protocol such as RIP [14] [43]

- With OSPF, there is no limitation on the hop count.
- The intelligent use of VLSM is very useful in IP address allocation.
- OSPF uses IP multicast to send link-state updates. This ensures less processing on routers that are not listening to OSPF packets.
- Updates are only sent when the state of a link changes instead of periodically. This ensures a better use of bandwidth.
- OSPF has better convergence than RIP. This is because routing changes are propagated instantaneously and not periodically.
- OSPF allows for better load balancing.
- OSPF allows for a logical definition of networks where routers can be divided into areas. This limits the explosion of link state updates over the whole network. This also provides a mechanism for aggregating routes and cutting down on the unnecessary propagation of subnet information.
- OSPF allows for routing authentication by using different methods of password authentication.
- OSPF allows for the transfer and tagging of external routes injected into an Autonomous System. This keeps track of external routes injected by exterior protocols such as BGP.
- The inherent structure of OSPF closely matches the structure of a large organization or large network.
- Each link is given a cost that reflects the desirability of choosing that link.
- Unresponsive routers, or processes, can be easily identified via periodic **hello** messages.

---

[13] If the network is working properly, I would suggest not changing the **hello** timeout.

17.9 Disadvantages of a OSPF Network

Like RIP, OSPF is far from perfect. There are additional complications with an advanced routing protocol such as OSPF.

- Requires an intelligent initial design.
- Open standard OSPF routers must be re–configured to interact with Cisco Systems routers[14].
- Difficult to merge two networks when two organizations merge[15].
- Significant, and pointless, overhead for very small networks.
- The overhead of OSPF is larger than the overhead of RIP due to significant utilization of the CPU[16] and may require an upgrade to some routers when moving from RIP to OSPF.

17.10 OSPF Advanced Topics

There are some advanced topics that will not be covered in detail for two main reasons. First, Cisco Systems and IBM have made some extensions to OSPF that are becoming part of the OSPF standards. Among these are stub areas, transit areas, and the ability to use virtual links to route to the backbone when the area border routers are not able to route to area zero.

Secondly, OSPF is a complex set of protocols and usually requires a number of class sessions to understand along with experience in the basics of OSPF design and configuration. The goal here is to get the basics of OSPF. If some of these topics would be of use in your network, there are many excellent classes in OSPF and Cisco Systems has put much of their information on line.

Lastly, the OSPF andOSPFv3[17] network on the Raspberry Pi computers can easily be extended to explore these advanced topics later.

---

[14] Cisco has developed a number of extensions to OSPF which must be taken into account if pure open standard OSPF routers are to be used with Cisco Systems equipment using extensions to OSPF. Fortunately, Quagga was designed to inter–operate with Cisco Systems routers and can easily be used in networks with stub areas, transit areas, and virtual links. These topics are beyond the scope of this book.

[15] This is due to a number of factors. It is often difficult to merge the different area 0's. Common area numbers require all the routers in one to be re-configured. Common functions in different areas *should* be merged and so on. Major re-configuration of a running network can expose weaknesses and can lead to severe, temporary outages. The more planning time spent on the merger, the fewer problems you can expect to encounter.

[16] Central Processing Unit

[17] Open Shortest Path First (IPv6)

17.11 OSPF Versions

For all intents and purposes, when talking about OSPF for IPv4 one is talking aboutOSPFv3 which is a very mature standard. Extensions to theOSPFv3 standard are being worked out, but at this point it is reasonable to expect any extensions to be backwards compatible with the current version.

The alternate version of OSPF isOSPFv3 for IPv6 addressing. Actually,OSPFv3 can advertise both IPv4 and IPv6 networks, but it would be a bit of an overkill to runOSPFv3 for a purely IPv4 network because IPv6 must be enabled forOSPFv3 processes to exchange information about LSAs asOSPFv3 does that using IPv6 link addresses.

17.12 OSPF on the Raspberry Pi

Quagga provides both OSPF and OSPFv3 for Raspbian. In order to run OSPF on the test–bed network, the network must reflect a two–level hierarchy. This is one of the reasons for the topology of both the Ring Laboratory and Star Laboratory networks. Below is a sample IPv4 routing table showing OSPF routes.

```
pi@router1-1:~\$ sudo vtysh 2601

Hello, this is Quagga (version 1.1.1).
Copyright 1996-2005 Kunihiro Ishiguro, et al.

router1-1# show ip route
Codes: K - kernel route, C - connected, S - static, R - RIP,
O - OSPF, I - IS-IS, B - BGP, P - PIM, A - Babel,
> - selected route, * - FIB route

K>* 0.0.0.0/0 via 192.168.1.250, eth1, src 192.168.1.1
O   10.1.0.0/24 [110/10] is directly connected, eth0, 00:41:22
C>* 10.1.0.0/24 is directly connected, eth0
O>* 10.2.2.0/24 [110/20] via 10.1.0.2, eth0, 00:41:12
O>* 10.3.0.0/24 [110/20] via 192.168.1.3, eth1, 00:41:11
C>* 127.0.0.0/8 is directly connected, lo
O>* 172.18.0.0/24 [110/20] via 10.1.0.2, eth0, 00:41:11
O   192.168.1.0/24 [110/10] is directly connected, eth1,
    00:41:22
C>* 192.168.1.0/24 is directly connected, eth1
router1-1#
```

17.13 OSPF Test–bed Network

To create the test–bed network for OSPF, care must be taken to insure the physical network is designed to support the logical topology of an OSPF network. That is, there must be a number of independent areas which connect to a common area 0. Area 0 will be built between the Pi#1 routers such that the Area Border Routers (ABR) will be the group's Pi#1 and other routers in the group will access other areas outside the group's local area via this ABR and area 0. The Laboratory Networks, see Figures 17.3 and 17.4, were designed with this in mind[18].

17.13.1 OSPF Ring Test–bed Network

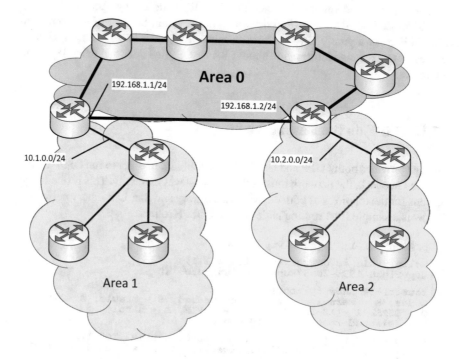

Fig. 17.3: OSPF Ring Network for Groups 1–6

If the groups are using the Ring Test–bed Network, all of the wiring and addressing can remain the same. The required **area** 0.0.0.0 is formed of the backbone connections using the various 192.168.g.0 networks. Each group will form its own area denoted as 0.0.0.g. Remember that OSPF requires networks to be completely contained in exactly one area.

---

[18] Actually, with OSPF and ISIS in mind. The other issue was to try and minimize the number of times the interface addresses needed to be changed.

17.13.2 OSPF Star Test–bed Network

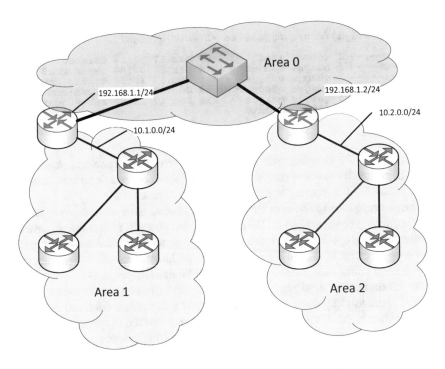

Fig. 17.4: OSPF Star Network for Groups 1–2

If the groups are using the Star Test–bed Network, all of the wiring and address-ing can remain the same. The required **area** 0.0.0.0 is formed of the backbone connections in the 192.168.1.0 network. Each group will form its own area denoted as 0.0.0.*g*. Remember that OSPF requires networks to be completely contained in exactly one area.

17.13.3 Contact and Configure the Router

Before the Pi router can be configured, the **daemons**, configuration, and log files must be created and their ownership set properly. A file "myfile.txt" is created by the command **sudo touch myfile.txt** or **sudo vi myfile.txt**, either is acceptable. "Touch" creates an empty file of size zero which can then be written or edited.

If the following files do not exist, create them with the **touch** command or by copying the sample files to the correct name. This would be a good time to copy the **ripd.conf** file to a different name if you wish to save your RIP configuration. It should be saved in the file **/etc/quagga/ripd.conf-save** but this might get over–written as time goes on. If you save it under another name, it will be kept as a small file you can always refer to when needed.

```
pi@router1-1:/etc/quagga\$ ls -l /etc/quagga/
total 20
-rw-r--r-- 1 root     root          57 Dec 17 12:26 daemons
-rw-r----- 1 quagga   quagga       184 Dec 20 07:44 ospfd.conf
-rw-r----- 1 quagga   quagga       201 Dec 17 12:56 ospf6d.conf
-rw-r----- 1 quagga   quaggavty      0 Dec 17 12:27 vtysh.conf
-rw-r----- 1 quagga   quagga       238 Dec 17 12:56 zebra.conf
pi@router1-1:/etc/quagga\$
```

Start the OSPF daemon with the command, **sudo systemctl restart ospfd**, and verify that it is running with **sudo systemctl status ospfd**. If the daemon is not running or if there is not a configuration file (**/etc/quagga/ospfd.conf**), you will be able to issue commands to configure the running configuration on Pi#1 and Pi#2 but you will not be able to save the configuration. Hopefully, any commands you issue without a the daemon running will have no effect at all, but if the commands do have some effect there may also be problems with running OSPF that will be difficult to diagnose.

To maximize the similarity with a Cisco Systems router, the Pi router is configured "remotely" by using **sudo vtysh 2601** to emulate configuring a router over a network. Also, the **vtysh** interface ensures that the proper commands are written to the correct configuration files without requiring the network administrator to contact each router daemon configuration individually[19].

17.14 Pi OSPF Configuration

The command set for OSPF shares some commands with the other command sets and also has additional commands and options. A list of the OSPF commands can be obtained by **router ospf ?** from the configuration menu.

17.14.1 Quagga OSPF Commands

From an **enable** prompt, start the OSPF process by entering **router ospf** which also opens the OSPF sub–menu and gives you access to the following set of commands:

---

[19] Based upon the author's experience, there is no guarantee the resulting configurations will produce the results you expect. If you use **sudo vtysh 2601**, things tend to work. You were warned.

area	OSPF area parameters
auto-cost	Calculate OSPF interface cost according to bandwidth
capability	Enable specific OSPF feature
compatible	OSPF compatibility list
default-information	Control distribution of default information
default-metric	Set metric of redistributed routes
distance	Define an administrative distance
distribute-list	Filter networks in routing updates
end	End current mode and change to enable mode
exit	Exit current mode and down to previous mode
list	Print command list
log-adjacency-changes	Log changes in adjacency state
max-metric	OSPF maximum/infinite-distance metric
mpls-te	MPLS-TE specific commands
neighbor	Specify neighbor router
network	Enable routing on an IP network
no	Negate a command or set its defaults
ospf	OSPF specific commands
passive-interface	Suppress routing updates on an interface
pce	PCE Router Information specific commands
quit	Exit current mode and down to previous mode
redistribute	Redistribute information from another routing protocol
refresh	Adjust refresh parameters
router-id	router-id for the OSPF process
router-info	OSPF Router Information specific commands
timers	Adjust routing timers

Most of these commands are beyond the scope of this book, but some are of interest to us even if we never plan to run OSPF in the real world. The order in which the commands are issued does not really matter as Quagga will write out the configuration from the running configuration in a specific order and the command control settings independently of each other. We will skip the commands that are more advanced than we need for our test–bed network.

area

The **area** command is the first command you should give Quagga after the **router ospf** command. This is one of the most critical commands for an OSPF router as it determines the area in which this process resides. Routers only form adjacencies with routers in the same area. The area number can be given as a single number or in the same format as an IPv4 address. Either way, Quagga stores the number in the IP address format. For **ospf6** the area *must* be given in the full IPv4 format. For example, 0.0.0.1 instead of 1.

An ABR must run two OSPF processes, one for the local area and one for area 0, and each is configured separately. If a router shares an interface with more routers in more than two distinct areas, your network should not be running OSPF.

auto–cost

When this is enabled the cost of a link will be automatically calculated by OSPF using the current default band–width. Unless your network has links whose cost does not depend upon speed, leave this alone[20]. **NOTE:** Any link with a higher speed than the default is set to a cost of zero.

default–information

This command controls whether the Pi will include its default route as part of all route announcements. The default setting is to *not* announce the default route. This can be set to **default--information originate** to cause this router to announce the default route or be left as the default at the group's discretion.

default–metric

This command can be used to increase the cost of routes learned from other proto-cols over those learned from OSPF. Leave this as the default of 1.

distance

By default, OSPF routes have an administrative distance of 110. This means that the router will prefer a route learned from OSPF over a route learned from another protocol with a higher administrative distance. For example, RIP has an administra-tive distance of 120; therefore, routes learned from OSPF are preferred over those learned from RIP.

log–adjacency–changes

When enabled this command will write a message into the log file any time a change in the adjacency of this router with a neighbor occurs. This can be very helpful during debugging.

---

[20] The network administrator can always cheat and assign the wrong speed to a link. For example, it would be a good idea to assign a very low speed to a leased link that is charged by the bit.

neighbor

In order to insure the router exchanges "hellos" with a directly connected router, each neighbor router should be noted with this command. For example, if this router is directly connected with a router whose IP address is 10.1.1.1/24 it should be noted with the command **neighbor 10.1.1.1/24**. Obviously this router must have an interface in the 10.1.1.0/24 network in order to have such a neighbor. When you configure your router, you should list each neighbor one at a time using this command.

network

The **ospfd** process must be told which interfaces will be allowed to receive and send routing announcements by the **network <interface>** command. The actual value for **<interface>** may be specified by either name or IP address. For example, the interface **eth0** with an address of 192.16.1.5 could be specified either as **interface eth0** or **interface 192.168.1.5** as both refer to the same interface and both are self–documenting. However, all interfaces should be entered either by name or address. Do not mix the two forms[21].

passive–interface

This command is used to inform **ospfd** not to send routing announcements out a specific interface. You could, for example, set interface **lo** to passive–interface.

redistribute

This command is used to force **ospfd** to add information learned from another protocol, say RIP, to outgoing route announcements. This allows a path through the network to utilize any open links regardless of what protocol announced that next link. In general, if this router is running multiple routing protocols, you may want to redistribute all known routes in order for other routers to know this router can reach networks that are not directly part of the OSPF area.

router–id

The **router-id** is used only during the election of the OSPF DR and BDR. The higher the binary value of the **router-id** the more likely the router is to be elected

---

[21] It is your router, but it seems easier to read if the two formats are not mixed. I prefer the **interface eth0** format as it is less typing.

DR or BDR. A possibility is to set this value to the IPv4 address of one of the router's interfaces. This should be unique in the area.

17.15 Configuration of OSPFv2 and IPv4 Lab Network

Before any configuration is done, check to see that the file **/etc/quagga/ospfd. conf** exists or simply create it. Remember that Quagga will allow someone to make all the changes they want to the configuration of a routing daemon such as **ospfd** if this file is missing, but *they will not work nor will they be kept.*

```
pi@router2-1:~$ sudo touch /etc/quagga/ospfd.conf
pi@router2-1:~$ sudo chown quagga:quagga /etc/quagga/ospfd.
    conf
pi@router2-1:~$ sudo systemctl restart ospfd
pi@router2-1:~$
```

The configuration of the backbone is different for the Ring Laboratory Network and the Star Laboratory Network. In reality, the backbone topology only matters for each group's Pi#1, so we will look at how to configure this router for each topology and then look at how to configure the other routers in the group's local area.

17.15.1 Ring Configuration, Pi#1

OSPF configuration requires a detailed knowledge of the network and of the OSPF protocol. Because this router will be an ABR, it should not be running RIP on any interfaces and **ripd** is not needed. Before we configure OSPF we will remove RIP from this router with **no router rip**. The local area, consisting of Group 1, is area 0.0.0.1 but the area could be area any unique number throughout the network[22]. It does not matter what the area number is, but the area number must be the same for all routers in the local area or the routers will *not* form full neighbor relationships. The area number *must* be unique in the OSPF network or there will be routing conflicts in area zero. If either of these two conditions are not met, the area will experience routing issues or not become part of the complete OSPF network.

```
1  pi@router2-1:~$ sudo vtysh
2
3  Hello, this is Quagga (version 1.2.4).
4  Copyright 1996-2005 Kunihiro Ishiguro, et al.
5
6  router2-1# configure terminal
7  router2-1(config)# router ospf
8  router2-1(config-router)# router-id 192.168.2.2
```

---

[22] At one point Area 0.0.0.51 was a common choice for the area computer people used to house servers and things.

```
 9    router2-1(config-router)# log-adjacency-changes
10    router2-1(config-router)# network 192.168.2.0/24 area 0
11    router2-1(config-router)# network 10.2.2.0/24 area 2
12    router2-1(config-router)# network 192.168.1.0/24 area 0
13    router2-1(config-router)# quit
14    router2-1(config)# interface eth0
15    router2-1(config-if)# ip ospf area 0
16    router2-1(config-if)# quit
17    router2-1(config)# interface eth1
18    router2-1(config-if)# ip ospf area 2
19    router2-1(config-if)# quit
20    router2-1(config)# interface eth2
21    router2-1(config-if)# ip ospf area 0
22    router2-1(config-if)# quit
23    router2-1(config)# exit
24    router2-1#write mem
25    router2-1#exit
26    pi@router2-1:~$
```

The first step in configuring the Ring Laboratory Network is to configure the routers that have interfaces in area 0. Each Group has one router, Pi#1, that is a border router and must have OSPF interfaces in both the local area and area zero (0.0.0.0); so first configure that router. Here is an example of how Pi#2.1 would be configured to be an ABR.

First the router process is enabled and configured with a **router-id**(lines 7 – 8), then it is set to log any changes in adjacency or neighbor formation (line 9), and thirdly each of the three interfaces is added by IPv4 address to exactly one OSPF area (lines 10 – 12).

The last step is to inform the process of which area each interface serves. This is done in lines 14 – 21 by extending the configuration of each interface. At this point the router is routing packets.

```
router2-1# show ip route
Codes: K - kernel route, C - connected, S - static, R - RIP,
       O - OSPF, I - IS-IS, B - BGP, P - PIM, A - Babel, N -
       NHRP,
       > - selected route, * - FIB route

K>* 0.0.0.0/0 via 192.168.2.2, eth0, src 192.168.2.2
O>* 10.1.1.0/24 [110/20] via 192.168.1.1, eth2, 00:02:12
R   10.1.1.0/24 [120/2] via 192.168.1.1, eth2, 1d22h09m
O   10.2.2.0/24 [110/10] is directly connected, eth1, 00:02:24
K * 10.2.2.0/24 is directly connected, eth1
C>* 10.2.2.0/24 is directly connected, eth1
O>* 10.3.3.0/24 [110/20] via 192.168.2.3, eth0, 00:02:20
R   10.3.3.0/24 [120/2] via 192.168.2.3, eth0, 1d04h48m
C>* 127.0.0.0/8 is directly connected, lo
R>* 172.17.0.0/16 [120/3] via 192.168.1.1, eth2, 1d04h48m
O   192.168.1.0/24 [110/10] is directly connected, eth2,
      00:02:12
K * 192.168.1.0/24 is directly connected, eth2
C>* 192.168.1.0/24 is directly connected, eth2
O   192.168.2.0/24 [110/10] is directly connected, eth0,
      00:02:30
K * 192.168.2.0/24 is directly connected, eth0
C>* 192.168.2.0/24 is directly connected, eth0
O>* 192.168.3.0/24 [110/20] via 192.168.2.3, eth0, 00:02:20
R   192.168.3.0/24 [120/2] via 192.168.2.3, eth0, 1d03h18m
```

```
O>* 192.168.5.0/24 [110/20] via 192.168.1.1, eth2, 00:02:12

router2-1# write mem
Building Configuration...
Configuration saved to /etc/quagga/zebra.conf
Configuration saved to /etc/quagga/ripd.conf
Configuration saved to /etc/quagga/ripngd.conf
Configuration saved to /etc/quagga/ospfd.conf
[OK]
router2-1#
```

At this point the configuration should be saved by **write memory**[23]. Notice
that Quagga is still reporting some routes learned from RIP. These routes are not
really a problem as Quagga follows the Cisco Systems convention to prefer OSPF
routes over RIP routes, but running RIP is no longer needed and can be turned off by
configuring Quagga to "forget" RIP via the command (in configuration mode) **no
router rip**. This will delete RIP from the running configuration and make this
change permanent when the configuration is written to memory (disk). After RIP is
deleted from the configuration, the route table now looks like this.

```
pi@router2-1:~$ sudo vtysh

Hello, this is Quagga (version 1.2.4).
Copyright 1996-2005 Kunihiro Ishiguro, et al.

router2-1# configure terminal
router2-1(config)# no router rip
router2-1(config)# exit
router2-1# show ip route
Codes: K - kernel route, C - connected, S - static, R - RIP,
       O - OSPF, I - IS-IS, B - BGP, P - PIM, A - Babel, N -
          NHRP,
       > - selected route, * - FIB route

K>* 0.0.0.0/0 via 192.168.2.2, eth0, src 192.168.2.2
O>* 10.1.1.0/24 [110/20] via 192.168.1.1, eth2, 00:21:33
O   10.2.2.0/24 [110/10] is directly connected, eth1, 00:21:45
K * 10.2.2.0/24 is directly connected, eth1
C>* 10.2.2.0/24 is directly connected, eth1
O>* 10.3.3.0/24 [110/20] via 192.168.2.3, eth0, 00:00:16
C>* 127.0.0.0/8 is directly connected, lo
O   192.168.1.0/24 [110/10] is directly connected, eth2,
       00:21:33
K * 192.168.1.0/24 is directly connected, eth2
C>* 192.168.1.0/24 is directly connected, eth2
O   192.168.2.0/24 [110/10] is directly connected, eth0,
       00:00:21
K * 192.168.2.0/24 is directly connected, eth0
C>* 192.168.2.0/24 is directly connected, eth0
O>* 192.168.3.0/24 [110/20] via 192.168.2.3, eth0, 00:00:16
O>* 192.168.5.0/24 [110/20] via 192.168.1.1, eth2, 00:21:33
router2-1#
```

---

[23] If you are an untrusting person like I am, you can log out of Quagga and restart **zebra**, **ripd**,
and **ospfd**. It is not really needed, but it might be a good habit to develop.

Notice all the RIP routes are gone and only networks that are directly connected or announced via OSPF are reachable, which is what we want as RIP takes too long to converge and creates too much overhead on a large network.

17.15.2 Star Configuration, Pi#1

If the backbone is configured using a star topology, the configuration of a group's Pi#1 is actually simpler. All of the same steps in Subsection 17.15.1 are followed *except* for two changes:

1. All commands referring to **eth2** are ignored completely.
2. The IPv4 address for **eth0** is 192.168.0.*g* where *g* is the group number.

```
 1    pi@router2-1:~$ sudo vtysh
 2
 3    Hello, this is Quagga (version 1.2.4).
 4    Copyright 1996-2005 Kunihiro Ishiguro, et al.
 5
 6    router2-1# configure terminal
 7    router2-1(config)# router ospf
 8    router2-1(config-router)# router-id 192.168.0.2
 9    router2-1(config-router)# log-adjacency-changes
10    router2-1(config-router)# network 192.168.0.0/24 area 0
11    router2-1(config-router)# network 10.2.2.0/24 area 2
12    router2-1(config-router)# quit
13    router2-1(config)# interface eth0
14    router2-1(config-if)# ip ospf area 0
15    router2-1(config-if)# quit
16    router2-1(config)# interface eth1
17    router2-1(config-if)# ip ospf area 2
18    router2-1(config-if)# quit
19    router2-1(config)# exit
20    router2-1#
```

17.15.3 Configuration, Pi#2

This router will be used to demonstrate how a router could have its own local network of routers that are not participating in OSPF. In this example, both **eth1** and **eth2** could be connecting to a RIP network that is fully contained inside the local OSPF area.

```
pi@router2-2:~\$ sudo vtysh 2601

Hello, this is Quagga (version 1.1.1).
Copyright 1996-2005 Kunihiro Ishiguro, et al.

router2-2# write term
Building configuration...

Current configuration:
!
```

```
hostname Router
log file /var/log/quagga/zebra.log
!
password zebra
enable password zebra
!
interface eth0
ip address 10.2.2.2/24
!
interface eth1
ip address 172.18.0.2/24
!
interface eth2
!
interface lo
!
interface wlan0
!
router rip
version 2
redistribute ospf
network eth1
network eth2
!
router ospf
ospf router-id 10.2.2.2
log-adjacency-changes
redistribute connected
redistribute rip
network 10.2.2.0/24 area 0.0.0.12
neighbor 10.2.2.1
!
ip forwarding
ipv6 forwarding
!
line vty
!
end
router2-2#
```

Remember: The order of the commands does not matter, but they should be done in some reasonable order to help the network administrator to enter all of the required information. Develop your own order that makes sense to you as Quagga does not require any particular order. It is not necessary to reboot the Pi, instead the following commands will restart the router with the new configurations as a final check for errors. Always start or restart zebra before any other routing processes.

```
pi@router2-2:~\$ sudo systemctl restart zebra
pi@router2-2:~\$ sudo systemctl restart ripd
pi@router2-2:~\$ sudo systemctl restart ospfd
```

17.16 OSPF Configuration for Pi#3 and Pi#4

For Pi#3 and Pi#4, the normal configuration for the group network can be followed. If the group has enough equipment, additional networks (with the first two octets of 10.*g.x.x*) can be created *within* the local area using RIP. Once the group is convinced all the networks are reachable, the steps for Pi#1 or Pi#2 should be used to migrate the networks to OSPF

17.17 Configuration of OSPFv3 and IPv6 Lab Network

Configuration of a working OSPF IPv4 network for IPv6 is relatively easy if the interfaces already have been assigned the proper addresses. As before, only Pi#1 needs to be configured differently for the star and ring network. The topology of the local area is the same for all the OSPF networks.

There is one minor irritation that needs to be cleared up with Quagga nomenclature concerning the naming of OSPFv3 for IPv6. Quagga uses the terms OSPF6 and **ospf6d** for OSPFv3 to emphasize the relationship with IPv6.

One interesting characteristic of OSPF in general is that it usually takes more effort to design the network than it does to configure the routers to implement that design. There are many, many settings that can be used to fine–tune the network and as usual changing these to the wrong values can lead to a broken network. When in doubt, take the defaults until you know what needs to be changed. All of our links are the same speed so the defaults work for either the ring network or the star network.

17.18 Configure Pi#1 for OSPFv3

For each group, Pi#1 will act as an ABR and therefore will have interfaces in the local area 0.0.0.*f* and area zero 0.0.0.0. For the Ring Laboratory Network, both **eth0** and **eth2** will be in area zero and **eth1** will be in area 0.0.0.*g*. The Star Laboratory Network is exactly the same except **eth2** either does not exist or is not in area 0.0.0.0. The following are the commands to issue to Quagga. Notice that the router–id is given in the format of an IPv4 address. While the router–id can be any unique number, it is customary to use the IPv4 address of one of the router's NICs.

```
Hello, this is Quagga (version 1.2.4).
Copyright 1996-2005 Kunihiro Ishiguro, et al.

router2-1# configure terminal
router2-1(config)# router ospf6
router2-1(config-ospf6)# router-id 192.168.2.2
router2-1(config-ospf6)# log-adjacency-changes
router2-1(config-ospf6)# interface eth0 area 0.0.0.0
router2-1(config-ospf6)# interface eth1 area 0.0.0.2
router2-1(config-ospf6)# interface eth2 area 0.0.0.0
```

```
router2-1(config-ospf6)# quit
router2-1(config)# exit
router2-1# write mem
Building Configuration...
Configuration saved to /etc/quagga/zebra.conf
Configuration saved to /etc/quagga/ripd.conf
Configuration saved to /etc/quagga/ripngd.conf
Configuration saved to /etc/quagga/ospfd.conf
Configuration saved to /etc/quagga/ospf6d.conf
Configuration saved to /etc/quagga/bgpd.conf
Configuration saved to /etc/quagga/isisd.conf
[OK]
router2-1#
```

Because IPv4 and IPv6 are incompatible, Quagga routers running both keep two separate route tables. To check IPv6 routes the command to Quagga is **show ipv6 route**. Once it has been determined that OSPF is working properly by examining the route table, Quagga can be configured to turn off RIPng if it is running.

```
router2-1(config)# no router ripng
router2-1(config)# exit
```

At this point the route table should show a number of OSPF routes for IPv6. In this example, Group 3 has not configured their routers to move from RIPng to OSPF. There are no routes into area 0.0.0.3 network fd86:9b29:e5e1:3310::/64.

```
Hello, this is Quagga (version 1.2.4).
Copyright 1996-2005 Kunihiro Ishiguro, et al.

router2-1# show ipv6 route
Codes: K - kernel route, C - connected, S - static, R - RIPng,
       O - OSPFv6, I - IS-IS, B - BGP, A - Babel, N - NHRP,
       > - selected route, * - FIB route

C>* ::1/128 is directly connected, lo
O>* fd86:9b29:e5e1:1000::/64 [110/20] via fe80::ebac:8872:fcbd
    :f1da, eth2, 00:01:32
O   fd86:9b29:e5e1:2000::/64 [110/10] is directly connected,
    eth1, 00:01:52
K * fd86:9b29:e5e1:2000::/64 is directly connected, eth1
C>* fd86:9b29:e5e1:2000::/64 is directly connected, eth1
O   fd86:9b29:e5e1:ff01::/64 [110/10] is directly connected,
    eth2, 00:01:37
K * fd86:9b29:e5e1:ff01::/64 is directly connected, eth2
C>* fd86:9b29:e5e1:ff01::/64 is directly connected, eth2
O   fd86:9b29:e5e1:ff02::/64 [110/10] is directly connected,
    eth0, 00:01:37
K * fd86:9b29:e5e1:ff02::/64 is directly connected, eth0
C>* fd86:9b29:e5e1:ff02::/64 is directly connected, eth0
O>* fd86:9b29:e5e1:ff05::/64 [110/20] via fe80::ebac:8872:fcbd
    :f1da, eth2, 00:01:32
C * fe80::/64 is directly connected, eth0
C * fe80::/64 is directly connected, eth2
C>* fe80::/64 is directly connected, eth1
router2-1#
```

17.19 Pi#2 OSPFv3 Configuration

For this example, let us assume Group 2 has chosen fd86:9b29:e5e1:2000::/64 for the network between local area routers and the ABR Pi#1 and fd86:9b29:e5e1:2001::/64 for a subnetwork connected to Pi#2 on **eth1**. Remember that each group can assign IPv6 networks with any prefix matching the network prefix Global ID (869b29e5e1in these examples) and ending in :gxxx:: which gives each group an enormous address space to use in assigning networks. If the interfaces have been properly assigned IPv6 addresses, the router can easily be configured to work on either Laboratory Networks.

```
 1  Hello, this is Quagga (version 1.2.4).
 2  Copyright 1996-2005 Kunihiro Ishiguro, et al.
 3
 4  router2-2# configure terminal
 5  router2-2(config)# router ospf6
 6  router2-2(config-ospf6)# router-id 10.2.2.2
 7  router2-2(config-ospf6)# log-adjacency-changes
 8  router2-2(config-ospf6)# interface eth0 area 0.0.0.2
 9  router2-2(config-ospf6)# interface eth1 area 0.0.0.2
10  router2-2(config-ospf6)# exit
11  router2-2(config)# exit
12  router2-2# show ipv6 route
13  Codes: K - kernel route, C - connected, S - static, R - RIPng,
14         O - OSPFv6, I - IS-IS, B - BGP, A - Babel, N - NHRP,
15         > - selected route, * - FIB route
16
17  C>* ::1/128 is directly connected, lo
18  O   fd86:9b29:e5e1:2000::/64 [110/10] is directly connected,
        eth0, 00:00:13
19  K * fd86:9b29:e5e1:2000::/64 is directly connected, eth0
20  C>* fd86:9b29:e5e1:2000::/64 is directly connected, eth0
21  O   fd86:9b29:e5e1:2001::/64 [110/10] is directly connected,
        eth1, 00:00:13
22  C>* fd86:9b29:e5e1:2001::/64 is directly connected, eth1
23  C * fe80::/64 is directly connected, eth1
24  C>* fe80::/64 is directly connected, eth0
25  router2-2# write mem
26  Building Configuration...
27  Configuration saved to /etc/quagga/zebra.conf
28  Configuration saved to /etc/quagga/ripd.conf
29  Configuration saved to /etc/quagga/ripngd.conf
30  Configuration saved to /etc/quagga/ospfd.conf
31  Configuration saved to /etc/quagga/ospf6d.conf
32  Configuration saved to /etc/quagga/bgpd.conf
33  Configuration saved to /etc/quagga/isisd.conf
34  [OK]
35  router2-2#
```

The other Pi devices, if they are routing, can be configured following this example and substituting the correct interfaces which will all be in area 0.0.0.*g*. If RIPng is running it should be turned off once OSPFv3 is running.

Projects

17.1 Start a ping of an address starting with 172 that is not owned by your group. Once everyone has a ping going, randomly disconnect an Ethernet cable between two groups and document the results. The results will vary from group to group.

17.2 Shutdown the switch that is acting as the backbone for area 0. What happens to the route tables?

17.3 Reboot the switch and watch the route tables. Can you see the networks come back one by one?

Exercises

17.1 Why would you shut down an interface?

17.2 Give an example of when routing all inter–area traffic over area 0.0.0.0 could cause a problem or a poor routing choice.

17.3 Given a network with the following three media, give a possible cost for each media to minimize the chance of a poor route being chosen.

- OC3 (155mbits/second)
- OC12 (622mbits/second)
- Ethernet (100mbit/second)

Further Reading

The RFC below provide further information about OSPF. This is *not* an exhaustive list and most RFC are typically dense and hard to read. Normally RFC are most useful when writing a process to implement a specific protocol.

RFCs Directly Related to This Chapter

OSPF	Title
RFC 1131	OSPF specification [82]
RFC 1245	OSPF protocol analysis [89]
RFC 1246	Experience with the OSPF Protocol [90]
RFC 1247	”OSPF Version 2” [91]
RFC 1364	BGP OSPF Interaction [94]
RFC 1370	Applicability Statement for OSPF [95]
RFC 1403	BGP OSPF Interaction [100]
RFC 1583	”OSPF Version 2” [116]
RFC 2178	OSPF Version 2 [164]
RFC 2328	”OSPF Version 2” [169]
RFC 3137	OSPF Stub Router Advertisement [196]
RFC 3509	Alternative Implementations of OSPF Area Border Routers [212]
RFC 5340	OSPF for IPv6 [259]
RFC 6549	OSPFv2 Multi-Instance Extensions [272]
RFC 7503	OSPFv3 Autoconfiguration [280]
RFC 8362	OSPFv3 Link State Advertisement (LSA) Extensibility [294]
RFC 8571	BGP - Link State (BGP-LS) Advertisement of IGP Traffic Engineering Performance Metric Extensions [305]

Other RFCs Related to OSPF

For a list of other RFCs related to the OSPF protocol but not closely referenced in this chapter, please see Appendix B–OSPF.

Chapter 18
Service Provider Protocols

18.1 Overview

Routing in a large ISP requires a different viewpoint than routing in an organization's network. Large networks need to make use of techniques to summarize routes and often view organization's networks as nothing more than nodes on their network. For these very large networks the details of their internal structure can become sensitive information that it would be best not to divulge to their competitors but routing between ISPs must still occur.

18.1.1 Autonomous Systems and ASNs

In order to summarize routes and start to hide internal structures, large networks can be viewed as ASs[1]. In order to be a true AS a network must have an internal IP structure such that all subnetworks have a common prefix and the AS must be assigned an ASN[2] by the IANA[3]. This allows all other ASs to view this network as a single pseudo–node with a network address equal to the summary prefix. While this greatly simplifies routing and route tables, this does require some discipline on the part of the network administrator and the local ISP. It is possible for an AS to have multiple summary prefixes, but this should be avoided unless there is a good reason to have more than one prefix[4].

---

[1] Autonomous Systems

[2] Autonomous System Number

[3] If the AS does not connect to the Internet it can use one of the public ASN much like the public IP address ranges.

[4] This is the author's opinion, but it seems like a good idea. The fewer summary prefixes the simpler everyone's routing will be. Remember: Simpler is usually faster.

© Springer Nature Switzerland AG 2020
G. Howser, *Computer Networks and the Internet*,
https://doi.org/10.1007/978-3-030-34496-2_18

18.1.2 RIP and OSPF Issues

While the protocols we have looked at on the Raspberry Pi can be used in very large networks, there are severe problems with both RIP and OSPF on a nation–wide level. Some of the expected problems are:

- RIP

  - RIPis very chatty. RIP updates will quickly overwhelm a large network re-gardless of the available bandwidth.
  - RIP determines the best route without taking bandwidth, lease costs, or other factors into account.
  - RIPhas a default maximum number of hops of 15 which gives a maximum horizon of 30 hops from end–to–end. This can be expanded, but nation–wide ISPs can be many router hops from end–to–end.
  - RIPdoes not have a mechanism for creating or using summary routes.

- OSPF

  - OSPF requires all traffic between areas to cross the unique area zero. Imple-menting OSPF on a nation–wide ISP would mean each client network could not take full advantage of OSPF because they would be limited to being a single OSPF area.
  - This would require all clients regardless of how small or simple the network to configure and run an instance of OSPF on their gateway router.
  - In order to correctly calculate the best route for long routes some rigid stan-dard for the cost of client links might be required. This would be difficult to implement and enforce.

- Many nation–wide ISPs were operating *before* OSPF was standardized and were using other protocols.

To address these issues, we will look at one possible solution that will run on the Raspberry Pi using two of the protocols provided by the Quagga package. These are protocols rarely used in an intranet, but they do serve to help one understand how messages cross the Internet.

18.2 ISIS Overview

While on the surface ISIS and OSPF are very similar, there are many conceptual differences which can impact the design and operation of the network itself. ISIS routes between routers while other routing protocols such as OSPF route between interfaces on routers. This means there is less emphasis on the details of how media connects to the device and more emphasis on the interactions between devices. As a result, the structure of an area's network or a backbone is more of a dynamic, logical structure and less constricted by the details of exactly where the media is physically connected to the hardware. In practice these conceptual differences are usually slight at the level of complexity encountered in a private network.

A network using ISIS actually consists of two types of networks. Areas use ISO-IGP internally to route between destinations and subnetworks that are completely contained in the local area. In order to route between devices which are not in the same area, a different type of routing called IS–IS[5] is used. Each of these types of routing interprets the OSI standard NSAP address in a different fashion but for our purposes this is only a minor issue[6].

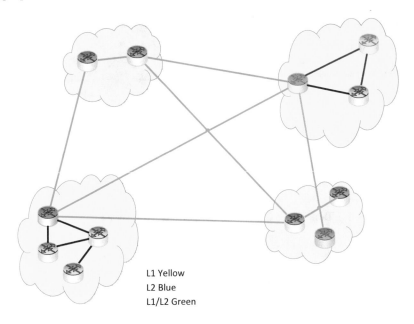

L1 Yellow
L2 Blue
L1/L2 Green

Fig. 18.1: A Small ISIS Network

Because ISIS can be used in a large number of different contexts as with OSPF, ISIS is a very flexible, although complex, set of protocols. The addressing schema is dependent upon the type of network and many of the fields are variable in length and their meanings can be locally defined to meet the needs of the organization. For this reason, this chapter will use a simplified version of ISIS that meets the Cisco Systems standards for addressing. However, the general form of the NSAP address is given in the next section along with the standardized Cisco Systems addressing.

---

[5] ISIS Inter–Area Routing

[6] This topic is intricate because so much of the form of an NSAP address depends upon the requirements of the network and services it provides. We will keep this to a more simplistic model as this chapter is intended to be an introduction to ISIS

Table 18.1: GOSIP Version 2 NSAP Structures

GOSIP NSAP: 47.0005.80.005a00.0000.1000.0020.00800a123456.00

← IDP →								
AFI	IDI	← DSP →						
47	0005	DFI	AA	Rsvd	RD	Area	ID	Sel
1*	2*	1*	3*	2*	2*	2*	6*	1*
47	0005	80	005a00	0000	1000	0020	00800a123456	00

*Length in octets.

IDP   Initial Domain Part
AFI   Authority and Format Identifier
IDI   Initial Domain Identifier
DSP   Domain Specific Part
DFI   DSP Format Identifier
AA   Administrative Authority
Rsvd   Reserved
RD   Routing Domain Identifier
Area   Area Identifier
ID   System Identifier
SEL   NSAP Selector

Table 18.2: Cisco Standard NSAP Addresses

Example: 47.0001.aaaa.bbbb.cccc.00

IS–IS Address Format:

← Area →		System ID	NSEL
47	0001	aaaa.bbbb.cccc	00

ISO–IGP Address Format:

Domain	Area	System ID	NSEL
47	0001	aaaa.bbbb.cccc	00

18.3 NSAP Addressing

NSAP addressing is extremely flexible and does not translate well to IPv4 or
IPv6 addressing. In fact, NSAP addressing is so complex and variable that the
US government was led to create a standard known as GOSIP for government systems [53] [85]. Australia and New Zealand used a modified GOSIP[7] standard and
the EU has its own standard. The ATM Forum adopted a different standard for ATM
addressing. Fortunately, private networks can develop their own standards or use
some existing format that meets their needs.

---

[7] One wonders if this standard was adopted because of the cute name.

The address is usually written as a variable length string of hexadecimal dig-its separated by "dots"[8] and the meaning of the fields of the address can be rede-fined to meet the needs of the network and devices. For our purposes, the standard Cisco Systems NSAP addressing schema can be used [55] [20].

Because the dots in an NSAP address can appear wherever they are needed for readability, various standards organizations have different rules for where to place the dots. For example consider the address 47.0001.aaaa.bbbb.cccc.00. This address will take two different forms for IS-IS and ISO-IGP as shown in Table 18.2.

Table 18.3: NSAP NET Conversion

Hexadecimal	Decimal	Hexadecimal	Decimal
0	00	8	08
1	01	9	09
2	02	a	10
3	03	b	11
4	04	c	12
5	05	d	13
6	06	e	14
7	07	f	15

Example:
MAC: b8:27:eb:7f:95:67
System ID: 1108.0207.1411.0715.0905.0607

18.4 ISIS Network

As stated before, ISIS networks are formed from two types of areas. Level 1 routers with the same area address form an ISO-IGP area and Level 2 routers with the same area address form an IS–IS backbone. Areas *must* be contiguous and all devices with NSAP addresses in the area must have the same area address or id. In fact, the area address *defines* the area for Level 1 or Level 2[9].

ISIS is a link–state protocol and only exchanges information when there is a change in the network such as a new link or router being added. Therefore, each router maintains a database of the state of the network and has a view of the network that is consistent with all other routers in the same area regardless of whether it is an IS-IS backbone or an ISO-IGP area. Level 1–2 routers must maintain two databases, one for the Level 1 area and one for the Level 2 backbone.

---

[8] The digits of the NSAP are often separated by dots for easier reading by humans. The network will ignore all dots in an NSAP address.

[9] ISIS does not allow for routing between two different Level 2 backbones. Some other protocol such as BGP must be used for this purpose.

18.5 Convergence of a ISIS Network

Like most modern routing protocols, ISIS is very quick to converge and frequently converges faster than a similar OSPF network. Once all the routers in an area have formed adjacencies, a router is elected to act as the DIS[10] for that area. Like most protocols, ISIS will elect a new DIS any time is it required. If at any time the DIS should fail to respond, the other routers will immediately hold a new election. Election depends upon the priority of the routers with any ties broken by the router ID. The routers then form a MST with the DIS as the root. This insures that all the routers in an area have the same view of the network. At this point the network area has converged.

If the network consists of different speed connections or connections that are preferred over others, the network administrator can configure each router connection with an *outgoing* weight. If no weight is set a default value is used[11]. These weights are used by the Dijkstra's algorithm to form a weighted shortest path from the DIS to each router to insure convergence.

18.5.1 Joining an Area

The easiest way to look at convergence in an ISIS area is to look at what happens when a new router boots up. When the router boots, it senses which interfaces have active links and sends an ISIS "hello" message. The router on the other end of the connection replies and an adjacency is formed[12]. Once the two routers have exchanged link–states via LSP[13], an election is held by the routers in the area to elect a new DIS. When this is finished, the routers all form a new view of the network by running Dijkstra's algorithm. Now the area has converged.

Notice that the presence of a new router may also lead to new subnetworks being part of the area in which case link–state updates may be required on the backbone. Like a local area, the backbone will also elect a DIS simply to act as root for a MST and to run Dijkstra's.

---

[10] Designated Intermediate System

[11] Typically 10 for Cisco Systems networks.

[12] While this is happening, the neighbor router is also informing its adjacent routers of a link–state update.

[13] Link State Packet Pseudonode

18.6 Advantages of a ISIS Network

One of the most important characteristics of ISIS is fast convergence. Link–state protocols react quickly to network changes and have minimal overhead when the network is stable.

Unlike most of the open standard routing protocols on the Internet, ISIS does not depend upon nor require, IP. ISIS can run over ATM, frame relay, and many other protocols as well as IP networks. Indeed, ISIS is frequently used as the protocol of choice by many of the largest ISPs precisely because it is so flexible.

Like OSPF, ISIS is a link–state protocol which converges very quickly. The overhead involved in convergence of the network is therefore minimal and is even less once convergence has been reached. The main goal of any network is to move packets as quickly as possible and any exchanges of information required by a routing protocol detract from moving user data.

ISIS is extremely flexible and scalable[14]. If some forethought is given to eventual growth, ISIS can grow from a single area network to a large, complex global network without the need to reconfigure many devices. While this may seem a minor issue, reconfiguring a large network is always stressful for everyone involved and can lead to unplanned outages in the network.

Like other routing protocols, ISIS can be implemented over a number of different physical connections, layer 2 networks, and in some cases even different layer 3 networks.

ISIS has complete protocol transparency which means it can easily carry any protocols that can be encapsulated. This is built into the protocol rather than added later which implies greater stability.

18.7 Disadvantages of a ISIS Network

The only serious disadvantage of ISIS is that is does not scale well to smaller networks. This is really not a serious problem, but more work is required to configure a very small ISIS network than a very small RIP network[15]. This disadvantage is insignificant in the light of the ability of ISIS to easily scale up to large or complex networks.

---

[14] Scalability is under appreciated. Networks grow in size and complexity and scalability is crucial in this growth.

[15] Even with all its problems, RIP is the protocol of choice for networks with few routers and subnetworks. Networks of less than 10–20 routers typically do fine with RIP until they begin to grow.

18.8 ISIS on the Pi

As of this writing, there is a problem with Quagga support of ISIS routing. The daemon **isisd** does *not* support multiple areas on the same router which means there is no support for inter-area routing even though Quagga allows for such a configuration. For the present, ISIS can run on the backbone of the network but does not allow interconnections between all devices. This will be addressed with BGP in Section 18.10.

The ISIS Test–Bed network in Figure 18.2 is shown wired as a ring, but a star will work as well. Each Group will run ISIS in two forms. On the set of Pi#1s (the backbone) the ISIS area will be 9999 and on all other routers in the Group the ISIS area will be 000*g* where *g* is the Group's number.

[htb]

ISIS Ring Test Bed Network

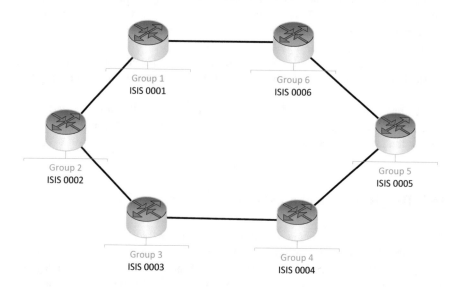

Fig. 18.2: IS–IS Test–bed Network

Like other routing protocols on the Pi, ISIS runs as a daemon (**isisd**) in cooperation with Zebra and the Linux kernel to populate the route table. The configuration is held in the file **/etc/quagga/isisd.conf** which can be created as follows:

```
pi@mail:~$ sudo touch /etc/quagga/isisd.conf
pi@mail:~$ ls -l /etc/quagga/
total 16
-rw-r--r-- 1 root     root          0 May 27 00:43 isisd.conf
-rw-r----- 1 quagga   quagga      250 Apr 14 12:13 ripd.conf
-rw-r----- 1 quagga   quagga       78 Apr 14 12:11 ripd.conf.
        sav
-rw-r----- 1 quagga   quaggavty     0 Apr 14 12:04 vtysh.conf
```

```
-rw-r----- 1 quagga quagga    235 Apr 14 12:13 zebra.conf
-rw-r----- 1 quagga quagga    200 Apr 14 12:11 zebra.conf.
     sav
```

After the file has been created, the daemon *must* be started before ISIS can be configured[16].

```
pi@mail:~$ sudo systemctl restart isisd
pi@mail:~$ sudo systemctl status isisd
isisd.service - IS-IS routing daemon
Loaded: loaded (/lib/systemd/system/isisd.service; enabled
     ; vendor preset: en
Active: active (running) since Mon 2019-05-27 00:49:29 EDT
     ; 9s ago
Docs: man:isisd
Process: 2249 ExecStart=/usr/sbin/isisd -d -A 127.0.0.1 -f
     /etc/quagga/isisd.c
Process: 2246 ExecStartPre=/bin/chown -f quagga:quagga /
     etc/quagga/isisd.conf
Process: 2243 ExecStartPre=/bin/chmod -f 640 /etc/quagga/
     isisd.conf (code=exit
Main PID: 2250 (isisd)
CGroup: /system.slice/isisd.service
2250 /usr/sbin/isisd -d -A 127.0.0.1 -f /etc/quagga/isisd.
     conf

May 27 00:49:29 mail.college.edu systemd[1]: Starting IS-
     IS routing daemon...
May 27 00:49:29 mail.college.edu systemd[1]: Started IS-IS
     routing daemon.
```

18.9 Quagga ISIS Commands

As might be expected, there is a wealth of commands to support ISIS routing on the
Pi. Only the bare minimum commands to set up the network will be covered here.
For more detail of the other options the reader is directed to the Quagga website [43].
The full command set appears to be implemented on the Pi and is available via **sudo
vtysh 2601** in the same manner as all other routing protocols.

18.9.1 Unique ISIS Router ID

Each router must have exactly one unique **System ID** which is entered via the
net command. The format used is very specific and the ID *must* be unique in
each area. Since it is hoped that Quagga will support multiple areas on one router
(**level-1-2**) in the future, it is best to insure that the ID is absolutely unique. To
do this we will follow the ATM standard and Cisco Systems's suggestion to use the

---

[16] Quagga will be very nice and let you configure the ISIS router all you want, but without the
daemon running all configuration commands are ignored.

MAC address of one of the device's interface[17]. In this case we will use the MAC address of the **eth0** interface which can be found by entering **sudo ifconfig eth0**.

```
1    pi@mail:~$ sudo ifconfig eth0
2     eth0: flags=4163<UP,BROADCAST,RUNNING,MULTICAST>  mtu 1500
3     inet 192.168.1.5  netmask 255.255.255.0  broadcast
         192.168.1.255
4     inet6 fe80::15cd:867b:ff1d:db20  prefixlen 64  scopeid 0
         x20<link>
5     inet6 fd51:42f8:caae:d92e::ff  prefixlen 64  scopeid 0x0<
         global>
6     ether b8:27:eb:7f:95:67  txqueuelen 1000   (Ethernet)
7     RX packets 1546  bytes 528627 (516.2 KiB)
8     RX errors 0  dropped 0  overruns 0  frame 0
9     TX packets 832  bytes 377150 (368.3 KiB)
10    TX errors 0  dropped 0 overruns 0  carrier 0  collisions 0
11   pi@mail:~$
```

The MAC address is on line 6 of the screen above and is given in hexadecimal as $b8 : 27 : eb : 7f : 95 : 67$. From this address can be formed the unique router address for the **net** command in the Cisco Systems format given in Table 18.2. Each hexadecimal digit must be entered as two decimal digits as in Table 18.3. The area must also be unique and is a four decimal digit number such as your group number. The AFI[18] for a private network must be 47 and for a router the SEL[19] must be 00. So if this router belongs to Group 1 in area 0001 the NET address should be: 47.0001.1108.0207.1411.0715.0905.0607.00. If the router connects to other groups, it is a backbone router and should be in the same area with all the other backbone routers, for example 9999[20] for a net identifier of 47.9999.1108.0207.1411.0715.0905.0607.00. Once this information has been placed in the correct format, the router can be configured using **vtysh**.

18.9.2 ISIS Area Configuration Steps

The following steps will properly configure this router to be in area 0001.

```
router1-1~$ sudo vtysh 2601

Hello, this is Quagga (version 1.1.1).
Copyright 1996-2005 Kunihiro Ishiguro, et al.

router1-1# config term
router1-1(config)# interface eth0
router1-1(config-if)# ip router isis 0001
router1-1(config-if)# isis circuit-type level-2-only
router1-1(config-if)# quit
```

---

[17] In reality, any unique numerical identifier could be used. Some organizations use an IPv4 address padded with zeroes. Some use the inventory number of the router. It can be any unique numeric string of 12 hex or 24 decimal digits and is only used to identify the router.

[18] Authority and Format Identifier (NSAP)

[19] NSAP Selector

[20] Remember that the backbone must be contiguous and can have any unique area number. It need not be area 0.0.0.0 or zero as in OSPF which means it should be easier to merge ISIS networks than OSPF networks.

```
router1-1(config)# router isis 0001
router1-1(config-router)# log-adjacency-changes
router1-1(config-router)# is-type level-2-only
router1-1(config-router)# net
    47.0001.1108.0207.1411.0715.0905.0607.00
router1-1(config-router)# quit
router1-1(config)# exit
router1-1# write term
Building configuration...

Current configuration:
!
log file /var/log/quagga/zebra.log
hostname router1-1
!
password raspberry
!
interface eth0
ip address 192.168.1.5/24
ip router isis 0001
isis circuit-type level-2-only
!
interface lo
!
interface wlan0
!
router isis 0001
net 47.0001.1108.0207.1411.0715.0905.0607.00
metric-style wide
is-type level-2-only
log-adjacency-changes
!
ip forwarding
ipv6 forwarding
!
line vty
!
end
router1# write mem
Building Configuration...
Configuration saved to /etc/quagga/zebra.conf
Configuration saved to /etc/quagga/ripd.conf
Configuration saved to /etc/quagga/isisd.conf
[OK]
router1# exit
router1-1~$
```

If the router is Pi number one for the group it should be configured to be part of the backbone which is designated as level 2 by ISIS. The steps below will configure the router to be a backbone router with the potential to connect to group 1 when, and if, Quagga supports a fully functional ISIS on the Raspberry Pi.

18.9.3 ISIS backbone Configuration Steps

The following steps will properly configure this router to be in area 9999.

```
router1-1~$ sudo vtysh 2601

   Hello, this is Quagga (version 1.1.1).
   Copyright 1996-2005 Kunihiro Ishiguro, et al.

   router1-1# config term
   router1-1(config)# interface eth0
```

```
router1-1(config-if)# ip router isis 9999
router1-1(config-if)# isis circuit-type level-2-only
router1-1(config-if)# quit
router1-1(config)# router isis 9999
router1-1(config-router)# log-adjacency-changes
router1-1(config-router)# is-type level-2-only
router1-1(config-router)# net
      47.9999.1108.0207.1411.0715.0905.0607.00
router1-1(config-router)# quit
router1-1(config)# exit
router1-1# write term
Building configuration...

Current configuration:
!
log file /var/log/quagga/zebra.log
hostname router1-1
!
password raspberry
!
interface eth0
ip address 192.168.1.5/24
ip router isis 9999
isis circuit-type level-2-only
!
interface lo
!
interface wlan0
!
router isis 9999
net 47.9999.1108.0207.1411.0715.0905.0607.00
metric-style wide
is-type level-2-only
log-adjacency-changes
!
ip forwarding
ipv6 forwarding
!
line vty
!
end
router1# write mem
Building Configuration...
Configuration saved to /etc/quagga/zebra.conf
Configuration saved to /etc/quagga/ripd.conf
Configuration saved to /etc/quagga/isisd.conf
[OK]
router1# exit
```

The last step is to verify that all routers on the backbone can **ping** each other. At this point the network is fragmented into a number of areas and can no longer function correctly. To allow the automatous areas to communicate accross area boundaries, a different protocol such as BGP must be used. However, BGP does not scale well to small and medium size networks, so for a work–around RIP and RIPng can be used to redistribute the routes between the Group AS and the backbone AS.

ISIS Projects

18.1 Set up either the Ring or Star Laboratory Network and insure that ISIS routing iw working properly.

18.2 Start a **`ping`** of an interface is another group. Experiment with disconnecting various connections and document the results.

ISIS Exercises

18.1 Draw a network diagram of your Group. Document each interface in your Group. You might find it more readable to use a table of the interfaces and NSAP addresses.

18.10 BGP Overview

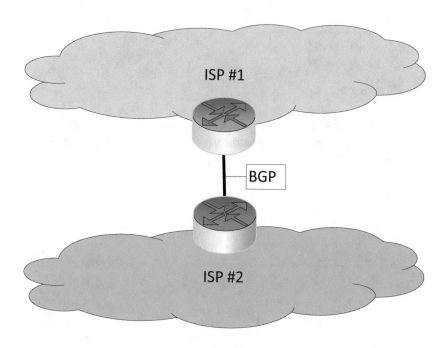

Fig. 18.3: BGP Connecting Two Large ISPs

BGP is an interdomain routing protocol designed to provide loop-free routing links between organizations. BGP is designed to run over a reliable transport protocol; it uses TCP (port 179) as the transport protocol because TCP is a connection-oriented protocol. The destination TCP port is assigned 179, and the local port is assigned a random port number. Cisco Systems software supports BGP version 4 and it is this version that has been used by Internet service providers (ISPs) to help build the Internet. RFC 1771 introduced and discussed a number of new BGP features to allow the protocol to scale for Internet use. RFC 2858 introduced multiprotocol extensions to allow BGP to carry routing information for IP multicast routes and multiple Layer 3 protocol address families, including IPv4, IPv6, and CLNS.

BGP is mainly used to connect a local network to an external network to gain access to the Internet or to connect to other organizations. When connecting to an external organization, external BGP (eBGP) peering sessions are created. Although BGP is referred to as an exterior gateway protocol (EGP), many networks within an organization are becoming so complex that BGP can be used to simplify the internal network used within the organization. BGP peers within the same organization exchange routing information through internal BGP (iBGP) peering sessions.

BGP uses a path-vector routing algorithm to exchange network reachability information with other BGP-speaking networking devices. Network reachability information is exchanged between BGP peers in routing updates. Network reachability information contains the network number, path-specific attributes, and the list of autonomous system numbers that a route must transit to reach a destination network. This list is contained in the AS-path attribute. BGP prevents routing loops by rejecting any routing update that contains the local autonomous system number because this indicates that the route has already traveled through that autonomous system and a loop would therefore be created. The BGP path-vector routing algorithm is a combination of the distance-vector routing algorithm and the AS-path loop detection.

BGP selects a single path, by default, as the best path to a destination host or network. The best path selection algorithm analyzes path attributes to determine which route is installed as the best path in the BGP routing table. Each path carries well-known mandatory, well-known discretionary, and optional transitive attributes that are used in BGP best path analysis. Quagga software provides the ability to influence BGP path selection by altering some of these attributes using the command-line interface (CLI.) BGP path selection can also be influenced through standard BGP policy configuration.

BGP uses the best-path selection algorithm to find a set of equally good routes. These routes are the potential multipaths. When there are more equally good multipaths available than the maximum permitted number, the oldest paths are selected as multipaths.

BGP can be used to help manage complex internal networks by interfacing with Interior Gateway Protocols (IGPs). Internal BGP can help with issues such as scaling the existing IGPs to match the traffic demands while maintaining network efficiency [17].

Currently the Internet consists of a number of extremely large ISP networks that are interconnected at what are known as Meet–points. Each of the large ISPs maintains at least one of these Meet–points where the US national ISPs and the international ISPs connect to a common high speed network to exchange packets that must travel between ISPs. There is a major problem with sharing routes between these ISPs as knowledge of the internal routing of an ISP could expose sensitive data about the network. The number of subnetworks that need to be exchanged would also lead to giant routing tables on all the routers which would seriously delay the movement of packets.

BGP can provide a solution to both of these issues by summarizing routes when possible, thereby limiting the number of detail routes exposed to outside networks[21]. Routers inside an ISP need only maintain a route to the BGP router for all the routes in foreign ISPs[22].

---

[21] It is assumed that an ISP would *not* use any detailed knowledge of a competitor's network to its advantage, nor would an ISP ever use this information to attack another network. Even so, it is better not to expose sensitive data outside your organization.

[22] Other protocols can provide summary routes to control the number of routes a router must maintain in the route table as well. However, the random manner in which IPv4 addresses were assigned means that summarization is not as efficient as it could be.

18.11 Policy Driven BGP Requirements

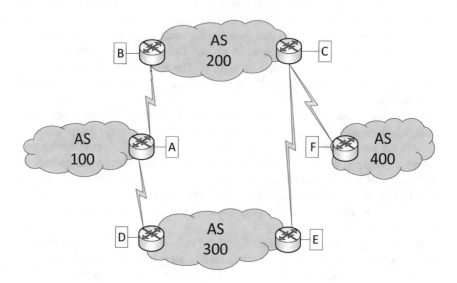

Fig. 18.4: Policy Driven BGP Requirements

Quite often the connection between ASs can be made using static routes. This is especially useful when the assignment of IP subnetworks allows for efficient router summarization[23]. However, Telco tariffs or other considerations might force an AS to prefer one route over another. Because these ASs are not advertising internal routes, a method such as used for OSPF or ISIS cannot always be used to determine the cheapest route.

Consider the example in Figure 18.4 [13] with a policy for AS 100 "**Always use AS 300 path to reach AS 400**". Suppose AS 400 is connected to AS 200 using static routes to router C. From the point of view of router A, the networks connected by router F have the same policies as the networks in the ISP autonomous system 200.

A new pathing policy that refers to AS 400 cannot be implemented in AS 100 because this requires being able to distinguish the networks in AS 400 from other autonomous systems. In this case, for autonomous system 100 to know of the existence of autonomous system 400, autonomous system 400 will have to be connected via BGP. Also, router F must be configured to announce the existence of AS 400.

---

[23] Remember: the goal of all routing is to move packets quickly and static routes are very fast, but cannot automagically change when the network changes.

18.12 BGP Operation

When multiple autonomous systems are interconnected, BGP supports two types of sessions between neighboring routers:

- EBGP[24]. Occurs between routers in two different autonomous systems. Typically these routers are adjacent to each other and share the same media and a subnetwork.
- IBGP[25]. Occurs between routers in the same autonomous system and is used to coordinate and synchronize routing policy within the autonomous system. Neighbors may be located anywhere in the autonomous system, even many hops away from one another.

Routes learned via BGP must be redistributed to some internal routing protocol used by the autonomous system such as OSPF or RIP. This implies that routers running BGP will also be running some other routing protocol and redistributing routes between the protocols.

18.13 Advantages of a BGP Network

- ISPs use BGP to control which routes within an AS are advertised to the outside world thereby protecting the details of the internal network.
- BGP can summarize routes where feasible to reduce the number of routes routers must keep in their route table while still allowing full access to all internal and external subnetworks.
- With BGP it is possible to prefer one interconnection between ISPs while keeping fall–back connections active.
- To a certain extent, BGP can be used to balance the load on multiple connections between large ISPs.

18.14 Disadvantages of a BGP Network

- BGP is complicated and intricate to configure.
- When running BGP, some other routing protocol must also be running for routing within the AS.
- To use BGP on the public Internet, the ASs must all have publicly registered ASNs[26].

---

[24] External BGP session

[25] Internal BGP session

[26] It is not difficult to register an ASN.

- All routes learned or advertised must at some point be redistributed via some other routing protocol. This means at least some of the routers running BGP must also run some other protocol.
- If the AS is not properly designed, BGP may not be able to take full advantage of route summarization. Therefore care must be taken in the design of networks which *might* eventually use BGP to avoid the necessity of changing large numbers of IP addresses[27] and re–configuring many routers.

18.15 BGP on the Pi

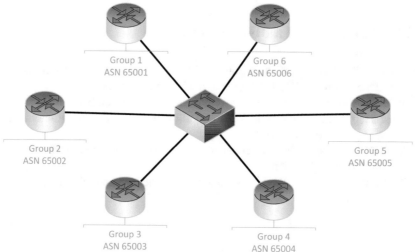

Fig. 18.5: BGP Lab Network

While it is perfectly possible to run BGP on a small controlled network of Pi's such as the Laboratory network shown in Figure 18.5, this is left as an experiment you may pursue on your own.

---

[27] Hopefully this can be done by simply reconfiguring a DHCP server. This is one of the advantages of dynamic assignment of IP addresses.

Further Reading (ISIS)

The RFCs below provide further information about Intermediate System to Intermediate System (ISIS), BGP, and Internet Service Providers. This is *not* an exhaustive list and most RFCs are typically dense and hard to read. Normally RFCs are most useful when writing a process to implement a specific protocol.

RFCs Directly Related to This Chapter

ISIS	Title
RFC 1142	OSI IS-IS Intra-domain Routing Protocol [84]
RFC 1169	Explaining the role of GOSIP [85]
RFC 1195	Use of OSI IS-IS for routing in TCP/IP and dual environments [87]
RFC 1637	DNS NSAP Resource Records [118]
RFC 2966	Domain-wide Prefix Distribution with Two-Level IS-IS [191]
RFC 2973	IS-IS Mesh Groups [192]
RFC 3277	Intermediate System to Intermediate System (IS-IS) Transient Blackhole Avoidance [198]
RFC 3719	Recommendations for Interoperable Networks using Intermediate System to Intermediate System (IS-IS) [221]
RFC 3787	Recommendations for Interoperable IP Networks using Intermediate System to Intermediate System (IS-IS) [222]
RFC 3847	Restart Signaling for Intermediate System to Intermediate System (IS-IS) [225]
RFC 4971	Intermediate System to Intermediate System (IS-IS) Extensions for Advertising Router Information [253]
RFC 5130	A Policy Control Mechanism in IS-IS Using Administrative Tags [256]
RFC 7775	IS-IS Route Preference for Extended IP and IPv6 Reachability [284]
RFC 8571	BGP - Link State (BGP-LS) Advertisement of IGP Traffic Engineering Performance Metric Extensions [305]

Other RFCs Related to ISIS

For a list of other RFCs related to the ISIS protocols but not closely referenced in this chapter, please see Appendices B–ISIS and B–ISP.

Further Reading (BGP)

The RFC below provide further information about BGP. This is a fairly exhaustive list and most RFC are typically dense and hard to read. Normally RFC are most useful when writing a process to implement a specific protocol.

RFCs Directly Related to BGP

BGP	Title
RFC 1266	Experience with the BGP Protocol [92]
RFC 1364	BGP OSPF Interaction [94]
RFC 1397	Default Route Advertisement In BGP2 and BGP3 Version of The Border Gateway Protocol [99]
RFC 1403	BGP OSPF Interaction [100]
RFC 1520	Exchanging Routing Information Across Provider Boundaries in the CIDR Environment [111]
RFC 2842	Capabilities Advertisement with BGP-4 [187]
RFC 2858	Multiprotocol Extensions for BGP-4 [188]
RFC 4893	BGP Support for Four-octet AS Number Space [249]
RFC 5396	Textual Representation of Autonomous System (AS) Numbers [260]
RFC 6811	BGP Prefix Origin Validation [275]
RFC 8388	Usage and Applicability of BGP MPLS-Based Ethernet VPN [295]
RFC 8503	BGP/MPLS Layer 3 VPN Multicast Management Information Base [303]
RFC 8571	BGP - Link State (BGP-LS) Advertisement of IGP Traffic Engineering Performance Metric Extensions [305]

Other RFCs Related to BGP

For a list of other RFCs related to BGP but not closely referenced in this chapter, please see Appendix B–BGP.

Further Reading (ISPs)

The RFC below provide further information about ISPs. This is a fairly exhaustive list and most RFC are typically dense and hard to read. Normally RFC are most useful when writing a process to implement a specific protocol.

RFCs Directly Related to ISPs

ISP	Title
RFC 3013	Recommended Internet Service Provider Security Services and Procedures [193]
RFC 3871	Operational Security Requirements for Large Internet Service Provider (ISP) IP Network Infrastructure [227]
RFC 4778	Operational Security Current Practices in Internet Service Provider Environments [248]
RFC 8501	Reverse DNS in IPv6 for Internet Service Providers [302]

Other RFCs Related to ISP

For a list of other RFCs related to ISPs but not closely referenced in this Chapter, please see Appendix B–ISP.

Chapter 19
Babel

Introduction

According to the most recent RFC, "Babel is a loop-avoiding distance-vector routing protocol that is designed to be robust and efficient both in networks using prefix-based routing and in networks using flat routing, and both in relatively stable wired networks and in highly dynamic wireless networks." [270].

While Babel is supported by Raspbian, it is not supported by the current version of Quagga. Also Babel is still in the process of becoming a standard open source routing protocol. As such, it can be expected to change and grow over time.

One of the distinct advantages of Babel is its support for networks comprising both wired and wireless links. While RIP, OSPF, and ISIS all run on any link, only Babel is designed to take the changing characteristics of a wireless network into consideration when determining the best route.

In this chapter we will examine the advantages and disadvantages of Babel. Because of the inherent security issues with running an unauthorized wireless network and some installation issues, the actual implementation of a Babel network is currently beyond the scope of this book.

19.1 Overview of Babel

Unlike most open routing standards, Babel was designed to answer the specific problem of unstable networks as is often encountered with wireless networks. Other routing protocols are designed to converge with a stable view of the network and only change when the network changes. This is fine for wired networks that rarely change, but as we saw with route flapping this can lead to increased overhead with an unstable network.

Babel concentrates on limiting the occurrence of routing problems such as routing loops and black holes. During convergence, Babel minimizes routing issues caused by network changes and produces a correct view of the network routes, but this view may be suboptimal. In short, convergence produces a correct network very quickly at the expense of not all routes being the best possible route.

If the solution is not optimal, why would one use Babel? The answer lies in the nature of wireless networks. In practice, full wireless coverage of an area leads to multiple networks serving any one mobile station at the same time. As the background conditions change, the relative strength of the wireless network sensed by a mobile station change as well. Frequently this leads to wireless interfaces connecting to different networks over a short period of time. Babel was designed to quickly converge as these wireless network connections change.

Babel attempts to find the best route based upon the quality of each link rather than the cost of each link. As networks appear and disappear, the quality of those links is less than the quality of links in networks that are stable. By using the quality of each link, Babel is able to avoid route flapping which can cause extensive overhead in RIP and OSPF.

When the overall network becomes stable, Babel slowly converges from a correct sub-optimal routing solution to a correct optimal solution. During this optimization of the network solution, Babel maintains correct routes to all known networks. This optimization can take many minutes to complete during which time packest are routed using less than optimal routes.

Because it was developed at a time when multiple network protocols were common, Babel was developed as a hybrid protocol to understand both IPv4 and IPv6 with a single process. This reduces the load on the router as the routing engine need not be tied to the underlying IP version.

For more on Babel there is an excellent overview given in RFC 6126 [270].

19.2 Babel on the Pi

While it is possible to run Babel on the Raspberry Pi, there are significant issues with installing the packages required[1]. It also appears there may be some significant conflicts with how **babeld** is started and stopped and how Quagga daemons are handled.

If there are problems with running Babel on the Raspberry Pi and if Babel is still being finalized as an open protocol, then why are we interested in Babel? There are a number of reasons. Babel will eventually be folded into Quagga or a fork of Quagga called FRR[2]. This means it will eventually have the same type of interface as the other routing protocols we have examined. Also Babel is specifically designed to work reasonably well for unstable networks. Babel seems to work well for medium–

---

[1] At the time of this writing, this is actually an understatement. Installing Babel requires a number of additional packages and a local compilation. While this is not as terrifying as the instructions might make it seem, Babel on the Pi will be left for a later edition or as a special project on the companion websites.

[2] Free Range Routing

sized networks and should provide an alternative between RIP for small networks and OSPF for large networks.

19.3 Babel Best Route

When a router comes up or senses the presence of a new link, it sends "hello" messages out each link to find its new neighbors. These routers respond with an IHU[3] message to form an association and to begin to exchange route information. Each router uses the information in the Hello message and IHU exchange to estimate the cost of the link between the two routers.

Once a router has information on the costs of each link in the network, it runs the Bellman–Ford[5] algorithm to find the best route to each known subnetwork[6]. This is a fairly fast algorithm and has a low impact upon the router CPU.

19.4 Convergence of a Babel Network

An interesting feature of Babel is how it converges in an unstable network. Babel has features to avoid recreating transient pathological routing issues such as routing loops and black holes. All routing protocols can experience poor routing choices during convergence as packet are sent over a more costly route, route in a loop, or even routed to a router that will drop the packets thus creating a black hole. If the connection is guaranteed delivery, these issues will cause the retransmission of packets which may follow the same useless path.

While Babel is also subject to some of these issues, it has unique features to avoid the same bad routes when seen multiple times as in route flapping. The resulting avoidance of reoccurring problems allows Babel to reach convergence much faster in unstable networks such as wireless or sensor networks. Some unstable networks go through the same issues in a cycle such as route flapping or wireless network interference. As these networks fluctuate they tend to past through the same states (link up/link down, wireless connection good/weak/bad/weak/good) over and over in a cycle. Other protocols are not designed to recognize cyclical bad behavior, even though such behavior is common in networks. Babel can protect itself from these cycles to a greater extent than most protocols.

---

[3] IHU[4]

[5] There is a detailed explanation of how Bellman–Ford helps Babel to avoid transient routing loops in RFC 6126. For an explanation of the Bellman–Ford algorithm see Cormen [25].

[6] Unlike Dijkstra's algorithm, Bellman–Ford can handle a network that includes links with a negative cost. A negative cost would mean we get paid for every packet we send across that link. When you find a situation where your network has a link with a negative cost, please let me know. We could get rich sending packets in a loop over that link. In short, a negative cost link will never occur but Bellman–Ford is ready when it does!

   Much like OSPF, Babel can be configured to force an update when a link changes rather than waiting for the next scheduled update. In a stable network with few changes, this can allow the network to begin convergence much sooner. Other distance–vector protocols such as RIP may not be able to do this.

   Unfortunately, routers are forced to send out updates at regular intervals which leads to unnecessary overhead. This is why Babel might not be a good choice for a stable network with few changes.

19.5 Advantages of a Babel Network

- Babel was designed to provide support for both IPv4 and IPv6 with a single process. In fact, updates for IPv4 and IPv6 networks can be transmitted in the same update message. Both RIP and OSPF require a separate routing protocol stack for IPv4 and IPv6. This should facilitate the transition from the older IPv4 to IPv6.
- Babel was designed to take the characteristics of wireless networks into account when determining the best route.
- Babel will dynamically change the cost of wireless connections if the quality of the connection changes.
- The issues with Babel actually point to the idea that new routing protocols can be developed and integrated with existing protocols.
- Babel uses a different algorithm, the Bellman–Ford algorithm, to determine the shortest path from the source to all other known networks.
- When (not if) Babel becomes integrated with existing Quagga protocols or the FRR platform, the EIGRP daemon will most likely follow[7]. This would bring another very useful protocol to the Pi router.
- Cisco Systems may support Babel in the future and it is always wise to keep up with Cisco Systems if you are interested in routing. Open source routing platforms are beginning to push back at the big name routers such as Cisco Systems and Juniper, but the larger networks are still running router hardware.
- Babel promises quick convergence and relatively low overhead.
- Babel was designed to be easily extended to allow for custom implementations. The only restriction is that extended versions of Babel must be backwards compatible with existing standard Babel.
- Some routing protocols are prone to routing pathologies during forced reconvergence. Babel was designed to proactively avoid routing loops and "black holes" when routes change, especially during route flapping.

---

[7] I look forward to adding a section on EIGRP at some point. It is an interesting protocol. It is simple and fast to converge.

19.6 Disadvantages of a Babel Network

- Babel is an emerging technology and as such is in a state of flux. The developers have made a commitment to being backwards compatible with all releases, but there may still be some less than desirable side–effects.
- Babel relies on periodic routing table updates rather than using a reliable transport; hence, in large, stable networks it generates more traffic than protocols that only send updates when the network topology changes. In such networks, protocols such as OSPF, IS-IS, or the Enhanced Internal Gateway Routing Protocol (EIGRP) might be more suitable [270].
- The overhead of inter–router messages generated by Babel may become a problem for extremely large networks.
- Babel does impose a hold time when a prefix is retracted. While this hold time does not apply to the exact prefix being retracted, and hence does not prevent fast reconvergence should it become available again, it does apply to any shorter prefix that covers it. This may make those implementations of Babel that do not implement the optional algorithm described in [RFC 6126] Section 3.5.5 unsuitable for use in networks that implement automatic prefix aggregation [270].
- Apparently Babel is not in widespread use and may never be. This usually leads to more bugs because fewer users are finding problems the hard way.

Advanced Projects

These projects will require compiling and running Babel on the Pi. There are tutorials on the web that are fairly clear but there are also horror stories about getting Babel to work on a wireless mesh. These two projects could require some knowledge of compiling C language programs using the the **make** command.

You might want to update Raspbian before you attempt these. These are optional projects.

19.1 Download the Babel routing daemon and compile it on your Pi. Can you get multiple Raspberry Pi's to route using Babel?

19.2 If you are in a controlled environment such as a computer lab or classroom, get permission to create wireless networks before using the on–board wireless of the Pi. Create an *ad hoc* mesh network of Pi Microcomputers connected only using wireless.

Further Reading

The RFCs below provide further information about the Babel routing protocols. This is a fairly exhaustive list and most RFCs are typically dense and hard to read. Normally RFCs are most useful when writing a process to implement a specific protocol.

RFCs Directly Related to This Chapter

Babel	Title
RFC 6126	The Babel Routing Protocol [270]
RFC 7298	Babel Hashed Message Authentication Code (HMAC) Cryptographic Authentication [278]
RFC 7557	Extension Mechanism for the Babel Routing Protocol [281]

Other RFCs Related to Babel

For a list of other RFCs related to the Babel routing protocol but not closely referenced in this chapter, please see Appendix B, Babel.

Part IV
Internet Services

Overview

"The time has come," the Walrus said,
"To talk of many things:
Of shoes — and ships — and sealing-wax —
Of cabbages — and kings —
And why the sea is boiling hot —
And whether pigs have wings."

Lewis Carroll, *Through the Looking–Glass, and What Alice Found There* [10]

/abstract*It is easy to become so wrapped up in building a network and running all kinds of interesting protocols and forget the goal of networking in the first place: to share resources among a set of devices.

The time has come to examine how resources are made available, found, and used across an internet. In the next few chapters we will examine some of the more common methods of sharing information such as the World–Wide Web, Email, and other services. Virtually all the services we will examine use the client/service model and use name services to relieve users of the need to know, and remember, esoteric strings of numbers. Much like contact lists on cellphones have made it possible to easily find people by name, Domain Name Service (DNS) has made it easier to find resource by name than to remember exactly what is the address at which those resources can be found.

It is easy to become so wrapped up in building a network and running all kinds of interesting protocols and forget the goal of networking in the first place: to share resources among a set of devices.

The time has come to examine how resources are made available, found, and used across an internet. In the next few chapters we will examine some of the more common methods of sharing information such as the World–Wide Web, Email, and other services. Virtually all the services we will examine use the client/service model and use name services to relieve users of the need to know, and remember, esoteric strings of numbers. Much like contact lists on cellphones have made it possible to easily find people by name, Domain Name Service (DNS) has made it easier to find resource by name than to remember exactly what is the address at which those resources can be found.

Chapter 20
Domain Name Service

Overview

While many people use Name Service and Domain Name Service (DNS) inter-changeably[1], there are major differences between the two. If this were not so, the Internet would be much less stable, more prone to system wide failures, and much less resilient to attack.

It is a fact that people can remember an alphanumeric name much easier than they can remember a string of digits. It is even easier to remember a mnemonic name such as **ford.com** or **YouTube.com**. The problem is that messages on the Internet must travel between IP addresses which are arbitrary strings of semi–random numbers. The solution is simple: assign every device a human–oriented name and devise a protocol to relate those names to network–oriented IP addresses. In this way, humans can contact resources without knowing their addresses. This leads to some interesting side–effects which we will discuss in this chapter.

This chapter deals with local name service, zone files (the database of names and IP addresses), primary name servers, secondary name servers, and providing domain name service for a registered domain. Much of DNS[2] works the same for both IPv4 and IPv6. Any differences will be explicitly pointed out in the text.

20.1 Fully Qualified Domain Name

Servers on the internet that are contacted by humans typically have a name con-sisting of a number of alphanumeric strings separated by "dots" or periods. These names form a hierarchy with the levels going from least significant to most sig-nificant as they are read left to right. The left–most level is the "hostname" and the right–most is the "top level domain." These top level domains are assigned by

---

[1] In my opinion *we* in the industry should be more careful about using these terms loosely.

[2] Domain Name Service

© Springer Nature Switzerland AG 2020
G. Howser, *Computer Networks and the Internet*,
https://doi.org/10.1007/978-3-030-34496-2_20

ICANN and administered by various nation wide ISPs and domain registrars. It is important to understand the role of ICANN and IANA as coordinating organizations with *no* control over content. As stated on the ICANN website:

> ICANN doesn't control content on the Internet. It cannot stop spam and it doesn't deal with access to the Internet. But through its coordination role of the Internet's naming system, it does have an important impact on the expansion and evolution of the Internet [38].

To begin we need to define some terms:

Fully Qualified Domain Name   A Fully Qualified Domain Name (FQDN[3]) consists of a hostname.[optional sub–domain name].domain.top–level–domain–name and is assigned by the local network administrator using a previously registered domain name and previously registered IP address range (or private IP address if the device does not connect to the Internet).

Top Level Domain Name   A top level domain name is the last part of an FQDN such as "com", "edu", "gov", or "uk". These are served by the root name servers.

Domain Name   A domain name is registered with ICANN and consists of an alphanumeric name followed immediately by a top level domain name. It is common to refer to a domain name such as "yahoo" as "yahoo.com" because a domain name can be registered with more than one top level domain. For example, "MyDomain" could be registered as "MyDomain.com", "MyDomain.net", or "MyDomain.info". This unfortunately leads to various simple, and unstoppable, domain phishing schemes such as registering domains like "whitehouse.com" and "white-house.gov" to trick traffic directed to "whitehouse.gov" but containing common mistakes and typos. These are sometimes called "typo–squating[4]."

Host Name   The first field of a Fully Qualified Domain Name is the hostname assigned to the actual device or an optional alias for the physical device. Common host names such as "www", "outlook", or "mail" are often aliases rather than host names. The reasoning behind using an alias will be discussed later.

alias   An alias is another name that can be used in place of the actual host name of a physical device. Most often this is done so that a human need not know the actual name of a device, but can instead refer to that device by a common alias that is the service desired[5].

Sub–Domain   A registered domain can be split into any number of sub–domains which can also be split into sub–domains and so on. For example, "mycollege.edu" might be split into "ce.mycollege.edu" for use by the Computer Engineering Department. The domain could be split even further into "masonhall.ce.mycollege.edu" and "research.ce.mycollege.edu". Splitting a domain does

---

[3] Fully Qualified Domain Name

[4] Typo–squatting is when hijackers register misspelled versions of your domain name to send the traffic to malicious sites. Registering all possible versions of your domain name including singular and plural versions, all common domain extensions and hyphenated and non hyphenated word compounds can allow a hacker to acquire traffic intended for your site [52]. In the early days of the Internet, an attempt to contact a non–existing FQDN would receive a generic "not found" message. Now the response is an ad asking you to register the domain.

[5] There is no theoretical limit to the number of aliases a device can have.

not mean the domain no longer exists, quite the contrary is true. All sub–domains belong to the owner of the domain and can easily be created as needed or desired. Sub–domains are not registered with ICANN.

Name Space   Name space is the logical structure constructed of all the FQDN reachable on the Internet. There is no structural connection between name space and geography or network topology.

All domain names and IP addresses are registered with ICANN which does not control these things except to insure their orderly registration. For example, if you wish to "squat" on a domain name such as "JohnDoeForPresident.com" ICANN will register that domain as yours if it is not in use. John Doe would have to work with you if he wanted that domain later.

20.1.1 A Typical FQDN

If we look at a possible name **www.ce.mycollege.edu**, we can break this name down into its components very easily. The Top Level Domain is obviously "edu" as that is the last part of the name. The domain can be thought of as "mycollege" or "mycollege.edu". Indeed, it is better to talk of the domain as "mycollege.edu" as someone else could own a domain "mycollege.org". For this reason, many organizations will register all possible versions of its domain; ".com", ".org", ".mil", and so on.

In the case of **www.mycollege.edu** there is no sub–domain, but **www.ce.mycollege.edu** in the same domain is actually in the sub–domain named "ce.mycollege.edu".

The maximum length of a FQDN is about 253 ASCII characters and is not case-sensitive.

20.2 Top Level Domain Names

Top level domain names[6] are controlled and maintained by the ICANN and serviced by 13 top level domain name servers. There are actually a large number of Top Level Domain name servers, but they share IP addresses that are so well known as to be built into standard operating systems as a special file. This can be over–ridden for an intranet.

This is why all domains *must* be registered. If you do not register your domain, DNS will not work for your domain and someone may register that domain before you do. The present system of domain registration is "first come, first served" and there is nothing to stop someone from registering the domain you want to use.

---

[6] Originally there were three main top level domains: edu(education), com(commercial), and mil(military).

20.3 Registered Domains

Technically all domain names are assigned as requested by the ICANN, but in reality most domains are now registered through an ISP instead. This was done to relieve the ICANN of some of the burden of administering all the domain names and, unfortunately, to allow some of the larger domain hosting companies another opportunity to advertise their services in return. That being said, registration is cheap and simple with a domain name costing on the order of $35 US per year.

Once an organization registers a domain, that domain and any sub–domains the organization cares to create belongs to them and no one else can put a FQDN in that domain on the Internet[7]

20.4 Sub–domains

An organization that owns a domain can further subdivide it into as many levels of sub–domains as it desires. Except for authoritative name servers, these sub–domains make little, or no, difference in how names are resolved. The main usefulness of sub–domains is local administration of hosts, aliases, and addresses. To a lesser extent, sub–domains are another way to promote important divisions of your organization. For example, a company could promote their R & D division by using a sub–domain "Research.mycompany.com" to show how important R & D is to the company.

20.5 Host Name

A host name refers directly to a device in a name space domain. Host names can be assigned to a device or to a service by using an alias such as **www** or **mail**. The advantage of assigning an alias to a service rather than using a device's hostname is that this relieves both the remote users *and* the domain of the need to know the device on which the service is running. If the domain uses an alias for the device running a service, such as a web server, the service can easily be moved to a different physical device or distributed among a number of physical devices without users being aware of the change. This makes administration of the services in a domain much more flexible.

A device may be configured with a hostname or may obtain one via DHCP or BOOTP. However the hostname is assigned, each device must have a host name that is unique in the domain or sub–domain. In other words, the FQDN formed for this host must be unique.

---

[7] However, a private intranet can violate this rule because *it cannot connect to the Internet.*

20.6 Types of Name Servers

The heart of DNS is the name server process, such as **bind9d** or **named** on Linux. There are a number of different types, but here we get into a small problem: most DNS name servers actually function as more than one type of server. Sometimes a name server functions as different types of name servers on different networks. This behavior is a bit problematic and sometimes can be a security risk [2].

20.6.1 Root DNS Servers

There are 13 root servers on the Internet in the domain **root-servers.net**, which is reserved. These are maintained by various large networking companies under contract with ICANN. Each of these servers is really a group of servers sharing an IP address in order balance the load created by the large number of requests made to the root servers.

The root servers provide authoritative responses to queries about TLD[8] servers. These servers are extremely well secured and are not directly visible on the Internet but only to each other. Should a device issue a ping[9] for **www.yahoo.com**[10] and not be able to resolve the name, the name server for that device might issue a query to a root server for the name server responsible for **yahoo.com**. One of the root servers will respond with the address of the TLD server responsible for the **com** domain. The name server can then query the TLD name server for a name server responsible for **yahoo.com** as the next step in resolving **www.yahoo.com**. Please refer to Figure 20.2.

20.6.2 Top Level Domain Name Servers

Each of the TLDs are serviced by at least one TLD name server. For example, if a query for **www.yahoo.com** is sent to a root server, the root server will respond with the address of a TLD name server for the TLD ".com" so that the inquiring name server can query the TLD name server for a name server for the **yahoo.com** domain. While this all seems a bit complicated there are two things to remember: it works very well and it distributes the load over many name servers instead of one or two[12]. Again, please refer to Figure 20.2.

---

[8] Top Level Domain

[9] Echo Request and Echo Response

[10] For the record, **www.yahoo.com** will always respond to a ping. Not all sites or hosts will answer a ping as this can lead to a DDOS[11] attack. Thank you, Yahoo.

[12] This distribution of name service also prevents a "single point of failure" in case of a problem with DNS or network connectivity.

20.6.3 Primary (Master) Name Server

This server is the authoritative name server for at least one domain and is registered with ICANN as the authoritative name server for this domain. The configuration of this server will contain the primary zone file for the domain and all configuration changes for this domain are made on this device. This server is responsible for resolving names within a domain into the corresponding addresses and also responsible for keeping the secondary, or slave, domain name servers up to date[13]. Changes made to the hosts and addresses in this file are disseminated to all secondary name servers to insure that all name servers provide the same response to a query[14].

In order to register a domain, a valid Primary Name Server with IPv4 address must be provided as part of the application for the domain along with the same information for a valid Secondary Name Server (usually in a different domain and IP address space).

20.6.4 Secondary (Slave) Name Server

As the name implies, this type of name server can respond with the proper address to match a FQDN. In this case the response is not marked as authoritative. The information about the domain is still kept in a zone file, but instead of being edited directly, the zone file is transferred from a Primary Name Server for the domain. It is typical for a domain to have more than one secondary name server to help balance the load so it is critical that all the secondary servers have the same information in their files. Later we will see how this is achieved and maintained when we discuss the zone file headers.

20.6.5 Resolving Name Server

A Resolving Name Server obtains its information from another name server that keeps zone files for a domain and caches (saves) the results. If a request can be met from the local cache, the Resolver does not contact any other name server but responds from its own cache as a non-authoritative name server. If the information is not in cache, the Resolver contacts the primary name server, caches the information, and responds as an authoritative name server for the domain.

Typically, a Resolver uses both Recursive and Iterative Queries to finally resolve a FQDN to an IP address. If not, somewhere in the chain of name service there needs to be a Resolver that does use both no matter how many name servers pass

---

[13] For the record, I prefer the terms "primary" and "secondary" because they are more descriptive and do not have a negative connotation. I am OK with "Master" and "Secondary" as an alternative.

[14] In my opinion, a domain should have *exactly one* primary name server. This is only an *opinion*.

the requests along. For simplicity, assume the Resolver is the name server passed to the Client via DHCP or manually configured at the NIC.

Resolving a FQDN to an IP address involves many steps as shown in Algorithm 10 and Figure 20.1. While the procedure looks complicated, it is actually a nested set of the same recursion over and over. In short the procedure is:

1. Check the cache for the information.
2. If it is there, quit and use the information from the cache. If not:

    a. Pass the query up the line to the next level.
    b. If the information is in the cache, return the information as a response and quit.
    c. If the information is not in the cache, repeat at the next level.

Algorithm 10 Resolving a FQDN

 1: **procedure** RESOLVE
 2:    **if** Not in local cache **then**
 3:        Pass query to Resolver process   ▷ Query 1
 4:        **if** Not in Resolver cache **then**
 5:            Pass query to area DNS Resolver   ▷ Query 2
 6:            **if** Not in DNS Resolver cache **then**
 7:                **if** name server for the domain is not in cache **then**
 8:                    **if** TLD name server not in cache **then**
 9:                        Check "hints" for root name server
10:                        Query root name server for TLD name server   ▷ Query 3
11:                        Enter TLD name server into cache   ▷ Response 3
12:                    **end if**
13:                    Query TLD name server for domain name server   ▷ Query 4
14:                    Enter name server for that domain into cache   ▷ Response 4
15:                **end if**
16:                Send query to name server for the domain   ▷ Query 5
17:                Enter response into cache   ▷ Response 5
18:            **end if**
19:            Respond with result from DNS Resolver cache   ▷ Response 2
20:            Enter result into Resolver cache
21:        **end if**
22:        Respond with result from Resolver cache   ▷ Response 1
23:        Enter response into Browser cache
24:    **end if**
25:    Use result from Browser cache
26: **end procedure**

It is easier to follow the query in Figure 20.1 for **www.mydomain.com** than to explain the steps as some generic theory, so let's look at how a web browser gets the address for **www.mydomain.com**[15]. Step 0 is to check the browser cache and use that information if it is there. If not:

1. The browser requests the IP address for **www.mydomain.com** from the resolver on the Client device (Query 1). If the resolver has the information in its cache, it responds with the information (Response 1) and quits because the query has been answered.
2. If the Resolver does not have the information, it requests it from the DNS Resolver address it learned from DHCP (Query 2). When a Response 2 is received from the DNS Resolver, it is relayed to the Web Browser and the request is completed.

Note: In Figure 20.1, query 2 is a Recursive Query where one question gives one *complete* answer; whereas queries 3, 4, and 5 are Iterative queries which may return either a Referral (to another DNS) or an answer [2].

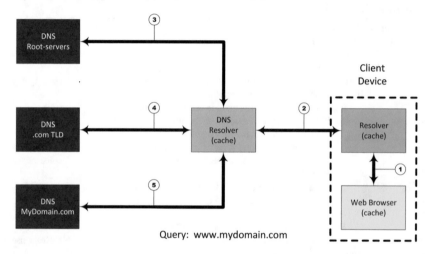

Fig. 20.1: Recursive and Iterative Query (www.mydomain.com). NOTE: Each arrow represents a pair of query and response.

As usual with a cache, the Resolver will age out any information that it has kept longer than the TTL (Time to live) specified by the primary name server.

By default, BIND acts as a Caching name server and therefore must support recursive DNS requests. BIND must also be configured with access to the root name servers ("hints" file), which presents a problem for an Intranet as we will see when we configure the Raspberry Pi name servers.

---

[15] If the browser points to **yahoo.com** and the zone file is set up correctly, the process returns the answer for **www.yahoo.com** after following the same steps. This is discussed in more detail in Chapter 21.

20.6.6 Forwarding Name Server

A NS[16] can also act simply as a DNS Forwarder, or a simpler version of a caching DNS, without the Client devices even being aware that it is a Forwarder. If the response for a query cannot be formed from the DNS Forwarder's own cache, the request is forwarded to some other NS to fill the request. The DNS Forwarder then caches the response and forwards it back to the Client. This is very little overhead for the DNS Forwarder but can help to relieve some of the DNS traffic on the network by taking advantage of the principle of **locality**. The assumption is that devices on the same network will tend to visit the same sites on the Internet and/or a single device will access the same hosts multiple times in a short time span. Think how many times Google.com is accessed in a few minutes in a research library. In practice, this is a fair assumption and it actually makes sense to have a number of DNS Forwarders to balance the load of DNS queries on a busy network.

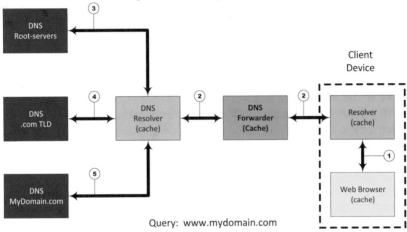

Fig. 20.2: A Typical Query to a DNS Forwarder

Again, in practice most DNS devices will function as different types of servers for the different domains they are configured to know. See Figure 20.2 as an example. Notice that the query to the DNS Forwarder, #2, is simply passed on and the final response, again #2, is cached and they passed back to the requester. In this way the DNS Forwarder populates its cache with all recent queries and acts in much the same way as an HTTP[17] proxy server[18] as discussed in Section 21.6 in Chapter 21.

---

[16] Name service

[17] Hyper–Text Transfer Protocol

[18] I have found many people semi-understand the concept of a web proxy. A DNS Forwarder and web proxy work much the same.

20.6.7 Stealth Name Server

Many Intranets face the problem of providing some services to the Internet at large while minimizing the exposure of the private network to outside attack. One solution is provide limited name service to public inquiry and full name service to those devices on the private network[19]. The main problem is how to export some, but not all, of the information about the private network to a fully exposed public name service as in Figure 20.3. The points of interest are noted on the figure as:

1. Public DNS queries from devices on the Internet and outside of the safe, private network. These queries must be handled by the public DNS directly. Recursive queries are not allowed as that could expose information about the private network and the devices on it.

2. The public DNS must only contain information about the public services and the devices which provide these services and *nothing* else. The zone files on this server are obtained from a hidden master that contains only information about the public servers in the DMZ[20]. These devices are considered to be at risk from outside attacks and should be protected as much as possible, but not trusted.

3. It would seem like a good idea to maintain the public DNS via zone file transfers from one or more hidden master DNS, but this would defeat the purpose of a stealth name server and DMZ. If such transfers were allowed, an adversary that compromised a public DNS would have critical information about the hidden master. Even a listing of the **named.conf.local** file would reveal the hidden master to the adversary.

4. Private name servers are configured to allow zone transfers *only* within the private part of the network. These servers provide authoritative responses to private devices and are configured to export recursive queries to name servers in the public Internet at large. This must be done using port 53 as this is the standard port for DNS queries on the Internet.

5. The private network is typically protected from the DMZ by a firewall. If private Internet addresses are used in the safe, private part of the network, this device must provide NAT services as discussed in Chapter 23.1.

6. Queries about the Internet are handled in the normal way (via port 53) either from the cache of the name server, local zone files, or by sending a recursive query to the Internet.

---

[19] The situation is more complicated if the private network used registered public IP addresses, but the difference is not great enough to worry about for this discussion of Stealth Name Servers.

[20] Demilitarized Zone

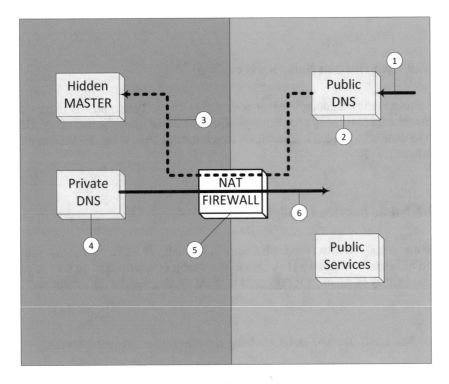

Fig. 20.3: DNS Stealth Server

A very good discussion of Stealth Name Servers by Ron Aitchison, along with sample configuration files, can be found on–line [3] [2].

20.6.8 Authoritative Only Name Server

Authoritative Only Name Servers are master or slave for any number of zones, but do not act as recursive resolvers. These name servers do not maintain a cache and never forward a query to another name server. Typically Authoritative Only Name Servers are used as public name servers in a DMZ or as high performance name servers.

BIND works well as an Authoritative Only Name Server in a DMZ with only minor issues, but it is not a good choice for a high performance name server. It is not possible to completely stop caching, but BIND allows the domain administrators to minimize caching by the option `recursion no;` in the `named.conf.option` file.

```
// options section fragment of named.conf
// recursion no = limits caching
options {
recursion no;
```

```
};
```

20.6.9 Split Horizon Name Server

In some specific situations, it might be desirable for local name servers to provide different responses based upon the type of query and the source of that query. This can be done with a Split Horizon Name server, but this topic is beyond the scope of this book [3] [2].

20.7 Name Service Configuration

The most common name service software in use today is **BIND** running on some form of Linux or Windows [11]. Due to the wealth of information on how to enhance security on Linux, it is suggested to avoid running **named** on Windows.

Table 20.1: Domain and Addressing Information for Example Network

Group	Domain Name	IP Network	Primary Name Server	Secondary
1	mycollege.edu	10.1.1.0	10.1.1.2	10.2.2.2
1	myhighschool.edu	172.17.0.0	172.17.0.1	10.1.1.1
2	birds.com	10.2.2.0	10.2.2.2	10.3.3.2
3	halo.mil	10.3.3.0	10.3.3.2	10.1.1.2
5	FakeISP.net[a]	192.168.1.0	192.168.1.1	

[a]Instructor's Pi is not really in any group. The Instructor will give you the IP address to use.

To begin with, the test–bed must be up and routing correctly[21]. Before the test–bed network can be configured to use name service, there must be some domains to work with. As the test–bed network will *not* ever be connected to the Internet at large, it makes not difference what names are chosen, but these names must be "registered" somewhere. Each group should chose a unique domain name to register and provide that information to the instructor. For purposes of this discussion, we will assume the domains listed in Table 20.1.

The actual mapping of FQDN to IP address is kept in files called zone files. For each domain there needs to be two zone files, one to map names to addresses and another to map addresses to names (known as an inverse zone file or "inverse"). While these files may eventually reside on a number of different servers, all the

---

[21] There is no real reason to use any one set of routing protocols over any other, but the network must be correctly routing to all networks or it does no good to be able to resolve names. The physical network must always be established first.

updating of these files is done on the primary, or "master", name server for a domain. The files are transferred as needed to secondary, or "slave", name servers whenever an update has been noted.

Inverse zone files are manually maintained in the same fashion as the normal zone files. Indeed, one of the most frustrating types of errors to track down is when the zone files are not properly updated. There are some very simple rules to follow to keep these files synchronized between multiple name servers which will be discussed in Section 20.13 in this Chapter.

20.8 **named** and Configuration Files

A name service daemon, **named** is typically installed either as part of the OS or as part of the BIND[22] package and is the process that handles DNS queries, zone file transfers, and caching[23]. Fortunately for everyone who uses the Internet, **named** works extremely well.

For a user, the primary use of the DNS system is to resolve domain names, such as **www.yahoo.com** to the appropriate IP address (either IPv4 or IPv6 [54]) in order to contact a service on a remote device[24].

20.8.1 Name Service Files

The most important configuration file for BIND is **/etc/BIND/named.conf** which loads three configuration files which can be modified by an administrator to control how this device provides name service. While it is possible to do all the configuration to the actual **named.conf** file, this is not the preferred method for current Linux distributions. These files are, more or less, self documenting.

Before any configuration is done, it is a good idea to create two directories for the zone files. Create the directories **/var/cache/bind/zone/masters** and **/var/cache/bind/zone/slaves**[25] and change the ownership of the directories[26] as follows:

---

[22] Berkeley Internet Name Domain service

[23] While an administrator can control caching to some extent, it appears that BIND does not allow you to completely turn off caching. Therefore, when a discussion talks about a name server not caching, take that with a grain of salt.

[24] Here the term "user" could apply to either a person or a process.

[25] Actually any directories will do or the files could even be created in **/etc/bind/**. If you prefer "primary" and "secondary", there will be no problems. For simplicity I will use "masters" and "slaves" because this is the most common usage.

[26] Zone transfers cannot happen if **bind** does not have ownership of these directories.

`sudo chown bind:bind /var/cace/bind/zone/masters`. Zone files will be created in these directories to keep things organized[27].

`named.conf.options`

This file contains the default options for **named** and in most cases should be left alone unless the administrator needs to add the DNS Resolvers provided by an ISP. This is where an administrator can change the default directory for the zone files along with configuring a forwarder. This file has good documentation within the file itself.

`named.conf.local`

This is the file that will be edited to point to the new zone files. Here is specified all the zones that this name server will answer as an authoritative name server; i.e., the zones that will be either master zones or slave zones. If other information is cached by this name server, it will still answer a request but not as authoritative. This file also contains the names of the inverse, or reverse, zone files for both IPv4 and IPv6.

`named.conf.default-zones`

Normally, the default zones should be included in a public name server on the Internet. Including this file directs the name server to reply with a negative response to any queries about the reserved private networks. On intranets, or private Internets, these reserved networks are used for IP addressing and cannot be seen from the public Internet. Blocking responses for some of these might be desired which is easily done by editing the `named.conf.default-zones` file to cause negative responses for those private networks not in use on this particular intranet.

Because intranets use private IP addresses, care must be taken that these zones are not accidentally blocked. For our purposes, none of the private IP addresses should be blocked. However, if this name server is to be on the public Internet, these zones should be blocked to enhance security.

20.8.2 Typical `named.conf.local` File

For most domains, the only file that needs to be edited to properly run a name server is the `named.conf.local` file. This file explicitly determines which domains this server can reply as an authoritative server; or in other words, the domains for

---

[27] The default directory can be changed in the "options" file as noted below.

which this server is either a primary NS or a secondary NS. Below is one possibility for the **named.conf.local** file for the name server on Group 1's Pi which is the primary for **mycollege.edu**.

```
1  pi@router1-1:~\$ sudo vi /etc/bind/named.conf.local
2  //
3  // Do any local configuration here
4  //
5
6  // Consider adding the 1918 zones here, if they are not used
      in your
7  // organization
8  //include "/etc/bind/zones.rfc1918";
9    zone "." in{
10    type hint;
11    file "root.servers";
12  };
13
14  zone "mycollege.edu" in{
15    type master;
16    file "/var/cache/bind/master/mycollege.edu";
17    allow-transfer {10.1.1.2; 172.17.0.1; };
18  };
19
20  zone "1.0.10.in-addr.arpa" in{
21    type master;
22    file "/var/cache/bind/master/mycollege.inv";
23    allow-transfer {10.1.1.2;};
24  };
25
26  zone "myhighschool.edu" in {
27    type slave;
28    file "/var/cache/bind/slave/myhighschool.edu";
29    masters {10.1.1.2; 172.17.0.1; };
30  };
31
32  zone "0.17.172.in-addr.arpa" in{
33    type slave;
34    file "/var/cache/bind/master/myhighschool.inv";
35    masters {10.1.1.2; 172.17.0.1; };
36  };
```

From this configuration file we can see that the name/IP information for the domain **mycollege.edu** is kept on this machine and can be edited, either directly or via some program, to reflect changing information for that domain. Also, reverse lookup, IP address to name, information is updated directly on this name server via the special zone file **1.0.10.in-addr.arpa**. More on this odd name a little later.

This name server will also respond authoritatively to queries about the domain **myhighschool.edu** and reverse lookup queries for the network 172.17.0.0. However, while these zones are kept in files on this name server, they must be updated on a different name server and then transferred to this one. While this seems like a manually intensive operation, **named** will automatically transfer the files whenever an update is noted.

Any DNS query sent to this name server will be answered as authoritative if, and only if, it is about **mycollege.edu**, **myhighschool.edu**, 10.1.0.0, or 172.17.0.0. All other queries will be answered with the help of other name servers and will receive a non-authoritative response from this name server.

Each zone statement defines how this name server will handle a query about that zone, but the first zone statement has a different purpose. This zone informs the name server how to handle a query about a zone for which it is not authoritative; i.e. a zone for which it is neither a primary or secondary name server. In this case, the query is answered with the help of a root name server, a top–level–domain server, and a name server that is authoritative for the domain or network in question, using the resolving name server method discussed previously in Figure 20.1 and Section 20.6.5. The special zone "." matches any query that is not matched by another zone and is typically the first zone in **named.conf.local** file even though the order of the zone statements has no meaning. Notice that this zone is neither "master" nor "slave" but instead is a "hints" zone and loads a file named **root.servers** which is installed as part of the BIND package. This file contains semi-current information on the 13 root servers and how to contact them. When this file is loaded, the name server contacts one of the root servers to determine if the information in the root.servers file is current. If not, the current information is transferred from the root server. If this file is missing, BIND has a number of the root servers "hard coded" and can contact at least one without any help from the "hints" zone.

The second zone in the example is a master zone for the **mycollege.edu** domain. This file has all the information needed to resolve a FQDN in the domain to the IP address assigned to the device with that name[28]. This zone statement identifies where the data file is located and exactly which secondary name servers can request a copy of the zone file[29].

The third zone is the master reverse, or inverse, lookup zone and contains the exact mapping of IP addresses to FQDN. The format of this zone name is a legacy of the original DNS on the ARPA.net and must be followed exactly. The name is formed of the network prefix (octets in *reverse* order) concatenated with "in-addr-arpa". Again, this zone is a master and the appropriate data file is named **/var/cache/bind/master/mycollege.inv**. The secondary name servers for this zone are listed in the allow-transfer statement to prevent unauthorized zone data transfer for security reasons.

The last two zones specify the domain, and inverse domain, for which this name server will act as a secondary, authoritative server. The zone statement lists the master name servers for these zones in order that the most current zone information may be downloaded when needed. How these master and slave zones are synchronized is handled by the "SERIAL" information in the appropriate zone files.

---

[28] This works even if the IP address is assigned via DHCP because when the address is assigned, the appropriate hostname is learned via DNS and assigned correctly. Without this cooperation between DNS and DHCP it would be very difficult to handle situations where large numbers of devices come and go on a network. A public network in a Starbucks coffee shop, for example, would be a headache to administer.

[29] If an adversary can get a copy of the zone file, they will learn much vital information about the network and domain.

20.8.3 Checking the named.conf.local File for Errors

It is possible to visually check this file for errors, but that is not very reliable. Fortunately, BIND provides a utility to check the file **named−checkconf**. Consider the following file 20.8.3, **named.conf.local−bad** which is missing a semicolon ("**;**") in the zone statement for "mycollege.edu" in the **allow-transfer** clause after 172.17.0.1. The correct clause should be:

allow−transfer {10.1.1.2; 172.17.0.1;};.

```
1   //
2   // Do any local configuration here
3   //
4
5   // Consider adding the 1918 zones here, if they are not used
        in your
6   // organization
7   //include "/etc/bind/zones.rfc1918";
8   zone "." in{
9     type hint;
10    file "root.servers";
11  };
12
13  zone "mycollege.edu" in{
14    type master;
15    file "/var/cache/bind/master/mycollege.edu";
16    allow-transfer {10.1.1.2; 172.17.0.1 };
17  };
18
19  zone "1.0.10.in-addr.arpa" in{
20    type master;
21    file "/var/cache/bind/master/mycollege.inv";
22    allow-transfer {10.1.1.2;};
23  };
```

If this is checked with the **named−checkconf** utility, the following messages are generated which give the line, 16, and a fairly good error message[30].

```
pi@router1-1:/etc/bind$ sudo named-checkconf named.conf.local-
    bad
named.conf.local-bad:16: missing ';' before '}'
pi@router1-1:/etc/bind$
```

It is a good idea to check any **named.conf** file after it is edited.

20.9 Primary and Secondary Name Servers

Data for name servers are kept in a number of zone files; one zone file for each domain serviced directly by this name server. The only operational difference between a primary (master) name server and a secondary name server is how the zone files are created and maintained. For a primary name server the zone files are manually updated while for a secondary name server the zone files are transferred from a master name server. Both primary and secondary name servers are considered to have the authoritative information for a domain; i.e. the information is correct and up to

---

[30] This utility does not return any messages if the file is syntactically correct. In my opinion this is a minor weakness.

date. Remember: The process generating a DNS query and receiving a response can tell if the response is authoritative, but not whether it came from a primary or secondary name server. It really doesn't matter and also provides a small measure of load balancing for a domain.

Table 20.2: Zone File Resource Record (RR) Types

TYPE	Value RFC	Meaning
A	1 RFC 1035 [77]	A host address to IPv4 mapping
AAAA	28 RFC 3596 [214]	A hostname to IPv6 mapping
NS	2 RFC 2874 [190]	An authoritative name server
CNAME	5 RFC 1035 [77]	Canonical Name. An alias name for a host. Causes redirection for a single RR at the owner-name.
DNSKEY	48 RFC 4034 [230]	DNSSEC. DNS public key RR.
DS	43 RFC 4034 [230]	DNSSEC. Delegated Signer RR.
HINFO	13 RFC 1035 [77]	Host Information - optional text data about a host.
KEY	25 RFC 2535 [177]	Public key associated with a DNS name.
MX	15 RFC 1035 [77]	Mail Exchanger. A preference value and the host name for a mail server/exchanger that will service this zone. [75] defines valid names.
NS	2 RFC 1035 [77]	Name Server. Defines the authoritative name server(s) for the domain (defined by the SOA record) or the subdomain.
NSEC	47 RFC 4034 [230]	DNSSEC. Next Secure record. Used to provide proof of non-existence of a name.
PTR	12 RFC 1035 [77]	IP address (IPv4 or IPv6) to host. Used in reverse maps.
RRSIG	46 RFC 4034 [230]	DNSSEC. Signed RRset.
SOA	6 RFC 1035 [77]	Start of Authority. Defines the zone name, an e-mail contact and various time and refresh values applicable to the zone.
SRV	33 RFC 2872 [189]	Defines services available in the zone, for example, ldap, http, sip etc.. Allows for discovery of domain servers providing specific services.
TXT	16 RFC 1035 [77]	Text information associated with a name. The SPF record should be defined using a TXT record and may (as of April 2006) be defined using an SPF RR. DKIM ([226] also makes use of the TXT RR for authenticating email. Related: How to define DKIM/ADSP RRs.

20.10 Zone Files

Zone files have a very specific format and BIND is *extraordinarily* picky about the format, but oddly enough is not very picky about "white space". A missing, or misplaced, period can lead to a disastrously broken name server that might appear to function perfectly until a query comes in that fails because of the format error. Such failures are not obvious as the typical response is "not found." Fortunately, BIND provides a couple of utilities that can spot errors and give a reasonable error message to help the administrator to correct the problem[31]

A zone file consists of comments, directives, and RRs[32] as given in Table 20.2. These will be discussed in detail later in Section 20.10.1.

Comments can, and should, be included by using the semicolon, ";", to end the processing of a line. The default domain part of each hostname that is not a FQDN is obtained from the zone statement in the **named.conf.local** file. In a zone file, every FQDN must end with a period[33]("."). Below is a sample zone file for the domain **mycollege.edu** and the network 10.1.0.0.

```
 1  pi@router1-1:~\$ sudo vi /var/cache/bind/master/mycollege.edu
 2  ;;;;;; Sample zone file
 3  \$TTL 172899; Time-to-live is 2 days
 4  ;
 5  ;            Authoritative data for:  mycollege.edu
 6  ;
 7  ;            CONTACT:  Gerry Howser 999/555-1212 (
        postmaster@mycollege.edu)
 8  ;
 9  @            IN SOA router1-1.mycollege.edu. postmaster.mycollege.
        edu. (
10            2019020700          ; SERIAL restarted on 02/07/2109
11            7200                ; REFRESH
12            600                 ; RETRY
13            3600000             ; EXPIRE
14            86400               ; MINIMUM
15            )
16                                   in      ns router1-1.mycollege
                                        .edu.
17                                   in      ns dns.myhighschool.
                                        edu.
18  router1-1.mycollege.edu.         in      a 10.1.0.1
19  dns.myhighschool.edu.            in      a 10.1.0.2
20  ;
21  ; end of header information
22  ;
23  \$origin mycollege.edu.                  ;redundant record, this is
        the default
24  ;
25  ; special names and records
26  mycollege.edu.                   in MX 0 router1-1.mycollege.
        edu.
27  ;
28  router1-1                        in      a        10.1.0.1
29  bongo.mycollege.edu.             in      a        10.1.0.1
30                                   in      MX       0 router1-1
```

---

[31] In my opinion, it would be nice if all error messages were as helpful as those from the BIND utilities.

[32] Resource Records

[33] This is actually the correct way to form a FQDN but many of us get lazy and leave it off. Zone files will correct that error even when we don't want it to do so.

31	mail	in	cname	router1-1.
	mycollege.edu.			
32	ftp	in	cname	router1-1.
	mycollege.edu.			
33	dns	in	cname	router1-1.
	mycollege.edu.			
34	www	in	cname	router1-1.
	mycollege.edu.			
35	r1-1-eth1	in	cname	router1-1.
	mycollege.edu.			
36	station25	in	a	10.1.0.25
37	station26	in	a	10.1.0.26
38	station27	in	a	10.1.0.27
39	laptop32	in	a	10.1.0.32
40	;			
41	\$origin cs.mycollege.edu.			
42	www	in	a	10.1.0.100
43	ftp	in	a	10.1.0.101

It is useful to think of a zone file as being composed of a header followed by a number of zone file statements even though this is not strictly true. What we will call the header is composed of:

- Semi-required comments that begin with a semi–colon.
- One required **TTL** record.
- One required @ **in SOA** record with clauses
- At least two **NS**, or name server records.
- One **A**, or authoritative record, for for each name server.

The mandatory TTL directive specifies the default value for TTL. Each RR can have a TTL directive to override this value.

It is a very good idea to have a comment in the header that documents the domain name for which this zone file provides authoritative data[34]. The other comment in this header provides contact information for the person responsible for this domain. This can be useful should the paper registration of the domain get lost or should someone report problems with name service for the domain[35]. This is typically name, voice phone number, and an email address. When provided, this information should be kept current.

The next part of the header is the SOA[36] record which contains information *critical* to the correct operation of DNS for this domain (or network). The first part of the record, "@ IN SOA" contains the name of a master name server for this domain and the email address of a person responsible for this domain's name service[37].

The five numbers inside the parenthesis control many things about the name server. The most important number that the administrator must keep track of is

---

[34] It is embarrassing to pick up a hard–copy or view a pdf of a zone file years later and wonder what domain it describes. Self-documenting files are nice when possible.

[35] For the test–bed network, feel free to fake this information if you wish. The correct name would still be nice to have.

[36] Start Of Authority

[37] Notice the white space. Formatting this file correctly can add to its readability and help you to spot errors.

the first one, the SERIAL number of the file[38]. This number controls the transfer mechanism between a primary and secondary name server and must be updated any time a change is made to the master copy of the zone file. When a secondary name service receives a **SERIAL** number from the primary name server that is larger than the one it has in its zone file, the secondary requests a zone file transfer from the master for that zone. For this reason, proper management of the **SERIAL** number for a zone file is critical[39].

Periodically, and at start up, a secondary name server will query the primary name server for the serial number of each zone file. If the secondary name server has a lower serial number than the master, it will initiate a zone transfer thus insuring both name servers have matching information about the zone. A very common issue is to update a zone file and then forget to update the serial number which leads to the secondary name servers responding with different, outdated information. The acknowledged best practice is to form the serial number from the date, i.e. "20190210nn" where *nn* starts at zero and increments. This documents the date of the last update to the zone files and allows for 100 updates to the zone in a single day. As it is perfectly acceptable to make many changes to a zone file before the serial number is updated, this should be more than adequate for any use. A good habit is to update the serial number when the file is saved and then immediately make the corresponding changes to the inverse zone file. If the inverse zone file does not need to be changed, update the serial number on the inverse file to match the regular zone file to document the fact that the two are still in agreement.

The other four values in the SOA record are for tuning purposes and these defaults should be fine but must be present.

- REFRESH This is the number of seconds the secondary will wait before trying to request an update from the primary name server. If the data in the zone is volatile this should be set to 1200 (20 minutes) and if the data is very stable this can be set as high as 43200 (12 hours) [149] [2]. If the BIND default NOTIFY is being used, this can be set much higher.
- RETRY This is the number of seconds the secondary will wait if the master does not respond before attempting again to contact the master. The best value depends upon a local knowledge of the network and typical traffic. Typical values are 180 (three minutes) to 900 (15 minutes).
- EXPIRE The time is seconds the secondary will continue to provide authoritative responses if it is unable to contact the primary. RFC 1912 recommends 1209600 to 2419200 (2 to 4 weeks) [2] [142].
- MINIMUM This was originally the equivalent to the `$TTL` directive. RFC 2308 made the `$TTL` directive mandatory and this field now is used as the time in seconds a NXDOMAIN record is cached. The maximum value allowed for BIND is 3 hours.

---

[38] Problems with the serial number will lead to inconsistent data between the master and slaves. Not only is this hard to track down, but it can be horribly difficult to correct.

[39] Most administrators one make a serial number mistake once. In my opinion, once is enough.

The next two records specify at least one master and at least one secondary name servers for this domain. Note that the **NS** records are followed by **A** records. Later we will see that this can, and usually does, create a phantom warning message when the zone file is checked for syntax errors.

The last record in the header is an origin directive, **$origin mycollege.edu.** which is redundant and only included for documentation purposes. The origin is assumed to match the domain name in the **named.conf.local** file for this zone.

The remainder of the zone file consists of the information required to perform forward look–up to resolve a fully–qualified name to an IP address (either IPv4 or IPv6).

20.10.1 DNS Resource Record Types

The actual zone data is made up of a number of RR statements. There are *many*[40] [3] different types of RR but most zone files will use only four or five of these. Some of the more typical ones are discussed below.

A

Authoritative Resource Records define the relationship between a single IP address and a single FQDN. This information should be the same in the zone file and inverse zone file. Hostnames normally associated with services such as "www" and "dns" should be avoided and pointed to by **CNAME**[41] instead.

AAAA

Authoritative record to provide a forward mapping from hostname to an IPv6 address.

NS

Name Service records point to the authoritative servers for a zone. For public IP addresses there should always be at least two name servers for a given zone.

---

[40] A full listing of all the different RR types takes many pages. There are quite a few "experimental" RR types as well.

[41] Canonical Name

356 20 Domain Name Service

CNAME

CNAME records allow one to create, and maintain, aliases for physical devices. For example, a web server should be defined as "www.mydomain.com" using a CNAME record that points to a physical device defined in an "A" record. This allows the web server to be moved to a different device without changing the name by which the server is known on the network.

SOA

This record is required in every zone and denotes the "Start Of Authority" for this zone. If this record is split over a number of lines, the values for SERIAL, RE-FRESH, RETRY, EXPIRE, and MINIMUM must be within parenthesis (as in all the examples in this chapter) and the opening parenthesis must be on the same line as "SOA". If the record is all on one line, the parenthesis are not allowed.

PTR

PTR Resource Records "point" to the public FQDN for a given IP address for inverse resolution of an address to hostname. This RR only appears in inverse zone files.

MX

These are mail exchange records. Email will be directed to the host with the lowest MX[42] number first. Hosts with the same MX number will attempt to load balance.

20.11 Inverse Zone Files

There should be one inverse zone file per network segment associated with a zone file on this name server. The format of the file is very similar to the normal zone file, but is much simpler and shorter[43]. The inverse zone file and regular zone file must be kept in agreement to a large extent to the point that it is normal to edit both files for any changes. Some sites see inverse zone files as a security risk and therefore do not have them. Any attempt to resolve an IP address to a hostname is answered with

---

[42] Mail Exchange Resource Record (DNS)

[43] I find it best to create the zone file, copy it to the required name for the inverse zone file, and then edit the resulting file.

the IP address and no additional information[44]. Below is a sample inverse zone file
to match one given before in Section 20.10.

```
 1  ;;;;;;  Sample Inverse Zone file
 2  $TTL 172899; Time-to-live is 2 days
 3  ;
 4  ;          Authoritative data for:  10.1.0.1 (mycollege.edu)
 5  ;
 6  ;          CONTACT:  Gerry Howser 999/555-1212 (
        postmaster@mycollege.edu)
 7  ;
 8  @          IN SOA router1-1.mycollege.edu. postmaster.mycollege.
        edu. (
 9    2019020700        ; SERIAL restarted on 02/07/2109
10    7200              ; REFRESH
11    600               ; RETRY
12    3600000           ; EXPIRE
13    86400             ; MINIMUM
14    )
15
16  ns router1-1.mycollege.edu.
17  ns dns.myhighschool.edu.
18
19  ;
20  ; end of header information
21  ;
22  $ORIGIN 0.1.10.in-addr.arpa.
23  1              PTR router1-1.mycollege.edu.
24
25  25             PTR station25.mycollege.edu.
26  26             PTR station26.mycollege.edu.
27  32             PTR laptop32.mycollege.edu.
28
29  ; cs.mycollege.edu.
30  100            PTR nowhere.cs.mycollege.edu.
31  101            PTR noname.cs.mycollege.edu.
```

Most of the information in this inverse zone file is self–explanatory, but there are
a few things to notice. First of all, this file was created by copying the forward zone
file and deleting all CNAME records. This file should have information about the
physical devices (or nothing to hide a device) but nothing at all about any aliases
for a device. The reason is very simple: an administrator might need to "move" a
service such as a web server to a new device. There is no need for anyone to know
exactly where the web server is located as this makes attacks easier and makes it
harder for the administrator to balance the workload on physical devices.

The **$ORIGIN 0.1.10.in-addr.arpa.** statement is required if a non–
standard inverse zone name is used, but should be included anyway as self documentation and to guard against typos in the **named.conf.local** file. This name
traces its roots back to the early days of the ARPANET which was the forerunner of
the Internet. The origin is formed from the network part of the IP address (in inverse
order) and the required suffix **.in-addr.arpa** to inform the name server that this
is an inverse zone file.

Notice that the names returned are not very helpful in learning much about the
device associated with an IP address. This is also intentional, but it might be helpful
to name devices library1, library2, or circulation, to give those who need to know a
good idea as to where the device is located.

---

[44] In this case, it might be wise to configure all devices to not answer pings from outside the
organization. This can also be done by configuring a firewall to not allow incoming pings. Before
DDOS attacks, every device answered pings.

All that is required for a device in an inverse zone is the host part of the address and a hostname. Any other information would not be helpful to those who need to know and could leak information about the domain such as which device to attack to cause the most damage to the network. For this reason, the last two devices, nowhere.cs.mycollege.edu and noname.cs.mycollege.edu, could have been given the names nowhere.mycollege.edu and noname.mycollege.edu to hide the existence of a sub–domain "cs".

This file should be edited each and every time the forward zone file is edited and the SERIAL number changed to match the forward zone SERIAL number. Errors in the SERIAL number can cause secondary name servers to not update the local copy of this file. This is one of the worst problems that can occur with DNS, in my opinion.

20.12 Checking Zone Files for Errors

Just as there is a BIND utility for checking **named.conf** files, BIND provides the utility **named–checkzone** to check zone files and inverse zone files. The following zone file has two errors, one on line 25 in the MX record (Missing number) and one on line 28 (trailing period in IP address 10.1.0.25. which should be 10.1.0.25 instead).

```
 1  pi@router1-1:/var/cache/bind/master$ sudo vi mycollege.edu.bad
 2  ;;;;;; Sample bad zone file
 3  ;
 4  ;       \ Authoritative data for:  mycollege.edu
 5  ;
 6  ;         CONTACT:  Gerry Howser 999/555-1212 (
        postmaster@mycollege.edu)
 7  ;
 8  $TTL 3600
 9  @          IN SOA router1-1.mycollege.edu. postmaster.mycollege.
        edu. (
10                          2019020700      ; SERIAL restarted on
                                02/07/2109
11                          7200            ; REFRESH
12                          600             ; RETRY
13                          3600000         ; EXPIRE
14                          86400           ; MINIMUM
15                          )
16                                          in      ns router1-1.mycollege
                                                .edu.
17                                          in      ns dns.myhighschool.
                                                edu.
18  router1-1.mycollege.edu.        in      a 10.1.0.1
19  dns.myhighschool.edu.           in      a 10.1.0.2
20  ;
21  ; end of header information
22  ;
23  $origin mycollege.edu.          ;redundant record, this is the
        default
24
25  ; special names and records
26  mycollege.edu.                  in MX   router1-1.mycollege.
        edu.
27  ;
28  router1-1                       in      a       10.1.0.1
29  station25                       in      a       10.1.0.25.
```

As expected, BIND is not very helpful in finding the problem, but the provided utility **named-checkzone** is. Note that the name of the domain for the zone file must be given as part of the command.

```
pi@router1-1:/var/cache/bind/master$ sudo named-checkzone
    mycollege.edu mycollege.edu.bad
mycollege.edu.bad:18: ignoring out-of-zone data (dns.
    myhighschool.edu)
dns_rdata_fromtext: mycollege.edu.bad:25: near 'router1-1.
    mycollege.edu.': not a valid number
dns_rdata_fromtext: mycollege.edu.bad:28: near '10.1.0.25.':
    bad dotted quad
zone mycollege.edu/IN: loading from master file mycollege.edu.
    bad failed: not a valid number
zone mycollege.edu/IN: not loaded due to errors.
pi@router1-1:/var/cache/bind/master$
```

After corrections, **named-checkzone** gives the following output[45].

```
pi@router1-1:/var/cache/bind/master$ sudo named-checkzone
    mycollege.edu mycollege.edu.good
mycollege.edu.good:18: ignoring out-of-zone data (dns.
    myhighschool.edu)
zone mycollege.edu/IN: loaded serial 2019020700
OK
pi@router1-1:/var/cache/bind/master
```

Inverse zone files can be, and should be, checked in exactly the same manner.

20.13 Zone File Transfers

It is customary to have multiple name servers for a domain; indeed, it is required if the domain is registered. It would be tedious and error–prone to manually update multiple name servers with the same information. To resolve this issue, zone files can be automatically transferred from master name servers to secondary name servers [151] either by transferring the complete zone file or by a process of incremental zone file transfer.

The first thing a secondary name server does is to contact the master for the zone to obtain the master's serial number[46]. If the master serial number is larger than the serial number in the secondary's copy of the zone file, a transfer is initiated. The master simply sends the zone file to the secondary which replaces its zone file with the new copy. If the serial number is the same or *less*[47] than the one from the secondary's copy of the zone, no transfer takes place.

For this reason changes to a master zone file will not be passed on to the secondary name servers until the serial number is incremented. Should you forget to increment the serial number, devices will get different responses to the same query depending upon which name server (master or slave) happens to respond. This can

---

[45] In my opinion, it should also print a message such as "No errors found." but at least the message: "zone mycollege.edu/IN: loaded serial 2019020700" gives you positive feedback.

[46] Remember, we have stated many times that the SERIAL number is critical in zone file transfers.

[47] Hopefully, the secondary name server copy of BIND will display an error message, but don't count on it.

lead to seemingly random failures of networked services because different *authoritative* responses will be received by devices for the same FQDN. Due to caching and other considerations, this leads to problems that are almost impossible to recreate or correct. When in doubt, it might be a good idea to simply increment the serial number for the master copy of the zone file.

20.14 Dynamic DNS and DHCP

DNS and DHCP can work together to provide a dynamic system to automatically update DNS zone files with the correct information for DHCP clients using DDNS[48] as shown in Figure 20.4. All of the work happens behind the scene after the DHCP Client on a device initiates the actions:

1. The Client process requests DHCP services.
2. The DHCP service checks the files for a free IP address, assigns a lease to the Client, and then updates the files to show the lease is assigned.
3. The DHCP service notifies the *master*[49] name service of the IP address and FQDN assignment.
4. The DDNS name service updates the zone and inverse zone files.

At this point, any queries to name service will return the proper information as updated by the dynamic requests. When the lease expires, the hostname and IP address will be available for reuse. This allows DDNS to return a not–found response for addresses not currently assigned.

```
 1  # global statements are applicable to all subnets
 2  authoritative; # assume this is the only DHCP server on
        network
 3  # global statements
 4  ...
 5  # DDNS statements
 6  ddns-updates on;               # default but good practice
 7  ddns-update-style interim;  # only supported active option
 8  allow client-updates;          # default but good practice
 9  do-forward-updates;            # default but good practice
10  ...
11  # zone clauses are optional and required
12  # only to define params for DDNS
13  # may be one or more zone clauses
14  zone example.com {
15  ...
16  primary ns1.example.com;
17  # uses name format could use IP address format
18  }
19  zone 2.168.192.in-addr.arpa {
20  ...
21  primary 192.168.2.15;
22  # the above can use a dns name, instead of an IP
23  # which is probably more flexible
24  # primary ns1.example.com.
25  }
```

---

[48] Dynamic Domain Name System

[49] Remember: Only the master name service zone files can be edited directly. Secondary name servers are updated by the master name service.

```
26
27  # must be at least one subnet clause
28  # in a dhcpd.conf file
29  subnet 192.168.2.254 netmask 255.255.255.0 {
30  # useable IP addresses in this subnet
31  range 192.168.2.0 192.168.2.32;
32  range 192.168.2.34 192.168.2.128;
33  ....
34  subnet statements
35  ....
36  # DDNS statements
37  ddns-domainname "example.com.";
38  # use this domain name to update A RR (forward map)
39  ddns-rev-domainname "in-addr.arpa.";
40  # use this domain name to update PTR RR (reverse map)
41  }
```

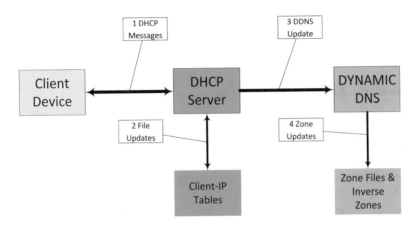

Fig. 20.4: A Client Request for DHCP and DDNS

20.15 Advanced Zone File Transfers

When a secondary name server requests a zone transfer, a complete transfer, or AFXR[50], of the zone is done by the master. If the zones are rather large or if DDNS is being used, full zone transfers can lead to a heavy overhead on the network. To resolve this problem, a secondary can request an IFXR[51] instead of a full AFXR[52].

In order for a IFXR to work, for each secondary the master must track the changes made since the last zone file transfer. When a IFXR is requested the master begins sending only the changes made since the last zone transfer to that secondary. When

---

[50] Asynchronous Full Transfer

[51] Incremental Zone Transfer

[52] Think of AFXR as a"all" transfer and IFXR as "incremental" transfer to help keep everything straight. [36]

the transfer is complete, the master notes the completed transfer and begins tracking any new updates as before. If a secondary requests an incremental zone transfer and the master does not support IFXR, a full zone transfer is done instead.

When a secondary receives the first incremental update, it makes a copy of the zone file and applies all update to that copy. When the IFXR is complete, the secondary replaces the zone file with the updated copy.

20.16 DNS in Isolation

There is a problem with DNS on an Intranet in that the root and TLD name servers are not available. In this section we will discuss a number of methods to solve this issue. Obviously, the best solution is to provide these services but it is possible to provide working DNS *without* root and TLD servers.

20.16.1 One Zone Solution

The simplest solution is often the best. It is hard to conceive of a situation where multiple domains, rather than multiple sub–domains, are required for an Intranet. Since Intranets are typically "owned" by a single organization, sub–domains of a single organization–wide domain will typically solve any name service problems. Should the domain require multiple name servers, these can be secondary name servers for the organization's chosen domain name. If needed, each sub–domain may be delegated to a separate name server to share the responsibility of maintaining the domain[53]

20.16.2 All Zones on Each NS Solution

The simplest solution is to provide one master name server for the entire Intranet with multiple secondary servers to balance the load. This is the simplest, and most straight forward, solution to the problem. Each name server is configured to provide either master or secondary service for each domain in the Intranet. This way each and every name server can provide an authoritative response for any Client request.

For example, consider an Intranet with two domains, `college.edu` and `bookstore.com`, and two name servers, `ns1.college.edu` (10.1.1.1) and `ns2.bookstore.com` (10.2.2.2). The following configuration files would allow

---

[53] In my experience, it is best to move the responsibility for maintaining data as close to those who actually produce the data as possible. Fewer mistakes will be made by those most familiar with the data.

each name server to respond correctly to any query about either **college.edu** or
bookstore.com.

Configuration File **/etc/BIND/named.conf.local** for
ns1.college.edu

```
 1  //
 2  // Do any local configuration here
 3  //
 4
 5  // Consider adding the 1918 zones here, if they are not used
        in your
 6  // organization
 7  //include "/etc/bind/zones.rfc1918";
 8
 9  zone "college.edu" in{
10  type master;
11  file "/var/cache/bind/master/college.edu";
12  allow-transfer {10.2.2.2; };
13  };
14
15  zone "1.1.10.in-addr.arpa" in{
16  type master;
17  file "/var/cache/bind/master/mycollege.inv";
18  allow-transfer {10.2.2.2;};
19  };
20
21  zone "bookstore.com" in {
22  type slave;
23  file "/var/cache/bind/slave/bookstore.com";
24  masters {10.2.2.2 };
25  };
26
27  zone "2.2.10.in-addr.arpa" in{
28  type slave;
29  file "/var/cache/bind/master/bookstore.inv";
30  masters {102.2.2;};
31  };
```

Configuration File **/etc/BIND/named.conf.local** for
ns2.example.org

Below is the same file for **ns2.example.org**.

```
 1  //
 2  // Do any local configuration here
 3  //
 4
 5  // Consider adding the 1918 zones here, if they are not used
        in your
 6  // organization
 7  //include "/etc/bind/zones.rfc1918";
 8
 9  zone "bookstore.com" in{
10  type master;
11  file "/var/cache/bind/master/bookstore.com";
12  allow-transfer {10.1.1.1; };
13  };
14
15  zone 2.2.10.in-addr.arpa" in{
16  type master;
17  file "/var/cache/bind/master/bookstore.inv";
18  allow-transfer {10.1.1.1;};
19  };
20
21  zone "college.edu" in {
```

```
22  type slave;
23  file "/var/cache/bind/slave/college.edu";
24  masters {10.1.1.1; };
25  };
26
27  zone "1.1.10.in-addr.arpa" in{
28  type slave;
29  file "/var/cache/bind/master/mycollege.inv";
30  masters {10.1.1.1;};
31  };
```

These configuration files would insure that each of the two name servers had complete knowledge of both of the domains expected on this Internet, **college.edu** and **bookstore.com**, and would insure that all queries for those domains received an authoritative response. This technique may be extended to as many domains as needed for a test–bed network or an Intranet.

20.17 All Services Solution

If someone is really interested in reproducing the actual, correct functioning of DNS, it is possible to reproduce both the root servers and the TLD servers using BIND on the Raspberry Pi. However, this requires some specialized configuration for BIND9. Sample configuration files and instructions are available from **www.gerryhowser.com/internet/**.

20.18 DNS Tools

Server–side tools have already been covered in Sections 20.8.3 and 20.12. Table 20.3 lists both server–side and Client DNS utilities. One topic that wasn't covered earlier is how to force **BIND9** to reload the configuration files. This is done by: **sudo systemctl restart bind9**.

Table 20.3: BIND9 Tools

Utility	Usage
named-checkconf	Checks the **named.conf.local** file for errors.
named-checkzone	Checks zone files for errors.
nslookup	Asks the DNS for information regarding a hostname or IP address.
dig	Asks the DNS for information with single inquiries or batch inquiries.

systemctl Command	Usage
start bind9	Starts[a] BIND and loads the configuration files.
restart bind9	Halts and restarts BIND.
status bind9	Displays the current status of the **daemon**.
reload bind9	Forces a reload of all the configuration files without restarting BIND.
stop bind9	Forces the **daemon** to gracefully halt[b].

a BIND will start automatically once it is installed and configured.

b This, like **start**, is rarely needed.

20.19 Client DNS tools

nslookup[54] is one of the most widely available client–side DNS tools and is available as part of virtually all network enabled OS. While **nslookup** is still very useful, it is not as useful as another very similar tool named DIG[55].

20.19.1 NSLookup and DNS

The **nslookup** command is available on the Raspbian, Linux, Unix, and Windows [48] and is apparently still supported but not being enhanced at this time. However, it still has its uses.

Format 1:   Lookup target using the device default name server.

---

[54] Name Service Lookup

[55] Domain Information Groper

```
nslookup [-opt] target
```

Format 2:   Lookup target using a specific name server.
```
nslookup [-opt] target dns
```

Format 3:   Enter interactive mode using the device default name server.
```
nslookup [-opt]
```

Format 4:   Enter interactive mode using a specific name server.
```
nslookup [-opt] - dns
```

20.19.2 NSLookup Examples

Format 1

Using the default name server to look up a host and an IP address.
```
pi@router1-1:~$ nslookup www.highschool.edu
Server:          192.168.1.1
Address:         192.168.1.1#53

www.highschool.edu        canonical name = dns.highschool.edu.
Name:   dns.highschool.edu
Address: 10.1.0.1

pi@router1-1:~$ nslookup 10.1.0.1
Server:          192.168.1.1
Address:         192.168.1.1#53

1.0.1.10.in-addr.arpa    name = dns.highschool.edu.

pi@router1-1:~$
```

Format 2

Using the specific name server to look up a host.
```
pi@router1-1:~$ nslookup www.highschool.edu 192.168.2.2
Server:          192.168.2.2
Address:         192.168.2.2#53

www.highschool.edu        canonical name = dns.highschool.edu.
Name:   dns.highschool.edu
Address: 10.1.0.1

pi@router1-1:~$
```

Format 3

Interactive mode.

```
pi@router1-1:~$ nslookup -all

Set options:
novc                    nodebug          nod2
search                  recurse
timeout = 0             retry = 3        port = 53        ndots =
    -1
querytype = A           class = IN
srchlist =
> set type=any
> set domain=highschool.edu
> mail
Server:         192.168.1.1
Address:        192.168.1.1#53

Name:   mail.highschool.edu
Address: 10.1.0.1
mail.highschool.edu     mail exchanger = 10 dns.highschool.edu
    .
> laptop
Server:         192.168.1.1
Address:        192.168.1.1#53

Name:   laptop.highschool.edu
Address: 10.1.0.25
> exit
```

Format 4

Interactive mode using a specific name server works, and looks, exactly like format 3 only using a specific name server specified by IP address or by hostname.

nslookup has many options which seem to change over time. For an explanation of some of these options, check the Raspbian **man** pages or see *Pro DNS and BIND* [2].

20.19.3 Dig and DNS

A more interesting tool for querying DNS is the DIG utility which can be found on Raspbian and most Linux/Unix distributions[56]. Like nslookup, DIG has many, many options that control where the query is sent and what amount of information is returned. In my opinion, this makes DIG a much more useful tool for debugging nasty DNS problems than nslookup. For casual inquiries there is not much difference between the two utilities. The best way to explain DIG is with a few examples. For more information on the various options, see *Pro DNS and BIND* [2].

A simple enquiry:

```
pi@router5-5:~$ dig www.highschool.edu
; <<>> DiG 9.11.5-P4-5.1-Raspbian <<>> www.highschool.edu
```

---

[56] Dig must be installed on Windows manually. For our purposes this is not worth the effort.

```
;; global options: +cmd
;; Got answer:
;; ->>HEADER<<- opcode: QUERY, status: NOERROR, id: 1340
;; flags: qr aa rd ra; QUERY: 1, ANSWER: 2, AUTHORITY: 3,
   ADDITIONAL: 2

;; OPT PSEUDOSECTION:
; EDNS: version: 0, flags:; udp: 4096
;; QUESTION SECTION:
;www.highschool.edu.              IN        A

;; ANSWER SECTION:
www.highschool.edu.      172899   IN        CNAME     dns.
   highschool.edu.
dns.highschool.edu.      172899   IN        A         10.1.0.1

;; AUTHORITY SECTION:
highschool.edu.          172899   IN        NS        dns.
   highschool.edu.
highschool.edu.          172899   IN        NS        dns.college.
   edu.
highschool.edu.          172899   IN        NS        dns.FakeISP.
   net.

;; ADDITIONAL SECTION:
dns.college.edu.         172899   IN        A         10.2.0.1

;; Query time: 1 msec
;; SERVER: 192.168.1.1#53(192.168.1.1)
;; WHEN: Mon Aug 19 20:34:37 EDT 2019
;; MSG SIZE   rcvd: 166

pi@router5-5:~$
```

An enquiry directed to a specific name server:

```
pi@router5-5:~$ dig @dns.college.edu mail.highschool.edu

; <<>> DiG 9.11.5-P4-5.1-Raspbian <<>> @dns.college.edu mail
    .highschool.edu
; (1 server found)
;; global options: +cmd
;; Got answer:
;; ->>HEADER<<- opcode: QUERY, status: SERVFAIL, id: 48821
;; flags: qr rd; QUERY: 1, ANSWER: 0, AUTHORITY: 0,
   ADDITIONAL: 1
;; WARNING: recursion requested but not available

;; OPT PSEUDOSECTION:
; EDNS: version: 0, flags:; udp: 4096
;; QUESTION SECTION:
;mail.highschool.edu.             IN        A

;; Query time: 1 msec
;; SERVER: 10.2.0.1#53(10.2.0.1)
;; WHEN: Mon Aug 19 20:36:22 EDT 2019
;; MSG SIZE   rcvd: 48

pi@router5-5:~$
```

As you can see, the information returned by a simple DIG inquiry is very detailed and can often help resolve complex issues. For that same reason, DIG also represents a security risk but that is beyond the scope of this text.

20.20 DNSSecurity

"Security is mostly a superstition. It does not exist in nature, nor do the children of men as a whole experience it."

Heller Keller

Complete security is a pipe dream and can never be achieved, however there are many things that should be done to secure DNS when connected to the public Internet[57].

First, and foremost, the devices running DNS must be physically secured. If an unauthorized person can gain access to your DNS server they can cut wires, pour electrolytes into the box (sugary soft drinks are great at destroying printed circuit boards), pound it with a sledge hammer, or steal the device[58].

20.20.1 General Software and OS Security

All software on the server must be kept up to date and this includes the operating system[59]. Users must only be allowed access that is required for their duties. All of the typical protections must be observed such as good passwords that are changed regularly, passwords are changed when key personnel leave, and so on.

20.20.2 DNSSEC

This still leaves the issue of outside attacks. DNS must be open to be of any use which makes these services unusually vulnerable to attackers and care must be taken to secure your name servers (and therefore your domain) from attacks. The most complete solution to securing a name server involves digital signatures and public key/private key security which is beyond the scope of this text. Fortunately, there are many good resources on line [3] and Ron Aitchison's book [2].

---

[57] Be aware of insider threats. Recent events in the past few years have shown that it is difficult to guard against an attack by someone whose job requires complete access to your stuff.

[58] Certain devices can be reset to "factory defaults" in seconds with a paperclip.

[59] I am not sure if this is the correct worm [22], but I did watch CNN "attacked" live on the air by a virus that caused all Windows 2000 devices to repeatedly reboot, including those at CNN. The problem was that if the OS had been kept up–to–date, the flaw the worm exploited was patched. Keep up to date.

Since the Raspberry Pi is capable of running the latest version of BIND, it will easily support DNSSEC[60]. Being somewhat paranoid[61], in my opinion even the DNS for an intranet must be secured as much as possible and I suggest running DNSSEC except in the case of a test–bed network.

If you have the time and interest, I direct you to Chapter 11 of Mr. Aitchison's book (or the on line equivalent) and RFCs 4033, 4034, and 4035. For our purposes, DNSSEC is definitely overkill, but then again our test–bed networks are not connecting to the Internet.

---

[60] Secure Domain Name Service

[61] Too many security failures have taught me three things:
1. You can never have too many backups.
2. You can never be too secure.
3. You can't trust anyone, especially yourself, to abide by security rules and procedures.

Projects

20.1 ping a device in another group *by FQDN.*

20.2 With permission, use **dig** to explore the zones for another domain.

20.3 Use **dig** to explore the **yahoo.com** domain.

20.4 (Optional) My personal experience has shown that it is often desirable to set up a Linux machine such as the Raspberry Pi as a caching name server even if no other device uses it for name service. Do this with each Pi in your group.

Exercises

20.1 Assuming that all processes do not have the relevant information cached and using Figure 20.1 as a guide, trace the steps to resolve **www.yahoo.com** then **mail.yahoo.com**?

20.2 What are the steps to resolve **bozo.bogus.com** if the domain **bogus.com** does not exist?

20.3 Can a hostname **bongo.bogus.com** be resolved if the domain is not registered? Explain.

Further Reading

Aitchison, Ron[62]. 2005. Pro DNS and BIND. Apress [2].

The RFC below provide further information about DNS[63]. This is a fairly exhaustive list and most RFC are typically dense and hard to read. Normally RFC are most useful when writing a process to implement a specific protocol.

Many of these RFC deal with IPv6 and security issues. DNS was fairly well settled long before IPv6 became widely used and so there was a flurry of RFC that needed to be produced to extend DNS to allow the same name server to work with both versions of IP and to enhance security.

---

[62] This is an excellent book for both the beginner and the network administrators in charge of DNS. An on–line version of this book can be found at: http://www.zytrax.com/books/dns/. Consider buying this book if you are heavily involved in DNS to support his work in the field.

[63] This list is extremely long. Only read those that interest you or apply to some problem you need to solve.

RFCs Directly Related to This Chapter

DNS	Title
RFC 0799	Internet Name Domains [64]
RFC 0882	DOMAIN NAMES - CONCEPTS and FACILITIES [69]
RFC 0883	DOMAIN NAMES - IMPLEMENTATION and SPECIFICATIONS [70]
RFC 0973	DOMAIN NAMES - IMPLEMENTATION and SPECIFICATIONS [74]
RFC 0974	Mail routing and the domain system [75]
RFC 1034	DOMAIN NAMES - CONCEPTS and FACILITIES [76]
RFC 1035	Domain names - implementation and specification [77]
RFC 1101	DNS Encoding of Network Names and Other Types [81]
RFC 1178	Choosing a Name for Your Computer [86]
RFC 1348	DNS NSAP RRs [93]
RFC 1591	Domain Name System Structure and Delegation [117]
RFC 1637	DNS NSAP Resource Records [118]
RFC 1706	DNS NSAP Resource Records [126]
RFC 1794	DNS Support for Load Balancing [129]
RFC 1912	Common DNS Operational and Configuration Errors [142]
RFC 1918	Address Allocation for Private Internets [144]
RFC 1982	Serial Number Arithmetic [149]
RFC 1995	Incremental Zone Transfer in DNS [151]
RFC 1996	A Mechanism for Prompt Notification of Zone Changes (DNS NOTIFY) [152]
RFC 2065	Domain Name System Security Extensions [156]
RFC 2136	Dynamic Updates in the Domain Name System (DNS UPDATE) [162]
RFC 2181	Clarifications to the DNS Specification [165]
RFC 2317	Classless IN-ADDR.ARPA delegation [168]
RFC 2606	Reserved Top Level DNS Names [179]
RFC 2671	Extension Mechanisms for DNS (EDNS0) [184]
RFC 2782	A DNS RR for specifying the location of services (DNS SRV) [185]
RFC 2874	DNS Extensions to Support IPv6 Address Aggregation and Renumbering [190]
RFC 3363	Representing Internet Protocol version 6 (IPv6) Addresses in the Domain Name System (DNS) [201]
RFC 3401	Dynamic Delegation Discovery System (DDDS) Part One: The Comprehensive DDDS [204]
RFC 3402	Dynamic Delegation Discovery System (DDDS) Part Two: The Algorithm [205]
RFC 3403	Dynamic Delegation Discovery System (DDDS) Part Three: The Domain Name System (DNS) Database [206]

DNS	Title
RFC 3404	Dynamic Delegation Discovery System (DDDS) Part Four: The Uniform Resource Identifiers (URI) [207]
RFC 3405	Dynamic Delegation Discovery System (DDDS) Part Five: URI.ARPA Assignment Procedures [208]
RFC 3425	Obsoleting IQUERY [209]
RFC 3484	Default Address Selection for Internet Protocol version 6 (IPv6) [211]
RFC 3596	DNS Extensions to Support IP Version 6 [214]
RFC 3597	Handling of Unknown DNS Resource Record (RR) Types [215]
RFC 3646	DNS Configuration options for Dynamic Host Configuration Protocol for IPv6 (DHCPv6) [220]
RFC 3833	Threat Analysis of the Domain Name System (DNS) [224]
RFC 4033	DNS Security Introduction and Requirements [229]
RFC 4343	Domain Name System (DNS) Case Insensitivity Clarification [234]
RFC 4367	What's in a Name: False Assumptions about DNS Names [236]
RFC 4472	Operational Considerations and Issues with IPv6 DNS [239]
RFC 4501	Domain Name System Uniform Resource Identifiers [240]
RFC 4592	The Role of Wildcards in the Domain Name System [241]
RFC 4697	Observed DNS Resolution Misbehavior [244]
RFC 4702	The Dynamic Host Configuration Protocol (DHCP) Client Fully Qualified Domain Name (FQDN) Option [245]
RFC 4703	Resolution of Fully Qualified Domain Name (FQDN) Conflicts among Dynamic Host Configuration Protocol (DHCP) Clients [246]
RFC 4871	DomainKeys Identified Mail (DKIM) Signatures [226]
RFC 4955	DNS Security (DNSSEC) Experiments [251]
RFC 5006	IPv6 Router Advertisement Option for DNS Configuration [254]
RFC 6106	IPv6 Router Advertisement Options for DNS Configuration [269]
RFC 7393	Using the Port Control Protocol (PCP) to Update Dynamic DNS [279]
RFC 7672	SMTP Security via Opportunistic DNS-Based Authentication of Named Entities (DANE) Transport Layer Security (TLS) [283]
RFC 8460	SMTP TLS Reporting [297]
RFC 8461	SMTP MTA Strict Transport Security (MTA-STS) [298]
RFC 8484	DNS Queries over HTTPS (DoH) [301]
RFC 8501	Reverse DNS in IPv6 for Internet Service Providers [302]

Other RFCs Related to DNS

For a list of other RFCs related to DNS but not closely referenced in this chapter, please see Appendix B–DNS.

Chapter 21
Hyper Text Transfer Protocol: The Web

Overview

The most common use of the Internet today is the World Wide Web, or "the web". Indeed, most people use the terms "the web" and "the Internet" interchangeably. This section will cover how to implement Apache web services and how to handle multiple virtual websites on the same server.

It is assumed that the reader has a passing knowledge of HTML[1] and can write simple web pages with tables, links, and images. If this is not the case, there are many tutorials on the web on building simple web sites[2]. All LAMP servers behave the same regardless of the underlying Linux distribution. Windows web servers, WAMP, work virtually the same as on Linux with the exception of the root directory for the server.

21.1 Apache Web Services

The Open Source web service, Apache, is one of the most common web services on the Internet and it is free to use[3].

---

[1] HyperText Markup Language

[2] This seems a bit circular, but it takes no background knowledge to search the web on how to built web sites.

[3] If Apache `httpd` is not pre-installed on your Pi, see Section 8.1.5

© Springer Nature Switzerland AG 2020   375
G. Howser, *Computer Networks and the Internet*,
https://doi.org/10.1007/978-3-030-34496-2_21

21.2 Installing a LAMP Server on the Raspberry Pi

At this point in the project, the various groups will begin to interact, so it is imperative that the date and time be accurate on all devices. Each time the Pi is booted, the date/time **must** be updated.

This section is a guide as to how to install the Apache part of a LAMP[4] server on your Pi to provide a web server. At this point, we will concentrate on installing Apache2 as our web server[5]. In this guide, replace **fineteas.co.uk** with your domain name. If you like, you can install web servers on all four Pi's in your group by naming them something like "www.fineteas.co.uk", "www2.fineteas.co.uk", and so on. The only changes should be the **ServerAlias** in the Apache configuration files.

21.3 Apache Resources Configuration

Until there is a reason to change the default server resources, it is best to leave them as is. However, it is important to know how Apache initializes and handles changing loads on the server. Apache handles load balancing and resource control mainly by controlling the number of child web servers that are running at any time. These are exact copies of the original service and are started and stopped as needed. Changing the configuration settings is quite complex and beyond the scope of this work. However, the Apache web site [24] [7]has excellent documentation and some tutorials to help with advanced uses of Apache. Most tuning is handled by the Apache MPM worker.

As the load on the server changes, Apache will adjust the number of running copies of the child servers by spawning new ones or allowing older ones to shutdown. The nice part of this is that no one has to administer this as Apache will take care of load-balancing for you.

21.4 Virtual Host: **fineteas.co.uk**

It is possible to have a single Apache server serve web pages for multiple virtual hosts simply by loading a configuration file for each virtual website and insuring the DNS zone files have the correct entries for **www.<site name>** and **<site name>**. To enable a new host, one needs only create a configuration file in the correct directory, **/etc/apache2/sites-available**, and instruct Apache to reload the con-

---

[4] LAMP - Linux Apache MySQL PHP server for active web pages.

[5] Later on you can install the rest of a LAMP[6] server. This is beyond the scope of this text, but can be found many places on the web [1].

figuration files. For example, to create the files for **www.fineteas.co.uk** one
needs to create the file **fineteas.co.uk.conf** from a copy of the default configuration file, **000-default.conf** in the **sites-available** directory.

```
pi@router5-1:~\$ cd /etc/apache2/sites-available/
pi@router5-1:/etc/apache2/sites-available$

pi@router5-1:/etc/apache2/sites-available$ sudo cp -p 000-
    default.conf fineteas.co.uk.conf
pi@router5-1:/etc/apache2/sites-available$ ls -l
total 16
-rw-r--r-- 1 root root 1332 Nov  3 07:34 000-default.conf
-rw-r--r-- 1 root root 6338 Nov  3 07:34 default-ssl.conf
-rw-r--r-- 1 root root 1332 Nov  3 07:34 fineteas.co.uk.conf
pi@router5-1:/etc/apache2/sites-available$
```

21.4.1 Virtual Host Configuration File Format

The format of the configuration file is not dependent upon "white space" but it helps
to indent each line to make the file more readable.

The configuration file specifies everything Apache needs to properly serve web
pages. The header, **<VirtualHost *:80>**, tells Apache to listen for requests
and send web pages using port 80 which is the standard port or socket for web
servers and browsers. The second field,
ServerAdmin webmaster@fineteas.co.uk, is an active email address
where Apache can send reports and some warnings via email. A human should
monitor this email periodically to insure the website is functioning normally. The
ServerName and **ServerAlias** identify this virtual website. These should be
set in DNS to point to the static IP address of this Pi. The final three entries are
the directories to hold the actual web pages and log files. These directories must
be created before Apache is restarted or reloaded, otherwise Apache will not start
properly.

Edit the configuration file to resemble the following:

```
<VirtualHost *:80>
    # The ServerName directive sets the request scheme, hostname
        and
    # port that
    # the server uses to identify itself. This is used when
        creating
    # redirection URLs. In the context of virtual hosts, the
        ServerName
    # specifies what hostname must appear in the request's Host:
        header to
    # match this virtual host.  You must set it explicitly for
        this virtual
    # host.
ServerName www.fineteas.co.uk
ServerAlias www.fineteas.co.uk
ServerAdmin webmaster@localhost
DocumentRoot /var/www/www.fineteas.co.uk/public_html

ErrorLog /var/www/fineteas.co.uk/logs/error.log
CustomLog /var/www/fineteas.co.uk/access.log combined
</VirtualHost>
```

The directory should now contain files such as:

```
pi@router5-1:/etc/apache2/sites-available$ ls -l
total 16
-rw-r--r-- 1 root root 1332 Nov  3 07:34 000-default.conf
-rw-r--r-- 1 root root 6338 Nov  3 07:34 default-ssl.conf
-rw-r--r-- 1 root root  624 Mar 13 12:02 fineteas.co.uk.conf
pi@router5-1:/etc/apache2/sites-available$
```

The log files must be created *before* Apache is started. Given the configuration file above, the files would be created by:

```
pi@router5-1:/etc/apache2/sites-available$ sudo mkdir /var/www
    /fineteas.co.uk
pi@router5-1:/etc/apache2/sites-available$ sudo mkdir /var/www
    /fineteas.co.uk/logs
pi@router5-1:/etc/apache2/sites-available$ sudo mkdir /var/www
    /fineteas.co.uk/public_html
pi@router5-1:/etc/apache2/sites-available$ cd /var/www/
    fineteas.co.uk/logs
pi@router5-1:/var/www/fineteas.co.uk/logs$ sudo touch error.
    log
pi@router5-1:/var/www/fineteas.co.uk/logs$ sudo touch access.
    log
pi@router5-1:/var/www/fineteas.co.uk/logs$ ls -l
total 0
-rw-r--r-- 1 root root 0 Mar 13 12:08 access.log
-rw-r--r-- 1 root root 0 Mar 13 12:08 error.log
pi@router5-1:/var/www/fineteas.co.uk/logs$
```

Page sources are placed in the directory:
`/var/www/fineteas.co.uk/public_html`.

21.5 Controlling the Apache2 `httpd` Daemon

Apache can be started, restarted, and reloaded using the same commands as most services via **sudo systemctl**.

```
pi@router5-1:/var/www/fineteas.co.uk/logs# sudo systemctl
    restart apache2
pi@router5-1:/var/www/fineteas.co.uk/logs# sudo systemctl
    status apache2
  apache2.service - The Apache HTTP Server
Loaded: loaded (/lib/systemd/system/apache2.service; enabled;
    vendor preset:
Active: active (running) since Wed 2019-03-13 12:23:24 EDT; 4s
    ago
Process: 1192 ExecStop=/usr/sbin/apachectl stop (code=exited,
    status=0/SUCCESS
Process: 1200 ExecStart=/usr/sbin/apachectl start (code=exited
    , status=0/SUCCE
Main PID: 1204 (apache2)
CGroup: /system.slice/apache2.service
 1204 /usr/sbin/apache2 -k start
 1207 /usr/sbin/apache2 -k start
 1208 /usr/sbin/apache2 -k start

Mar 13 12:23:23 router5-1 systemd[1]: Starting The Apache HTTP
    Server...
Mar 13 12:23:24 router5-1 apachectl[1200]: AH00558: apache2:
    Could not reliably
Mar 13 12:23:24 router5-1 systemd[1]: Started The Apache HTTP
    Server.
pi@router5-1:/var/www/fineteas.co.uk/logs
```

The warning is not an issue. It only warns you that Apache will use the IPv4 loopback address internally for **localname**.

21.5.1 Enable/Disable Virtual Website

After the directories have been created the website can be enabled. The following commands will enable/disable a website after which Apache must be reloaded **sudo service apache2 reload** or restarted **sudo service apache2 restart**.

Enable

```
sudo a2ensite fineteas.co.uk.conf
  pi@router5-1:~\$ sudo a2ensite fineteas.co.uk.conf
  Enabling site fineteas.co.uk.
  To activate the new configuration, you need to run:
  systemctl reload apache2
  pi@router5-1:~\$ sudo systemctl reload apache2
  pi@router5-1:~\$ sudo systemctl status apache2
   apache2.service - The Apache HTTP Server
  Loaded: loaded (/lib/systemd/system/apache2.service; enabled;
     vendor preset:
  Active: active (running) since Wed 2019-03-13 12:31:39 EDT; 2
     min 4s ago
  Process: 945 ExecReload=/usr/sbin/apachectl graceful (code=
     exited, status=0/SU
  Process: 420 ExecStart=/usr/sbin/apachectl start (code=exited,
     status=0/SUCCES
  Main PID: 519 (apache2)
  CGroup: /system.slice/apache2.service
    519 /usr/sbin/apache2 -k start
    953 /usr/sbin/apache2 -k start
    954 /usr/sbin/apache2 -k start

  Mar 13 12:31:36 router5-1 systemd[1]: Starting The Apache HTTP
     Server...
  Mar 13 12:31:39 router5-1 apachectl[420]: AH00558: apache2:
     Could not reliably d
  Mar 13 12:31:39 router5-1 systemd[1]: Started The Apache HTTP
     Server.
  Mar 13 12:33:07 router5-1 systemd[1]: Reloading The Apache
     HTTP Server.
  Mar 13 12:33:08 router5-1 apachectl[851]: AH00558: apache2:
     Could not reliably d
  Mar 13 12:33:08 router5-1 systemd[1]: Reloaded The Apache HTTP
     Server.
  Mar 13 12:33:38 router5-1 systemd[1]: Reloading The Apache
     HTTP Server.
  Mar 13 12:33:39 router5-1 apachectl[945]: AH00558: apache2:
     Could not reliably d
  Mar 13 12:33:39 router5-1 systemd[1]: Reloaded The Apache HTTP
     Server.
  pi@router5-1:~\$
```

Disable

```
sudo a2dissite fineteas.co.uk.conf
    i@router5-1:~\$ sudo a2dissite fineteas.co.uk.conf
    Site fineteas.co.uk disabled.
    To activate the new configuration, you need to run:
    systemctl reload apache2
    pi@router5-1:~\$ sudo systemctl reload apache2
    pi@router5-1:~\$ sudo systemctl status apache2
       apache2.service - The Apache HTTP Server
    Loaded: loaded (/lib/systemd/system/apache2.service; enabled;
       vendor preset:
    Active: active (running) since Wed 2019-03-13 12:31:39 EDT; 4
       min 9s ago
    Process: 1040 ExecReload=/usr/sbin/apachectl graceful (code=
       exited, status=0/S
    Process: 420 ExecStart=/usr/sbin/apachectl start (code=exited,
       status=0/SUCCES
    Main PID: 519 (apache2)
    CGroup: /system.slice/apache2.service
       519 /usr/sbin/apache2 -k start
      1048 /usr/sbin/apache2 -k start
      1049 /usr/sbin/apache2 -k start

    Mar 13 12:31:39 router5-1 systemd[1]: Started The Apache HTTP
       Server.
    Mar 13 12:33:07 router5-1 systemd[1]: Reloading The Apache
       HTTP Server.
    Mar 13 12:33:08 router5-1 apachectl[851]: AH00558: apache2:
       Could not reliably d
    Mar 13 12:33:08 router5-1 systemd[1]: Reloaded The Apache HTTP
       Server.
    Mar 13 12:33:38 router5-1 systemd[1]: Reloading The Apache
       HTTP Server.
    Mar 13 12:33:39 router5-1 apachectl[945]: AH00558: apache2:
       Could not reliably d
    Mar 13 12:33:39 router5-1 systemd[1]: Reloaded The Apache HTTP
       Server.
    Mar 13 12:35:44 router5-1 systemd[1]: Reloading The Apache
       HTTP Server.
    Mar 13 12:35:44 router5-1 apachectl[1040]: AH00558: apache2:
       Could not reliably
    Mar 13 12:35:44 router5-1 systemd[1]: Reloaded The Apache HTTP
       Server.
    pi@router5-1:~\$
```

Because there are no files in the directory for this website, if you point a browser at the site, you will see something like Figure 21.1.

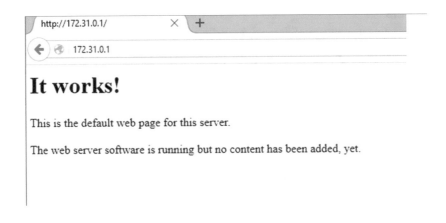

Fig. 21.1: New Website (default) for `http/:www.fineteas.co.uk`

21.6 Web Proxy

Apache also allows the Pi to act as a web proxy for the network. This is normally done on a device that has direct access to the Internet and intercepts all browser requests for web pages on port 80. Implementing a web proxy is not covered in this text.

Web Proxies provide:

- A method to filter web content that might be offensive[7].
- Caching of frequently accessed web pages. This can lead to faster access to common sites such as Google.
- A method to limit access to sites with possible malware.

---

[7] Schools and Libraries typically use proxies to filter adult content. I understand and agree to an extent, but there is a fine line between filtering and censoring.

Projects

21.1 Create a web page for your domain. If you know PHP[8] or some other way to create dynamic web pages, create a dynamic page that shows a different image each time the page is accessed.

21.2 Create a web page that links to another page on a different web server in your domain.

Exercises

21.1 HTML: Create a file named **index.html** in the correct directory to display a custom page introducing your web site to users. The answer to this exercise is a screen shot of the page.

21.2 Modify your first web page to include

    a. A link to another page on this server.
    b. An image related to the page.

21.3 What are some of the other file types a web server must understand for the default **index** web page?

21.4 Explain at least one way a web page can be hidden from some users but not others.

---

[8] PHP: Hypertext Preprocessor

Further Reading

The RFC below provide further information about HTTP. This is a fairly exhaustive list and most RFC are typically dense and hard to read. Normally RFC are most useful when writing a process to implement a specific protocol.

RFCs Directly Related to This Chapter

HTTP	Title
RFC 1945	Hypertext Transfer Protocol – HTTP/1.0 [147]
RFC 2068	Hypertext Transfer Protocol – HTTP/1.1 [157]
RFC 2145	Use and Interpretation of HTTP Version Numbers [163]
RFC 2295	Transparent Content Negotiation in HTTP [167]
RFC 2518	HTTP Extensions for Distributed Authoring – WEBDAV [176]
RFC 2616	Hypertext Transfer Protocol – HTTP/1.1 [180]
RFC 2660	The Secure HyperText Transfer Protocol [183]
RFC 3143	Known HTTP Proxy/Caching Problems [197]
RFC 4969	IANA Registration for vCard Enumservice [252]
RFC 7804	Salted Challenge Response HTTP Authentication Mechanism [285]
RFC 8336	The ORIGIN HTTP/2 Frame [292]
RFC 8470	Using Early Data in HTTP [300]
RFC 8484	DNS Queries over HTTPS (DoH) [301]

Other RFCs Related to HTTP

For a list of other RFCs related to HTTP but not closely referenced in this chapter, please see Appendix B–HTTP.

Chapter 22
Simple Mail Transfer Protocol: Email

Overview

Modern email grew out of a number of older protocols that were often vendor specific. While many early network OS were able to send short messages, almost none were designed to interoperate with all of the products from the same vendor[1], let alone between different vendors. By far, the most successful solution to this problem is SMTP, or email, and its later enhancements. However, email administrators and spammers are involved in a constantly escalating battle over the delivery of unwanted electronic mail.

In reality, email consists of two different types of protocols: server–to–server protocols and client/server protocols. The electronic mail service between servers, or server–to–server email, is SMTP. Servers exchange email based mainly upon the domain with little regard to the user names involved. Early on it was common for a single email to traverse many, many servers before it reached its destination. As connectivity between domains on the Internet has improved, it is more and more common for an email to be transferred directly from the source server to the destination without the need for intermediary servers.

The email protocols that most users are familiar with are the second type or client/server protocols such as Outlook or webmail. These protocols are tasked with transferring email between a user to the server. This is true even if the client, such as webmail, is running on the same physical device as the SMTP service.

This chapter will cover configuring an email server and email client support to run on the Pi. Email clients will be discussed only briefly.

---

[1] For example, early messaging on the Burroughs line of products did not interoperate well between BTOS and the various operating system versions of the Master Control Program (MCP). Burroughs used the term MCP long before TRON was ever thought of. MCP goes back to the early Burroughs mainframes and was first written around 1961. Surely there were some attempts to produce specialty products to interact, but the author is not familiar with these.

© Springer Nature Switzerland AG 2020
G. Howser, *Computer Networks and the Internet*,
https://doi.org/10.1007/978-3-030-34496-2_22

22.1 Early Attempts at Email

Email evolved from early messaging applications built into UNIX and mainframe operating systems that supported time–sharing terminals[2]. It was an obvious extension of these applications to create a method to send longer messages between devices on a network.

Before a working electronic method of email was developed, users struggled with solutions such as "sneaker net". The information was copied to a floppy disk[3] and then handed to the intended recipient. If it was not possible to walk the disk to the recipient, it might need to be sent via normal mail. Obviously this is very inefficient.

22.1.1 File Sharing as a Work–around

One of the early attempts to work around the lack of an application to send the equivalent of an office memo between devices was to use network file sharing applications. A user would create a memo as a file and then share that file with the recipient user. This worked well as long as everyone in the organization could share files and used the same office applications which was not always the case. The need for a standard method of creating files and moving those files between networks in a standard format was obvious.

Another inventive solution was used for a number of years at Lincoln University in Jefferson City, Missouri. The administration building was networked together using proprietary hardware that ran BTOS to allow file sharing. Each administrative assistant had a microcomputer running BTOS and a high quality printer that used a daisy wheel to produce typed pages. The network allowed for the sharing of resources on the central file server, each microcomputer, and all printers[4].

Users are very inventive. Before very many users were given microcomputers, the users discovered that when they printed a document they were presented a choice of *all* the printers on the network and the computer to which each printer was attached. This meant that a user could produce a document and send it to another user by printing the document on the recipient's printer. A memo could be sent to 20 people by issuing print commands for 20 printers.

This was so effective that it was difficult to convince these users to start using email for inter–office communications when it became available a few years later.

---

[2] These applications were much like Twitter in that the messages were typically limited to a small number of characters.

[3] Even floppy disks created a minor problem due to the number of sizes and formats. It was difficult to explain to a user that a soft–sector 5.25 diskette was not compatible with a 5.25 hard sector diskette drive. Explaining the difference today would be a good student research question.

[4] Printers were shared so that a user could still print important documents if their local printer was out of service.

22.1.2 BITNET

In the early 1990s there were two competing email systems running on the Internet, one of which was BITNET[5] and the other was SMTP. BITNET was a relay system for delivering email. A university, or other organization, would buy services from a telecommunications provider to connect to an established BITNET site and agree to allow other sites to connect to them for free. All email was sent to the established BITNET site including those sent to users connected to some other organization. Email then bounced from BITNET site to BITNET site until it reached the destination.

Messages could take days to traverse the BITNET and make many stops along the way. Some email could be held up at an intermediate site for days if a site was experiencing connectivity problems or if a site was simply down. This was problematic, but at least email almost always got through.

Until connectivity improved in the late 1990s, SMTP on the Internet acted much like BITNET. Both were widely used and there were a number of sites that ran both BITNET and SMTP and could act as a gateway between the two different types of email[6].

As high–speed connectivity became more common, and as most sites began to have all SMTP servers available 24 hours, seven days a week, SMTP became the email method of choice[7].

---

[5] Because It's Time Network

[6] NODAK, University of North Dakota, was one of the BITNET gateways used by SMTP servers in the Midwest. It was very common to see NODAK show up in the email headers if you knew where to look.

[7] Connectivity and speed were vastly increased after Senator Al Gore secured funding for the expansion of the Internet through the National Science Foundation.

22.2 The SMTP server

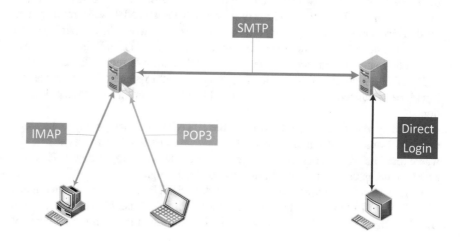

Fig. 22.1: Server–to–Server Email

The backbone of any email network is the SMTP server as in Figure 22.1. These servers, also known as MTAs[8], transfer mail between domains and also connect with the users of the email in a client/server relationship. Users tend to think of their local email client (such as Outlook or Gmail) as the entire package, but this is not the case. First we will look at the server view of email and then discuss some common client protocols.

One of the most common SMTP servers is sendmail. While sendmail is extremely flexible, it can be daunting to configure for a small and uncomplicated network [26]. Fortunately, the default configuration, with one modification, is adequate for many implementations.

Another SMTP server that is very popular on the Internet is Postfix. While postfix is somewhat easier to configure, almost any MTA will work for small, simplistic email installations as long as it is able to deal with spam and the most common attacks from spammers and hackers.

22.2.1 Email from a Server Point of View

Server–to–server email is completely different from the user experience, or client–to–server email. The server simply verifies and delivers incoming email and delivers out–going email. The issues arise from the "verifies" action. Entire careers can be made from verifying that incoming email is valued email and not spam. We will

---

[8] Mail Transfer Agents

focus on receiving and delivering email, but the idea of accepting all incoming email is not viable on the internet. Great care must be taken in placing an SMTP server on the internet as spammers are always on the lookout for an unprotected server through which they can spread spam. Unfortunately, this is beyond the scope of this book.

By convention and standards, all email servers must forward certain types of problematic email, such as a nonexistent address, to a special email account called "postmaster" [66].

22.3 SMTP Relay and Spam

Electronic mail is one of the most useful and most contentious uses of the Internet, or it would be except for spam. While controlling spam is actually the responsibility of the client protocols, there is one important issue on the server side that helps to control spam and that is SMTP Relay. During the early days of email, it was not always possible to send email directly from the source SMTP server to the destination SMTP server. An email might be forwarded, or relayed, from server to server over a large number of hops due to network topology and server availability. This required an SMTP server to receive email from a different domain that was destined for a third domain. This relay was, and still is, a prime target for spammers. If a spammer can find an open relay, they can generate large numbers of email messages for any domain they want[9]. The worst part about this is that the server performing the SMTP Relay becomes the last known server when the spam is tracked back to the source. If a domain relays enough spam, it will be "black listed" and servers will refuse any and all email *from all email addresses in that domain* as being spam.

When in doubt, turn off SMTP Relay to close this obvious target for spammers[10].

22.4 DNS MX Records and SMTP

One of the most critical services to ensure the proper operation of email (SMTP) is DNS, especially the proper use of the MX records. The MX records control how, and where, email gets processed which means the proper configuration of the MX records is quite complex. Most of these complexities are beyond the scope of this text[11].

---

[9] This can easily be done using nothing more sophisticated than `telnet`.

[10] It may seem trivial to point this out, but many, many attacks successfully utilize security issues that have been resolved for so long that administrators overlook them.

[11] These complexities are often covered in a good `sendmail` or DNS text.

22.4.1 MX Records

A correct MX record consists of four parts. The first part is the hostname, or alias, of the email server as it is seen by the external network. If this part is blank, the server will appear as the domain name *without a hostname.* For example, if you wish to receive email of the sort **user@example.com** and **user@mail.example.com**, the DNS zone file for **example.com** would contain MX records such as:

```
1          in MX 0 mail.example.com.
2   mail   in MX 5 mail.example.com.
```

The second part of the **MX** record, simply identifies the record as an MX record. This should not point to a FQDN with a **cname** clause[12], but directly to a **A** record.

The third part of the record is the server priority. The MTA will attempt to deliver the email to the *lowest numbered* server. This server could be in the domain or not as dictated by the needs of the organization. If multiple servers have the same priority, the email will be load–balanced by sending to the servers in a round–robin fashion. Should an attempt to hand off the email to a server fail, the next higher numbered server will be attempted and so on until the email is handed to a server. If no viable server can be found, the email fails and a message is sent back that the email failed.

The final part of the MX record is a FQDN which should appear later in the zone file in an **A** record. If instead the FQDN points to a CNAME record, the email system may fail.

22.4.2 Advantages of Using an Email Alias

There are two main advantages to using an alias for an email server rather than the actual FQDN (these advantages apply to most servers advertised via DNS):

- This allows the administrator to easily move services from one server to another without impacting the users. The administrator can simply change the CNAME and automatically *move* the service to a different service. In the above example, the records could be changed to move the SMTP service from **mail.example. com** to **mail2.example.com** as below:

  ```
  1          in MX 0 mail.example.com.
  2   mail   in MX 5 mail.example.com.
  ```

  ```
  1          in MX 0 mail2.example.com.
  2   mail   in MX 5 mail2.example.com.
  ```

- Users, and processes, expect certain services to be found with common names in any domain. Using aliases allows the administrator to provide email as **user@example.com** instead of forcing **user@mail2.example.com** which is not the common practice on the Internet.

---

[12] Failing to do this may result in working email, non-working email, or email that seems to work/not work at random. This can be very difficult to find and correct.

22.5 Configuring SMTP and Sendmail

Sendmail is one of the oldest and most common MTA found on the Internet. While Sendmail is sometimes difficult to configure, for our purposes the configuration is relatively easy[13].

22.5.1 Pre-configuration Tasks

Before one can even attempt to configure any MTA, there are a few tasks that must be accomplished. First, and most important, the test–bed network must be fully functional including DNS. If all future MTA cannot be "pinged" by name, email will surely fail.

The second task is to change the hostname of the Pi running email to be a FQDN by running **sudo raspi-config** and changing the hostname to include the domain. For example, **dumbo** for a domain **circus.org** must be changed to **dumbo.circus.org** before **sendmail** is configured.

At this time, the **ping** command should be used to verify that all planned SMTP servers can contact each other by name.

22.5.2 Configure Sendmail for Email Exchange

Because many processes send email messages to **root** for warnings or errors, the default configuration for **sendmail** is set up for local host mail *only* which means that no sessions with other SMTP servers will be accepted and email cannot be exchanged. To allow sessions with other servers, the file **/etc/sendmail.cf** must be edited and a new Sendmail configuration generated[14].

Change the line

```
DAEMON_OPTIONS ('Family=inet,  Name=MTA-v4, Port=smtp, Addr
    =127.0.0.1') dnl
```

to

```
DAEMON_OPTIONS ('Family=inet,  Name=MTA-v4, Port=smtp) dnl
```

to allow sessions from any host. After you save the file, you *must* run the configuration utility to generate a new configuration file[15]. For our purposes, simply take all the defaults as below:

```
sudo sendmailconfig
```

---

[13] The configuration we will be using is *not* suitable for use on the Internet or even for a home network. This configuration will not attempt to control spam at all. You were warned!

[14] For some reason, the Sendmail package was changed to overwrite any direct configuration changes to the file **sendmail.mc**.

[15] There are other methods that require more work and thought to generate the configuration. Those methods are explained in the **/etc/mail/sendmail.mc** file comments.

22.5.3 Testing Sendmail

The most effective way to test the configuration of DNS and SMTP is to send an email from the command line. Replace the **<stuff>** below with whatever text you want. When all the text has been entered, press *Ctrl* and *d* at the same time to end the message. If all is well, **sendmail** will send your email to **pi@mail.example.com** or whatever user you wish to contact.

```
sendmail -v pi{@}mail.example.com
<stuff>
<Ctrl-d>
```

22.6 Postfix MTA

If you wish to install postfix instead of sendmail, you will need to re-run the configuration script **sudo dpkg-reconfigure postfix**. Remember, you can have as many SMTP servers in your domain as you want, but you should not attempt to run multiple SMTP servers on the same Pi. Before you start, check to insure that your master DNS zone file for your domain has an entry such as **group1.edu. A 10.1.0.3** or whatever is appropriate for your group.

In this example we will configure postfix for a domain **FakeISP.net** using **sudo dpkg-reconfigure postfix**.

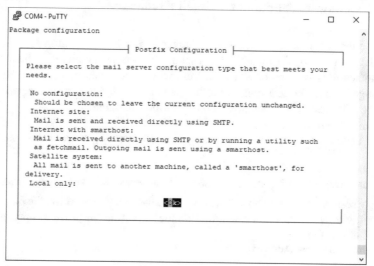

Fig. 22.2: Configuring **postfix**, Screen 1

1. The first screen, Figure 22.2, allows the administrator to choose between a number of standard uses for postfix. This screen is mainly documentation but you should read it the first time. Press **tab** to highlight **<Ok>** and then press **enter**.

The Pi should now display a screen like 22.3. Briefly, these choices have the following meanings:

- **No configuration** This choice will leave the current configuration intact and not allow changes. This might be useful in manually verifying the current configuration.
- **Internet Site** This is an email MTA that sends and accepts email from the Internet or an Intranet and processes it for local users. This is one of the most common types of postfix MTA and the one we will configure.
- **Internet site with smarthost** This is an email MTA that accepts email from the Internet, but sends outgoing email by forwarding it to another email host. This type of postfix MTA is designed for a site with many users and a high volume of email.
- **Satellite system** A satellite system interacts with the users, but all mail is relayed through another postfix MTA which sends and receives on the Internet.
- **Local only** Such a system does not send or receive email over a network, but only from local users (or processes) to be sent to other local users. This is useful if all the users are on the same closed system that does *not* connect to any other host.

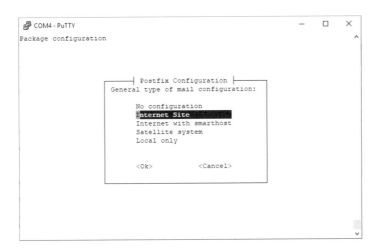

Fig. 22.3: Configuring **postfix**, Screen 2

2. The next step is to choose which type of MTA is best for your application. Use the keyboard arrows to move up or down until **Internet site** is highlighted, as in Figure 22.3 and then tab to **<Ok>** and press **enter**.

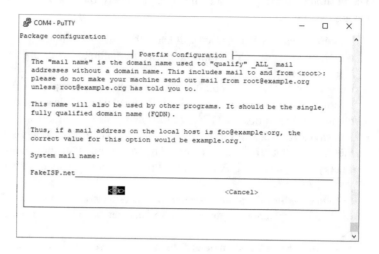

Fig. 22.4: Configuring **postfix**, Screen 3

3. Now you must enter your domain name. This is not the FQDN, but only the domain name in order to allow email addresses of the type **user1@group1.com**
instead of **user1@mail.group1.com**. The correct entry for the postfix
SMTP server **mail.FakeISP.net** is shown in Figure 22.4.

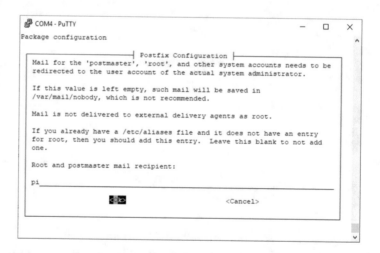

Fig. 22.5: Configuring **postfix**, Screen 4

4. By convention [119], all domains *must* support a standard email address of
postmaster to act as a point of contact and a dead letter address. Like other
MTAs, postfix will send any undelivered email (with the correct domain name)

to **postmaster** if the problem is that the user name is not defined and therefore has no mail box. Some system messages that are sent to the user **root** will also be forwarded to **postmaster** so this mail box needs to be checked regularly[16]

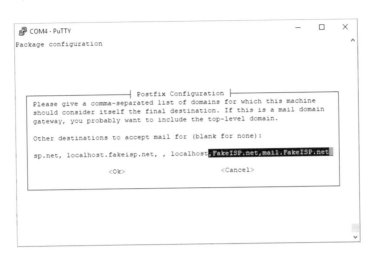

Fig. 22.6: Configuring **postfix**, Screen 5

5. A postfix MTA is often a terminal server, receives, and handles email for users using more than one form of the domain name. For example, **mail.example.org** may need to receive mail as **user@example.org** as well as **user@mail.example.org**. On this screen, see Figure 22.6, the administrator can add additional versions of the domain name (comma delimited)[17]. Add any versions of the domain part of an email address that you feel you need. All the versions listed, except for **localhost** and "blank", *must* have the proper DNS zone file entries.

---

[16] It is not uncommon for postmaster to also receive inquiries about the domain because this is the one mail box that should always exist. Care must be taken with inquires such as "Please reply with the current email address for my good friend John Doe" as these can be phishing attacks. Such messages should be forwarded to "John Doe" and he should make the decision to contact the sender or not. This is a good thing as he is most likely to know if this is a legitimate inquiry or not. This should be a standard procedure for all domains and the users should be educated as to how to deal with these inquiries and the dangers they can present.

[17] Apparently it does not matter if the same form is entered more than once.

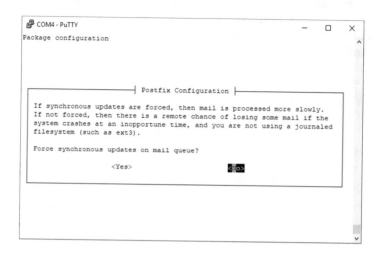

Fig. 22.7: Configuring **postfix**, Screen 6

6. Forcing synchronous updates for the mail queues causes email to be processed slower by postfix, but also helps prevent the loss of email in the event of a system crash, see Figure 22.7. If this server handles important email, this should be changed to **yes**. For less important servers, this can be set to **no** with the understanding that some email could be lost in the event of a hardware, service process, or network failure. Unless an extremely large volume of email is expected, this should be set to **yes**.

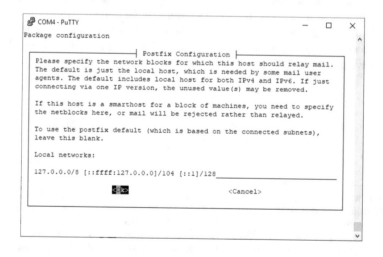

Fig. 22.8: Configuring **postfix**, Screen 7

7. Postfix is designed to facilitate using multiple physical devices to handle large email systems. If this is the case, this screen needs to be configured with the networks for which this Pi is the **smarthost**. This screen can be left as it is to allow Postfix to use all locally connected interfaces by default.

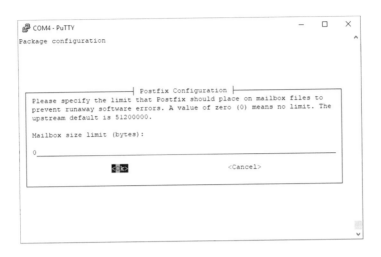

Fig. 22.9: Configuring **postfix**, Screen 8

8. Postfix is capable of enforcing quotas on all mail users. If a non–zero value is entered on this screen, the value is taken to be the maximum mailbox size in bytes. For testing, it is best to leave this blank.

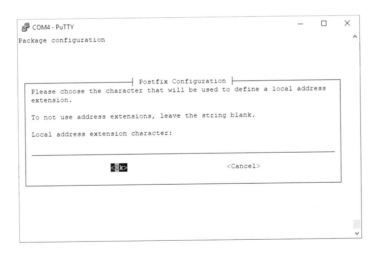

Fig. 22.10: Configuring **postfix**, Screen 9

9. There is no need to use address extensions, so press <enter>.

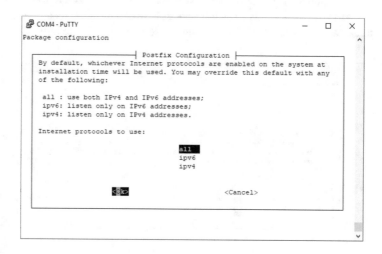

Fig. 22.11: Configuring **postfix**, Screen 10

10. Postfix can use either IPv4 or IPv6. Unless there is some reason to eliminate one of the protocols, choose **all** to use both versions of IP.

As soon as the last screen is finished, **postfix** will automatically run a script to reconfigure itself with the desired values.

```
setting synchronous mail queue updates: false
Adding sqlite map entry to /etc/postfix/dynamicmaps.cf
changing /etc/mailname to FakeISP.net
setting myorigin
setting destinations: router5-1.fakeisp.net, router5-1.fakeisp
    .net,localhost.fakeisp.net, , localhost,FakeISP.net,mail.
    FakeISP.net
setting relayhost:
setting mynetworks: 127.0.0.0/8 [::ffff:127.0.0.0]/104
    [::1]/128
setting mailbox_size_limit: 0
setting recipient_delimiter:
setting inet_interfaces: all
setting inet_protocols: all
WARNING: /etc/aliases exists, but does not have a root alias.

Postfix (main.cf) is now set up with the changes above.  If
    you need to make
changes, edit /etc/postfix/main.cf (and others) as needed.  To
    view Postfix
configuration values, see postconf(1).

After modifying main.cf, be sure to run 'service postfix
    reload'.

Running newaliases
```

After postfix has been reconfigured, the configuration files *must* be reloaded via either **sudo service postfix reload, sudo service postfix restart**, or by rebooting the Pi. If you check the status of postfix, you should see a screen similar to the one below[18].

```
postfix.service - Postfix Mail Transport Agent
Loaded: loaded (/lib/systemd/system/postfix.service; enabled
    ; vendor preset:
Active: active (exited) since Sun 2019-04-28 12:45:56 EDT;
    19s ago
Process: 4827 ExecReload=/bin/true (code=exited, status=0/
    SUCCESS)
Process: 4799 ExecStart=/bin/true (code=exited, status=0/
    SUCCESS)
Main PID: 4799 (code=exited, status=0/SUCCESS)

Apr 28 12:45:56 mail.FakeISP.net systemd[1]: Starting
    Postfix Mail Transport Age
Apr 28 12:45:56 mail.FakeISP.net systemd[1]: Started Postfix
    Mail Transport Agen
Apr 28 12:46:10 mail.FakeISP.net systemd[1]: Reloading
    Postfix Mail Transport Ag
Apr 28 12:46:10 mail.FakeISP.net systemd[1]: Reloaded
    Postfix Mail Transport Age
```

22.7 Client Support Services

To this point, we have been concerned with server–to–server email or SMTP, but if the users cannot send/receive email this is all pointless. Users interact with email via services which fall into two classes which we will call local services and client services. Local services are used by users who are logged into Linux machines that are running server–to–server email. For machines that are not constantly part of the email network, such as laptops or desktop machines, there are two very popular client services available. Which client service is chosen depends upon whether the user must log into the SMTP server or not and whether messages are to be available from multiple client machines.

22.8 Alpine

Before discussing client email services, it is useful to create a number of user accounts on your Raspberry Pi SMTP server using the built–in Linux script, **adduser**. For example, to add "Jane D. Student" with a sign–on user name of **jstudent** you would see:

```
pi@mail:~ sudo adduser jstudent
Adding user `jstudent' ...
Adding new group `jstudent' (1003) ...
Adding new user `jstudent' (1003) with group `jstudent' ...
Creating home directory `/home/jstudent' ...
Copying files from `/etc/skel' ...
Enter new UNIX password:
```

---

[18] For some reason text that should wrap around on the Pi console is not displayed properly. Fortunately, this is not usually a problem.

```
Retype new UNIX password:
passwd: password updated successfully
Changing the user information for jstudent
Enter the new value, or press ENTER for the default
Full Name []: Jane D. Student
Room Number []: 123
Work Phone []: (999)999-9999
Home Phone []: (123)555-1212
Other []: Majoring in Computer Science
Is the information correct? [Y/n] Y
```

While there are a number of options when adding a user, for examining email the
defaults will work[19].

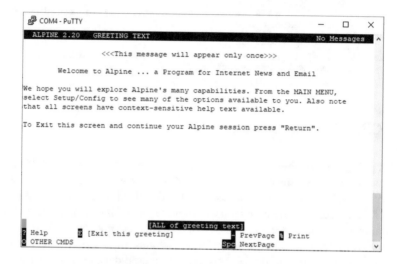

Fig. 22.12: Alpine Welcome Screen

The first time a user starts Alpine with the command **Alpine**, Alpine will create
mail directories and then display a welcome screen, as in Figure 22.12. Alpine,
unlike some menu–driven software, is very intuitive and easy to use. Commands are
entered by simply pressing the letter key unless they are preceded by the ^ character,
such as **^E** to exit, in which case the control key and character are pressed at the
same time. Alpine will also highlight the expected action for each screen in which
case you can press the enter key to take that action.

The main Alpine screen is displayed each time a user starts Alpine or presses **M**
as a command selection. From this screen a user can compose a new email message,
view messages in the current folder, list the mail folders, update the address book
(contacts), configure Alpine options, or quit Alpine and exit the program as shown
in Figure 22.13.

---

[19] In practice, it is recommended that a semi–standard password be used, such as the student
number, and users be forced to change their password the first time they sign in.

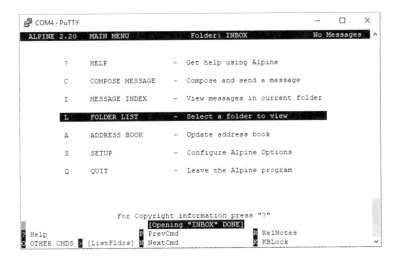

Fig. 22.13: Alpine Main Screen

Alpine will configure all options correctly upon the first execution, but a user might want to change how their name is displayed or add a signature by going to setup from the main screen.

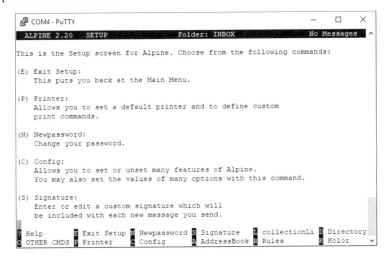

Fig. 22.14: Alpine Setup and Configuration Screen

22.8.1 Sending Email From Alpine

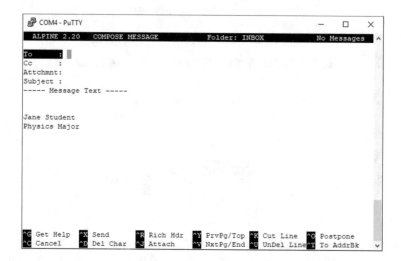

Fig. 22.15: Alpine Compose (Send) Email Screen

Email is sent from the Compose menu, see Figure 22.15 in Alpine which can be reached from any screen by pressing "c". This screen is intuitive and needs little explanation[20]. All email must have a destination in the "To:" field and should have a subject as some clients may interpret a blank subject line as a sign of spam. When the email has been composed, it is sent by pressing "control+x" as noted at the bottom of the menu. When email is sent a copy is placed in the user's **sent-mail** folder and the email is passed to the SMTP server for processing. Alpine does *not* actually send the email as that is done by either sendmail or postfix.

---

[20] Notice that the signature create by Jane Student is placed at the bottom of the text as plain text.

22.8.2 Reading Email From Alpine

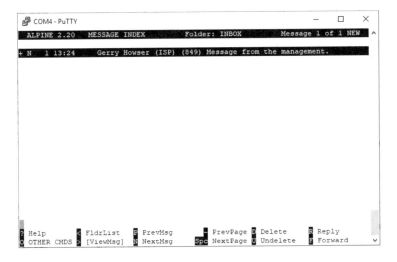

Fig. 22.16: Alpine INBOX with One Message

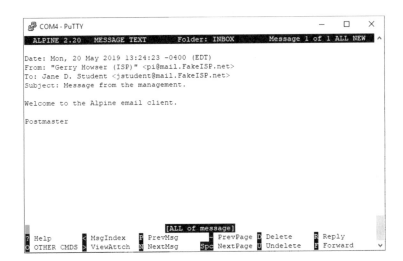

Fig. 22.17: Reading an Email with Alpine

22.9 POP3 Services

Most email users do not have accounts on email servers, so there must be some other interface to allow normal users to use email without opening large numbers of accounts on servers which presents a severe security risk. One such solution is Post Office Protocol or POP3[21]. When email is received for a user, the SMTP server and MTA store the email in the same manner it would be stored for a user with an account on the server. If no such account exists, the email is typically forwarded to the account **postmaster** to be either deleted or forwarded to the correct user. It is common for this to be done without notifying the sender that the user does not exist although some servers might be configured to send notifications.

When a user starts an email client, such as Thunderbird [50] or Pegasus email [35], the client immediately contacts the server and logs on. Email is then transferred from the server to the client device where it is held until the user opens it for reading. Typically the email is deleted from the server after it is downloaded to the client machine. This makes POP3 ideal for someone who typically accesses their email from a single machine such as an office desktop for a work related account. This also frees up space on the SMTP server. Because of this, POP3 can be thought of as a "store and forward" service.

While deleting the email from the server to free up space on the server might be one of POP3's greatest advantages, it became obvious early on that this is a severe disadvantage if the user checks email from a number of different client devices. Once the email is downloaded to the client device and removed from the server, it cannot be accessed from a different client device. While there are configurations that allow a user to work around this disadvantage, there is a better solution for users that access their email from a number of different machines and that is IMAP[22].

22.10 IMAP Services

For users that access email using a number of different client machines, a different service was developed called Internet Message Access Protocol or IMAP. For sending email, IMAP works the same as POP3 but the philosophy of how incoming email is handled is entirely different. Email is still downloaded to the client machine, but it is not deleted from the server until the user manually deletes it via the email client. When a user does delete email it is deleted from the client device's file system and the client then sends a message to the server to delete the email from the server as well. This means the user has access to all their email from any client that supports IMAP. In the early days of email some servers only supported POP3 or IMAP but not both. Today virtually all SMTP servers and email clients support

---

[21] Post Office Protocol
[22] Internet Message Access Protocol

both. Some users see IMAP as a potential security and privacy risk, but for most users IMAP is the better choice of protocols.

22.11 Web–based Services

There are too many webmail sites with too many different protocols to discuss here, but there are some hybrid sites that allow both IMAP and web access. Gmail, Hotmail, and Yahoo are three of the most popular but there may easily be hundreds of sites that allow a user to access email via a web–based client or an IMAP client. There may even be some proprietary email services that use other methods to deal with reading and storing email.

22.12 Automatic Email Lists

In the early days of email, it became obvious that there was a need for some automated way to handle email lists. One of the best email list handlers was, and most likely still is, Listserv. While it should be possible to run the free version of Listserv on Raspbian, there is not a package to directly support installation. This creates a problem for the scope of this text.

However, there are other email list managers that are both free and have a package for Raspbian such as Mailman Email List Service[23]. Mailman Email List Service provides all the expected functionality of an automatic email list manager which includes:

- Users can add/remove themselves from supported lists without assistance by sending an email or via a web interface.
- Users can manage their preferences via email or a web interface.
- Through-the-web list creation and removal (with automatic support depending on the MTA).
- Multi-lingual support: list web pages and email notices can be in any of nearly two dozen supported languages, configurable per-site, per-list, and per-user.
- "Real name" support for members.
- Support for personalized deliveries and VERP[24] such as message delivery for foolproof bounce detection. An additional header in the email "envelope" contains the *exact* email used to subscribe. Many of us manage to acquire a large number of email addresses and tend to forget which one was used for which email list. If all these are automagically collected into a single inbox, things can get dicey trying to manage all those lists.
- Emergency moderation.

---

[23] http://www.list.org/features.html
[24] Variable Envelope Return Paths

- Through the web membership management to make administering the members of a list easier for those who are web–oriented.
- Moderated newsgroup support.
- Flexible moderation and privacy controls.
- Subscription invitations[25].
- Auto-response controls to help control the proliferation of "I am out of the office until Monday." messages.
- User controllable delivery options.
- Urgent: header support (bypasses digests to reach all users immediately).
- Users can specify a preferred address to use by default.
- Flexible architecture for integration with your own site.

First check for the presence of **mailman** files: **ls -l /etc/mailman/**. If there are no files here, the Pi must be connected to the Internet and the package installed as in Section 8.1. Once it is known that **mailman** is properly installed and email is working, **mailman** can be configured to interact with the correct MTA [315].

22.12.1 Mailman With Apache

In order to use the web–interface, Apache must be aware of Mailman Email List Service. The proper configuration file is already in the **/etc/mailman/** directory and can be copied directly to the appropriate Apache directory and the new virtual host enabled.

```
sudo cp -p /etc/mailman/apache.conf /etc/apache2/sites-
    available/mailman.conf
sudo a2ensite mailman.conf
sudo a2enmod cgid
sudo systemctl restart apache2
```

22.12.2 Mailman With Posfix

There are a few steps that need to be done to configure postfix and Mailman Email List Service to work together. These steps are fairly simple if you use "cut and paste" whenever possible.

Step 1:

Activate the correct MTA by editing the configuration file **/etc/mailman/mm_cfg. py** to uncomment, or enter, the line **MTA = 'Postfix'** being very careful to insure

---

[25] Is this a good idea? In my opinion, this could be easily misused during an election year . . . and they are all election years in the US.

there are no spaces or tabs at the beginning of the line. It might be at the very end of the file.

```
 1  #-------------------------------------------------------
 2  # Uncomment if you use Postfix virtual domains (but not
 3  # postfix-to-mailman.py), but be sure to see
 4  # /usr/share/doc/mailman/README.Debian first.
 5  MTA='Postfix'
 6
 7  #-------------------------------------------------------
 8  # Uncomment if you want to filter mail with SpamAssassin. For
 9  # more information please visit this website:
10  # http://www.jamesh.id.au/articles/mailman-spamassassin/
11  # GLOBAL_PIPELINE.insert(1, 'SpamAssassin')
12
13  # Note - if you're looking for something that is imported from
          mm_cfg, but you
14  # didn't find it above, it's probably in /usr/lib/mailman/
          Mailman/Defaults.py.
```

Step 2:

Now the aliases used by **mailman** must be generated.

```
sudo /usr/lib/mailman/bin/genaliases
```

Step 3:

Now use the **postconf** command to add the necessary entries to **/etc/postfix/**
/etc/postfix/main.cf[26]:

```
sudo postconf -e 'relay_domains = lists.example.com'
sudo postconf -e 'transport_maps = hash:/etc/postfix/transport
     '
sudo postconf -e 'mailman_destination_recipient_limit = 1'
sudo postconf -e 'alias_maps = hash:/etc/aliases, hash:/var/
     lib/mailman/data/aliases'
```

Step 4:

In /etc/postfix/master.cf double check that you have the following transport (again, towards the end of a long file):

```
mailman    unix  -     n    n    -    -    pipe
   flags=FR user=list argv=/usr/lib/mailman/bin/postfix-to-
       mailman.py
   ${nexthop} ${user}
```

This calls the **postfix-to-mailman.py** script when a mail is delivered to a list.

---

[26] Like many critical Linux commands, these do not return *any* feedback if all is well. In my opinion, a command should always give feedback even if it is nothing more than "done".

Step 5:

Associate the domain **lists.example.com** to the mailman transport with the
transport map. Remember to replace **example.com** with your group's domain.
Edit the file **/etc/postfix/transport**:

```
lists.example.com        mailman:
```

Step 6:

Now have Postfix build the transport map by entering the following from a terminal
prompt:

```
sudo postmap -v /etc/postfix/transport
postmap: name_mask: all
postmap: inet_addr_local: configured 4 IPv4 addresses
postmap: inet_addr_local: configured 7 IPv6 addresses
postmap: open hash /etc/postfix/transport
postmap: Compiled against Berkeley DB: 5.3.28?
postmap: Run-time linked against Berkeley DB: 5.3.28?
```

Step 7:

Then add Mailman Email List Service aliases in /etc/aliases

```
mailman:                 "|/var/lib/mailman/mail/mailman post
    mailman"
mailman-admin:           "|/var/lib/mailman/mail/mailman admin
    mailman"
mailman-bounces:         "|/var/lib/mailman/mail/mailman bounces
    mailman"
mailman-confirm:         "|/var/lib/mailman/mail/mailman confirm
    mailman"
mailman-join:            "|/var/lib/mailman/mail/mailman join
    mailman"
mailman-leave:           "|/var/lib/mailman/mail/mailman leave
    mailman"
mailman-owner:           "|/var/lib/mailman/mail/mailman owner
    mailman"
mailman-request:         "|/var/lib/mailman/mail/mailman request
    mailman"
mailman-subscribe:       "|/var/lib/mailman/mail/mailman
    subscribe mailman"
mailman-unsubscribe:     "|/var/lib/mailman/mail/mailman
    unsubscribe mailman"
```

Step 8:

This build (compilation and packaging) of Mailman Email List Service runs as user
list and it must have permission to read **/etc/aliases** and read and write
the file **/var/lib/mailman/data/aliases**. Give **list** ownership of these

files and run the utility **newaliases** to fix everything up in the **/etc/aliases** file:

```
sudo chown root:list /var/lib/mailman/data/aliases
sudo chown root:list /etc/aliases
sudo newaliases
```

Step 9:

Now all that is needed is to restart postfix with the command: **sudo systemctl restart postfix**

22.12.3 Mailman Command Line Interface

NOTE: all commands must be issued from the **/usr/lib/mailman/bin** directory, so start with **cd /usr/lib/mailman/bin**. This is a small subset of the available commands. For more information, see Site Administrator Documentation [34] at url **http://www.gnu.org/software/mailman/site.html**. For now, simply create a list for testing.

```
pi@router1-1:/usr/lib/mailman/bin$ sudo newlist
Enter the name of the list: tester
Enter the email of the person running the list: pi@highschool.
    edu
Initial tester password:
Hit enter to notify tester owner...

pi@router1-1:/usr/lib/mailman/bin$
```

Users can now use Alpine to join the list and use it.

Projects

22.1 Create a user and send email to the user **pi** using **Alpine**.

22.2 Create a second user and send email to the first user.

22.3 Send mail to **pi** at some other group's domain.

22.4 Using **dovecat**, set up your email server to provide IMAP or POP3 services. Using a PC or Linux box connected to your Group's network, attempt to use the PC to send/receive email on your local network.

> **Warning:** This will seriously mess up any existing IMAP or POP3 mail configuration on the PC, so be very sure you know what you are doing.

Exercises

22.1 Although this is beyond the scope of this text, why would you not want to bounce email to a non–existant user for your domain?

22.2 Suppose a user, "Jane Doe", has an email account on your network. What are some common user names you might expect for this user? What about aliases?

Further Reading

RFCs Directly Related to Simple Mail Transfer Protocol

Except in limited circumstances, modern SMTP should really be considered as ESMTP[27], especially when talking about RFCs, as most recent RFCs dealing with spam, security, or enhanced attachments really apply to ESMTP. For our purposes, the two terms are interchangeable.

MTA	Title
RFC 1894	An Extensible Message Format for Delivery Status Notifications [141]
RFC 2505	Anti-Spam Recommendations for SMTP MTAs [175]
RFC 5321	Simple Mail Transfer Protocol [258]
RFC 7672	SMTP Security via Opportunistic DNS-Based Authentication of Named Entities (DANE) Transport Layer Security (TLS) [283]
RFC 8460	SMTP TLS Reporting [297]
RFC 8461	SMTP MTA Strict Transport Security (MTA-STS) [298]

SMTP	Title
RFC 0773	Comments on NCP/TCP mail service transition strategy [62]
RFC 0821	Simple Mail Transfer Protocol [65]
RFC 0822	STANDARD FOR THE FORMAT OF ARPA INTERNET TEXT MESSAGES [66]
RFC 0876	Survey of SMTP implementations [68]
RFC 0974	Mail routing and the domain system [75]
RFC 1047	Duplicate messages and SMTP [78]
RFC 1425	SMTP Service Extensions [101]
RFC 1426	SMTP Service Extension for 8bit-MIMEtransport [102]
RFC 1427	SMTP Service Extension for Message Size Declaration [103]
RFC 1428	Transition of Internet Mail from Just-Send-8 to 8bit-SMTP/MIME [104]
RFC 1460	Post Office Protocol - Version 3 [105]
RFC 1648	Postmaster Convention for X.400 Operations [119]
RFC 1651	SMTP Service Extensions [120]
RFC 1652	SMTP Service Extension for 8bit-MIMEtransport [121]
RFC 1653	SMTP Service Extension for Message Size Declaration [122]
RFC 1725	Post Office Protocol - Version 3 [128]
RFC 1830	SMTP Service Extensions for Transmission of Large and Binary MIME Messages [132]
RFC 1845	SMTP Service Extension for Checkpoint/Restart [133]

---

[27] Enhanced Simple Mail Transfer Protocol

SMTP	Title
RFC 1846	SMTP 521 Reply Code [134]
RFC 1854	SMTP Service Extension for Command Pipelining [135]
RFC 1869	SMTP Service Extensions [136]
RFC 1870	SMTP Service Extension for Message Size Declaration [137]
RFC 1891	SMTP Service Extension for Delivery Status Notifications [140]
RFC 1939	Post Office Protocol - Version 3 [146]
RFC 1985	SMTP Service Extension for Remote Message Queue Starting [150]
RFC 2034	SMTP Service Extension for Returning Enhanced Error Codes [154]
RFC 2197	SMTP Service Extension for Command Pipelining [166]
RFC 2487	SMTP Service Extension for Secure SMTP over TLS [174]
RFC 2505	Anti-Spam Recommendations for SMTP MTAs [175]
RFC 2554	SMTP Service Extension for Authentication [178]
RFC 2635	DON'T SPEW A Set of Guidelines for Mass Unsolicited Mailings and Postings (spam*) [181]
RFC 2645	ON-DEMAND MAIL RELAY (ODMR) SMTP with Dynamic IP Addresses [182]
RFC 2821	Simple Mail Transfer Protocol [186]
RFC 3030	SMTP Service Extensions for Transmission of Large and Binary MIME Messages [194]
RFC 3461	Simple Mail Transfer Protocol (SMTP) Service Extension for Delivery Status Notifications (DSNs) [210]
RFC 3974	SMTP Operational Experience in Mixed IPv4/v6 Environments [228]
RFC 4141	SMTP and MIME Extensions for Content Conversion [231]
RFC 4408	Sender Policy Framework (SPF) for Authorizing Use of Domains in E-Mail, Version 1 [237]
RFC 4871	DomainKeys Identified Mail (DKIM) Signatures [226]
RFC 4954	SMTP Service Extension for Authentication [250]
RFC 5039	The Session Initiation Protocol (SIP) and Spam [255]
RFC 5068	Email Submission Operations: Access and Accountability Requirements [37]
RFC 5235	Sieve Email Filtering: Spamtest and Virustest Extensions [257]
RFC 5321	Simple Mail Transfer Protocol [258]
RFC 6430	Email Feedback Report Type Value: not-spam [271]
RFC 6647	Email Greylisting: An Applicability Statement for SMTP [273]
RFC 6858	Simplified POP and IMAP Downgrading for Internationalized Email [277]

SMTP	Title
RFC 7672	SMTP Security via Opportunistic DNS-Based Authentication of Named Entities (DANE) Transport Layer Security (TLS) [283]
RFC 8314	Cleartext Considered Obsolete: Use of Transport Layer Security (TLS) for Email Submission and Access [291]
RFC 8460	SMTP TLS Reporting [297]
RFC 8461	SMTP MTA Strict Transport Security (MTA-STS) [298]

ESMTP	Title
RFC 3030	SMTP Service Extensions for Transmission of Large and Binary MIME Messages [194]
RFC 4141	SMTP and MIME Extensions for Content Conversion [231]

RFCs Related to Client Services

POP	Title
RFC 1460	Post Office Protocol - Version 3
RFC 1725	Post Office Protocol - Version 3
RFC 1734	POP3 AUTHentication command
RFC 1939	Post Office Protocol - Version 3
RFC 1957	Some Observations on Implementations of the Post Office Protocol (POP3)
RFC 2095	IMAP/POP AUTHorize Extension for Simple Challenge/Response
RFC 2195	IMAP/POP AUTHorize Extension for Simple Challenge/Response
RFC 2384	POP URL Scheme
RFC 2449	POP3 Extension Mechanism
RFC 2595	Using TLS with IMAP, POP3 and ACAP
RFC 3206	The SYS and AUTH POP Response Codes
RFC 5034	The Post Office Protocol (POP3) Simple Authentication and Security Layer (SASL) Authentication Mechanism
RFC 5721	POP3 Support for UTF-8
RFC 6856	Post Office Protocol Version 3 (POP3) Support for UTF-8
RFC 6858	Simplified POP and IMAP Downgrading for Internationalized Email
RFC 7817	Updated Transport Layer Security (TLS) Server Identity Check Procedure for Email-Related Protocols

IMAP	Title
RFC 1730	Internet Message Access Protocol - Version 4
RFC 1731	IMAP4 Authentication Mechanisms
RFC 1732	IMAP4 Compatibility with IMAP2 and IMAP2bis
RFC 1733	Distributed Electronic Mail Models in IMAP4
RFC 2060	Internet Message Access Protocol - Version 4rev1
RFC 2061	IMAP4 Compatibility with IMAP2bis
RFC 2062	Internet Message Access Protocol - Obsolete Syntax
RFC 2086	IMAP4 ACL extension
RFC 2087	IMAP4 QUOTA extension
RFC 2088	IMAP4 non-synchronizing literals
RFC 2095	IMAP/POP AUTHorize Extension for Simple Challenge/Response
RFC 2177	IMAP4 IDLE command
RFC 2180	IMAP4 Multi-Accessed Mailbox Practice
RFC 2192	IMAP URL Scheme
RFC 2193	IMAP4 Mailbox Referrals

IMAP	Title
RFC 2195	IMAP/POP AUTHorize Extension for Simple Challenge/Response
RFC 2221	IMAP4 Login Referrals
RFC 2342	IMAP4 Namespace
RFC 2359	IMAP4 UIDPLUS extension
RFC 2595	Using TLS with IMAP, POP3 and ACAP
RFC 2683	IMAP4 Implementation Recommendations
RFC 2971	IMAP4 ID extension
RFC 3348	The Internet Message Action Protocol (IMAP4) Child Mailbox Extension
RFC 3501	INTERNET MESSAGE ACCESS PROTOCOL - VERSION 4rev1
RFC 3502	Internet Message Access Protocol (IMAP) - MULTIAPPEND Extension
RFC 3503	Message Disposition Notification (MDN) profile for Internet Message Access Protocol (IMAP)
RFC 3516	IMAP4 Binary Content Extension
RFC 3691	Internet Message Access Protocol (IMAP) UNSELECT command
RFC 4314	IMAP4 Access Control List (ACL) Extension
RFC 4315	Internet Message Access Protocol (IMAP) - UIDPLUS extension
RFC 4466	Collected Extensions to IMAP4 ABNF
RFC 4467	Internet Message Access Protocol (IMAP) - URLAUTH Extension
RFC 4469	Internet Message Access Protocol (IMAP) CATENATE Extension
RFC 4549	Synchronization Operations for Disconnected IMAP4 Clients
RFC 4551	IMAP Extension for Conditional STORE Operation or Quick Flag Changes Resynchronization
RFC 4731	IMAP4 Extension to SEARCH Command for Controlling What Kind of Information Is Returned
RFC 4959	IMAP Extension for Simple Authentication and Security Layer (SASL) Initial Client Response
RFC 4978	The IMAP COMPRESS Extension
RFC 5032	WITHIN Search Extension to the IMAP Protocol
RFC 5092	IMAP URL Scheme
RFC 5161	The IMAP ENABLE Extension
RFC 5162	IMAP4 Extensions for Quick Mailbox Resynchronization
RFC 5182	IMAP Extension for Referencing the Last SEARCH Result
RFC 5232	Sieve Email Filtering: Imap4flags Extension
RFC 5255	Internet Message Access Protocol Internationalization
RFC 5256	Internet Message Access Protocol - SORT and THREAD Extensions

IMAP	Title
RFC 5257	Internet Message Access Protocol - ANNOTATE Extension
RFC 5258	Internet Message Access Protocol version 4 - LIST Command Extensions
RFC 5267	Contexts for IMAP4
RFC 5464	The IMAP METADATA Extension
RFC 5465	The IMAP NOTIFY Extension
RFC 5466	IMAP4 Extension for Named Searches (Filters)
RFC 5490	The Sieve Mail-Filtering Language – Extensions for Checking Mailbox Status and Accessing Mailbox Metadata
RFC 5530	IMAP Response Codes
RFC 5593	Internet Message Access Protocol (IMAP) - URL Access Identifier Extension
RFC 5738	IMAP Support for UTF-8
RFC 5788	IMAP4 Keyword Registry
RFC 5819	IMAP4 Extension for Returning STATUS Information in Extended LIST
RFC 5957	Display-Based Address Sorting for the IMAP4 SORT Extension
RFC 6154	IMAP LIST Extension for Special-Use Mailboxes
RFC 6203	IMAP4 Extension for Fuzzy Search
RFC 6237	IMAP4 Multimailbox SEARCH Extension
RFC 6785	Support for Internet Message Access Protocol (IMAP) Events in Sieve
RFC 6851	Internet Message Access Protocol (IMAP) - MOVE Extension
RFC 6855	IMAP Support for UTF-8
RFC 6857	Post-Delivery Message Downgrading for Internationalized Email Messages
RFC 6858	Simplified POP and IMAP Downgrading for Internationalized Email
RFC 7017	IMAP Access to IETF Email List Archives
RFC 7162	IMAP Extensions: Quick Flag Changes Resynchronization (CONDSTORE) and Quick Mailbox Resynchronization (QRESYNC)
RFC 7377	IMAP4 Multimailbox SEARCH Extension
RFC 7817	Updated Transport Layer Security (TLS) Server Identity Check Procedure for Email-Related Protocols
RFC 7888	IMAP4 Non-synchronizing Literals
RFC 7889	The IMAP APPENDLIMIT Extension
RFC 8437	IMAP UNAUTHENTICATE Extension for Connection Reuse
RFC 8438	IMAP Extension for STATUS=SIZE
RFC 8440	IMAP4 Extension for Returning MYRIGHTS Information in Extended LIST
RFC 8457	IMAP "$Important" Keyword and "\Important" Special-Use Attribute

IMAP	Title
RFC 8474	IMAP Extension for Object Identifiers
RFC 8508	IMAP REPLACE Extension
RFC 8514	Internet Message Access Protocol (IMAP) - SAVEDATE Extension

Other RFCs Related to SMTP/Email

For a list of other RFCs related to SMTP/Email but not closely referenced in this chapter, please see Appendix B–SMTP, B–ESMTP, B–IMAP, B–POP, and B–MIME.

Chapter 23
Other Services

Overview

There are many useful services provided on the Internet: too many to cover in one book. Some of these are critical and some are just nice to have when they are needed. This chapter covers a few that can be implemented on the Raspberry Pi.

23.1 Network Address Translation

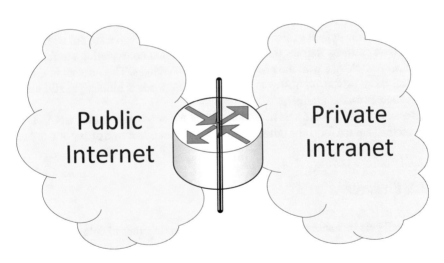

Fig. 23.1: NAT (Network Address Translation)

In the late 1990's, it became apparent that the Internet was growing too fast and would soon run out of usable IPv4 addresses. While many people were scrambling

© Springer Nature Switzerland AG 2020
G. Howser, *Computer Networks and the Internet*,
https://doi.org/10.1007/978-3-030-34496-2_23

to perfect a new addressing scheme and revamping all of the IPv4 protocols, it was obvious that the "IP next generation" known now as IPv6 could not be launched in time to prevent a meltdown. Where could an almost unlimited supply of IPv4 addresses be found?

Currently all public networks connect to the Internet via some form of router that segregates traffic on either side yet forwards packets that need to cross from one network to the other. In reality, a router forms a barrier between two networks with rules and protocols to determine what traffic can cross that barrier and what traffic cannot. We know how to do this very well.

Private IP addresses cannot be accessed from the public Internet nor can these addresses access the Internet so it does not matter how many devices have an address of 192.168.1.1/24 because they will never be accessible to each other. As a matter of fact, the beauty of private addressing is that absolute privacy. Yet a SOHO network can access any site on the public Internet while remaining completely hidden and private.

If a way could be found to allow a router to connect a private network to a public network while preserving all the benefits of private addresses, then as many private networks as needed could access the public Internet while preserving their privacy and allowing private addresses to be reused millions of times. The solution is NAT.

23.1.1 NAT Explained

In essence, NAT is very simple. NAT devices act as a proxy for all of the private IPv4 addresses "behind" the barrier between private and public networks. When a packet is received for a public address such as a website, the NAT router, which is also acting as the default gateway for the network, sends the packet out with its own public IPv4 address. Replies that come back from that conversation are forwarded to the original device with the proper private IPv4 address. The Internet is shielded from all those private addresses and the private network is hidden yet still able to use resources on the Internet.

Because a NAT device acts as a choke–point for out–bound traffic, a NAT box is an excellent choice for some other critical services such as a web proxy or a firewall.

23.2 Firewall

A firewall gets its name from good building design. In order to delay the spread of fires to allow people to escape, most large buildings are designed as smaller areas that are walled off from the rest of the building by fire resistant walls. We say there is a firewall between these areas. Unfortunately, there are services such as water, electrical services, and network cabling that must be allowed to pierce firewalls.

Every time a firewall is pierced, its effectiveness is compromised. A perfect firewall would not even have doors.

Network firewalls, such as the one built into Quagga, act the same way. A router acting as a firewall does not allow any traffic on one side to ever get over to the other side. The two networks the firewall connects cannot exchange packets and both are absolutely protected. Of course, if this is the case there is no need to even connect the two networks so some traffic must pierce the firewall.

Any firewall can be configured to allow traffic to cross based upon a set of rules. Traffic can be restricted based upon any number of characteristics including source/destination network, source/destination MAC address, socket, or traffic flow.

Most firewalls manage traffic by an access list. If the traffic matches an entry in the list, it is allowed or denied based upon the entry. In my opinion, Cisco Systems has it right in their training. First block *all* traffic and then open up the allowed traffic one rule at a time. It is equally possible to allow all traffic and then close off traffic that should not cross the firewall. This is a dangerous thing to do as the administrator may easily miss a possibly malicious set of traffic. It is far better to have a user call to ask why their traffic is blocked than to explain that you never thought anyone would attack the network in such an odd way. For example, who would ever expect someone from outside your network to "spoof" an address that should only occur *inside* your network? It happens.

23.3 The Telnet Service

Many would question why Telnet is still useful when **ssh** is more common and much more secure. The answer is quite simple: You can use Telnet to explore the messaging between clients and servers. You can even use Telnet to send spam email[1]

For example, suppose you wish to understand the conversation between a web browser and a web server. If you have **httpd** running on your Raspberry Pi, you could telnet to 127.0.0.180 and contact the web server as if you were a browser. You would see a response such as:

```
pi@router1-1:~$ telnet 127.0.0.1 80
Trying 127.0.0.1...
Connected to 127.0.0.1.
Escape character is '^]'.
HEAD
HTTP/1.1 400 Bad Request
Date: Wed, 21 Aug 2019 20:43:50 GMT
Server: Apache/2.4.38 (Raspbian)
Connection: close
Content-Type: text/html; charset=iso-8859-1

Connection closed by foreign host.
pi@router1-1:~$
```

---

[1] Do not try this at home. This technique is easily blocked and may be illegal where you are located. Use these techniques only with prior permission or on your own network.

If you know the correct messages to type, it is very easy to send spam email using this technique if the email administrator has not properly secured the server. Telnet is useful for troubleshooting some protocols.

Further Reading

The RFC below provide further information about NAT. This is a fairly exhaustive list and most RFC are typically dense and hard to read. Normally RFC are most useful when writing a process to implement a specific protocol.

RFCs Directly Related to NAT and Telnet

NAT	Title
RFC 3519	Mobile IP Traversal of Network Address Translation (NAT) Devices [213]
RFC 5902	IAB Thoughts on IPv6 Network Address Translation [263]
RFC 7393	Using the Port Control Protocol (PCP) to Update Dynamic DNS [279]
RFC 7857	Updates to Network Address Translation (NAT) Behavioral Requirements [286]

TELNET	Title
RFC 0097	First Cut at a Proposed Telnet Protocol [59]
RFC 0137	Telnet Protocol - a proposed document [60]
RFC 0779	Telnet send-location option [63]

Other RFCs Related to NAT and Telnet

For a list of other RFCs related to NAT or Telnet, please see Appendix B.

Appendix A
Solutions to Selected Exercises

Overview

There are a few things you should keep in mind:

1. Networking problems sometimes have multiple correct solutions. Given is *one* solution and there may be others that are equally correct.
2. If there is a mistake in the solution, please check the website for errata, double–check your work and discuss it with a local expert.
3. If you have need of an additional solution and the steps above have not helped (or you can't find a solution anywhere) you may contact me for a solution as long as you keep in mind I may not respond quickly or at all. Do not take this as a rejection, but rather as a sign I am having a delay getting to your request. I will eventually respond, I hope.

This appendix will give the solutions to some exercises in this book. At some point there may be additional solutions on line at https://www.springer.com/us/book/9783030344955 and www.gerryhowser.com/book/9783030344955/solutions, both of which are under construction at the time of this writing.

© Springer Nature Switzerland AG 2020
G. Howser, *Computer Networks and the Internet*,
https://doi.org/10.1007/978-3-030-34496-2

Appendix B
Request For Comments

The RFCs below are relevant to this book. This is *not* an exhaustive list and most RFCs are typically dense and hard to read. Normally RFCs are most useful when writing a process to implement a specific protocol.

AppleTalk

AppleTalk	Title
RFC 1378	The PPP AppleTalk Control Protocol (ATCP)
RFC 1504	Appletalk Update-Based Routing Protocol: Enhanced Appletalk Routing
RFC 6281	Understanding Apple's Back to My Mac (BTMM) Service
RFC 6760	Requirements for a Protocol to Replace the AppleTalk Name Binding Protocol (NBP)

ARP

ARP	Title
RFC 1027	Using ARP to implement transparent subnet gateways
RFC 1293	Inverse Address Resolution Protocol
RFC 1390	Transmission of IP and ARP over FDDI Networks
RFC 1577	Classical IP and ARP over ATM
RFC 1931	Dynamic RARP Extensions for Automatic Network Address Acquisition
RFC 5494	IANA Allocation Guidelines for the Address Resolution Protocol (ARP)

© Springer Nature Switzerland AG 2020
G. Howser, *Computer Networks and the Internet*,
https://doi.org/10.1007/978-3-030-34496-2

ATM

ATM	Title
RFC 1483	Multiprotocol Encapsulation over ATM Adaptation Layer 5
RFC 1577	Classical IP and ARP over ATM
RFC 1680	IPng Support for ATM Services
RFC 5828	Generalized Multiprotocol Label Switching (GMPLS) Ethernet Label Switching Architecture and Framework

Babel

Babel	Title
RFC 6126	The Babel Routing Protocol
RFC 7298	Babel Hashed Message Authentication Code (HMAC) Cryptographic Authentication
RFC 7557	Extension Mechanism for the Babel Routing Protocol

BGP

BGP	Title
RFC 1266	Experience with the BGP Protocol
RFC 1267	Border Gateway Protocol 3 (BGP-3)
RFC 1322	A Unified Approach to Inter-Domain Routing
RFC 1364	BGP OSPF Interaction
RFC 1397	Default Route Advertisement In BGP2 and BGP3 Version of The Border Gateway Protocol
RFC 1403	BGP OSPF Interaction
RFC 1520	Exchanging Routing Information Across Provider Boundaries in the CIDR Environment
RFC 2842	Capabilities Advertisement with BGP-4
RFC 2858	Multiprotocol Extensions for BGP-4
RFC 4893	BGP Support for Four-octet AS Number Space
RFC 5396	Textual Representation of Autonomous System (AS) Numbers
RFC 6811	BGP Prefix Origin Validation
RFC 8388	Usage and Applicability of BGP MPLS-Based Ethernet VPN
RFC 8416	Simplified Local Internet Number Resource Management with the RPKI (SLURM)
RFC 8503	BGP/MPLS Layer 3 VPN Multicast Management Information Base
RFC 8571	BGP - Link State (BGP-LS) Advertisement of IGP Traffic Engineering Performance Metric Extensions

BOOTP

BOOTP	Title
RFC 0906	Bootstrap loading using TFTP
RFC 0951	BOOTSTRAP PROTOCOL (BOOTP)
RFC 1395	BOOTP Vendor Information Extensions
RFC 1497	BOOTP Vendor Information Extensions
RFC 1532	Clarifications and Extensions for the Bootstrap Protocol
RFC 1533	DHCP Options and BOOTP Vendor Extensions
RFC 1542	Clarifications and Extensions for the Bootstrap Protocol
RFC 2132	DHCP Options and BOOTP Vendor Extensions

CIDR

CIDR	Title
RFC 1517	Applicability Statement for the Implementation of Classless Inter-Domain Routing (CIDR)
RFC 1518	An Architecture for IP Address Allocation with CIDR
RFC 1519	Classless Inter-Domain Routing (CIDR): an Address Assignment and Aggregation Strategy
RFC 1520	Exchanging Routing Information Across Provider Boundaries in the CIDR Environment
RFC 1812	Requirements for IP Version 4 Routers
RFC 1817	CIDR and Classful Routing
RFC 1917	An Appeal to the Internet Community to Return Unused IP Networks (Prefixes) to the IANA
RFC 2036	Observations on the use of Components of the Class A Address Space within the Internet
RFC 4632	Classless Inter-domain Routing (CIDR): The Internet Address Assignment and Aggregation Plan
RFC 7608	IPv6 Prefix Length Recommendation for Forwarding

DDNS

DDNS	Title
RFC 4703	Resolution of Fully Qualified Domain Name (FQDN) Conflicts among Dynamic Host Configuration Protocol (DHCP) Clients
RFC 7393	Using the Port Control Protocol (PCP) to Update Dynamic DNS

Default route

default	Title
RFC 1397	Default Route Advertisement In BGP2 and BGP3 Version of The Border Gateway Protocol
RFC 2461	Neighbor Discovery for IP Version 6 (IPv6)
RFC 4191	Default Router Preferences and More-Specific Routes

DHCP

DHCP	Title
RFC 0951	BOOTSTRAP PROTOCOL (BOOTP)
RFC 1497	BOOTP Vendor Information Extensions
RFC 1532	Clarifications and Extensions for the Bootstrap Protocol
RFC 1533	DHCP Options and BOOTP Vendor Extensions
RFC 1541	Dynamic Host Configuration Protocol
RFC 1542	Clarifications and Extensions for the Bootstrap Protocol
RFC 2131	Dynamic Host Configuration Protocol
RFC 2132	DHCP Options and BOOTP Vendor Extensions
RFC 3046	DHCP Relay Agent Information Option
RFC 3315	Dynamic Host Configuration Protocol for IPv6 (DHCPv6)
RFC 3396	Encoding Long Options in the Dynamic Host Configuration Protocol (DHCPv4)
RFC 3646	DNS Configuration options for Dynamic Host Configuration Protocol for IPv6 (DHCPv6)
RFC 4075	Simple Network Time Protocol (SNTP) Configuration Option for DHCPv6
RFC 4361	Node-specific Client Identifiers for Dynamic Host Configuration Protocol Version Four (DHCPv4)
RFC 4701	A DNS Resource Record (RR) for Encoding Dynamic Host Configuration Protocol (DHCP) Information (DHCID RR)
RFC 4702	The Dynamic Host Configuration Protocol (DHCP) Client Fully Qualified Domain Name (FQDN) Option
RFC 4703	Resolution of Fully Qualified Domain Name (FQDN) Conflicts among Dynamic Host Configuration Protocol (DHCP) Clients
RFC 6842	Client Identifier Option in DHCP Server Replies
RFC 7969	Customizing DHCP Configuration on the Basis of Network Topology
RFC 8026	Unified IPv4-in-IPv6 Softwire Customer Premises Equipment (CPE): A DHCPv6-Based Prioritization Mechanism
RFC 8115	DHCPv6 Option for IPv4-Embedded Multicast and Unicast IPv6 Prefixes

DHCP	Title
RFC 8353	Generic Security Service API Version 2: Java Bindings Update
RFC 8415	Dynamic Host Configuration Protocol for IPv6 (DHCPv6)
RFC 8539	Softwire Provisioning Using DHCPv4 over DHCPv6

DNS

DNS	Title
RFC 0799	Internet Name Domains
RFC 0882	DOMAIN NAMES - CONCEPTS and FACILITIES
RFC 0883	DOMAIN NAMES - IMPLEMENTATION and SPECIFICATIONS
RFC 0973	DOMAIN NAMES - IMPLEMENTATION and SPECIFICATIONS
RFC 0974	Mail routing and the domain system
RFC 1034	DOMAIN NAMES - CONCEPTS and FACILITIES
RFC 1035	Domain names - implementation and specification
RFC 1101	DNS Encoding of Network Names and Other Types
RFC 1178	Choosing a Name for Your Computer
RFC 1183	New DNS RR Definitions
RFC 1348	DNS NSAP RRs
RFC 1591	Domain Name System Structure and Delegation
RFC 1637	DNS NSAP Resource Records
RFC 1706	DNS NSAP Resource Records
RFC 1794	DNS Support for Load Balancing
RFC 1876	A Means for Expressing Location Information in the Domain Name System
RFC 1912	Common DNS Operational and Configuration Errors
RFC 1918	Address Allocation for Private Internets
RFC 1982	Serial Number Arithmetic
RFC 1995	Incremental Zone Transfer in DNS
RFC 1996	A Mechanism for Prompt Notification of Zone Changes (DNS NOTIFY)
RFC 2065	Domain Name System Security Extensions
RFC 2136	Dynamic Updates in the Domain Name System (DNS UPDATE)
RFC 2137	Secure Domain Name System Dynamic Update
RFC 2181	Clarifications to the DNS Specification
RFC 2230	Key Exchange Delegation Record for the DNS
RFC 2308	Negative Caching of DNS Queries (DNS NCACHE)
RFC 2317	Classless IN-ADDR.ARPA delegation
RFC 2535	Domain Name System Security Extensions
RFC 2536	DSA KEYs and SIGs in the Domain Name System (DNS)

DNS	Title
RFC 2539	Storage of Diffie-Hellman Keys in the Domain Name System (DNS)
RFC 2540	Detached Domain Name System (DNS) Information
RFC 2606	Reserved Top Level DNS Names
RFC 2671	Extension Mechanisms for DNS (EDNS0)
RFC 2672	Non-Terminal DNS Name Redirection
RFC 2673	Binary Labels in the Domain Name System
RFC 2694	DNS extensions to Network Address Translators (DNS_ALG)
RFC 2782	A DNS RR for specifying the location of services (DNS SRV)
RFC 2845	Secret Key Transaction Authentication for DNS (TSIG)
RFC 2874	DNS Extensions to Support IPv6 Address Aggregation and Renumbering
RFC 2930	Secret Key Establishment for DNS (TKEY RR)
RFC 3007	Secure Domain Name System (DNS) Dynamic Update
RFC 3110	RSA/SHA-1 SIGs and RSA KEYs in the Domain Name System (DNS)
RFC 3123	A DNS RR Type for Lists of Address Prefixes (APL RR)
RFC 3363	Representing Internet Protocol version 6 (IPv6) Addresses in the Domain Name System (DNS)
RFC 3401	Dynamic Delegation Discovery System (DDDS) Part One: The Comprehensive DDDS
RFC 3402	Dynamic Delegation Discovery System (DDDS) Part Two: The Algorithm
RFC 3403	Dynamic Delegation Discovery System (DDDS) Part Three: The Domain Name System (DNS) Database
RFC 3404	Dynamic Delegation Discovery System (DDDS) Part Four: The Uniform Resource Identifiers (URI)
RFC 3405	Dynamic Delegation Discovery System (DDDS) Part Five: URI.ARPA Assignment Procedures
RFC 3425	Obsoleting IQUERY
RFC 3484	Default Address Selection for Internet Protocol version 6 (IPv6)
RFC 3492	Punycode: A Bootstring encoding of Unicode for Internationalized Domain Names in Applications (IDNA)
RFC 3596	DNS Extensions to Support IP Version 6
RFC 3597	Handling of Unknown DNS Resource Record (RR) Types
RFC 3645	Generic Security Service Algorithm for Secret Key Transaction Authentication for DNS (GSS-TSIG)
RFC 3646	DNS Configuration options for Dynamic Host Configuration Protocol for IPv6 (DHCPv6)
RFC 3761	The E.164 to Uniform Resource Identifiers (URI) Dynamic Delegation Discovery System (DDDS) Application (ENUM)
RFC 3833	Threat Analysis of the Domain Name System (DNS)

DNS	Title
RFC 3958	Domain-Based Application Service Location Using SRV RRs and the Dynamic Delegation Discovery Service (DDDS)
RFC 4025	A Method for Storing IPsec Keying Material in DNS
RFC 4033	DNS Security Introduction and Requirements
RFC 4034	Resource Records for the DNS Security Extensions
RFC 4035	Protocol Modifications for the DNS Security Extensions
RFC 4255	Using DNS to Securely Publish Secure Shell (SSH) Key Fingerprints
RFC 4343	Domain Name System (DNS) Case Insensitivity Clarification
RFC 4367	What's in a Name: False Assumptions about DNS Names
RFC 4398	Storing Certificates in the Domain Name System (DNS)
RFC 4431	The DNSSEC Lookaside Validation (DLV) DNS Resource Record
RFC 4470	Minimally Covering NSEC Records and DNSSEC On-line Signing
RFC 4472	Operational Considerations and Issues with IPv6 DNS
RFC 4501	Domain Name System Uniform Resource Identifiers
RFC 4509	Use of SHA-256 in DNSSEC Delegation Signer (DS) Resource Records (RRs)
RFC 4592	The Role of Wildcards in the Domain Name System
RFC 4635	HMAC SHA (Hashed Message Authentication Code, Secure Hash Algorithm) TSIG Algorithm Identifiers
RFC 4641	DNSSEC Operational Practices
RFC 4648	The Base16, Base32, and Base64 Data Encodings
RFC 4697	Observed DNS Resolution Misbehavior
RFC 4701	A DNS Resource Record (RR) for Encoding Dynamic Host Configuration Protocol (DHCP) Information (DHCID RR)
RFC 4702	The Dynamic Host Configuration Protocol (DHCP) Client Fully Qualified Domain Name (FQDN) Option
RFC 4703	Resolution of Fully Qualified Domain Name (FQDN) Conflicts among Dynamic Host Configuration Protocol (DHCP) Clients
RFC 4725	ENUM Validation Architecture
RFC 4759	The ENUM Dip Indicator Parameter for the "tel" URI
RFC 4848	Domain-Based Application Service Location Using URIs and the Dynamic Delegation Discovery Service (DDDS)
RFC 4871	DomainKeys Identified Mail (DKIM) Signatures
RFC 4955	DNS Security (DNSSEC) Experiments
RFC 4956	DNS Security (DNSSEC) Opt-In
RFC 4979	IANA Registration for Enumservice 'XMPP'
RFC 4986	Requirements Related to DNS Security (DNSSEC) Trust Anchor Rollover
RFC 5001	DNS Name Server Identifier (NSID) Option
RFC 5006	IPv6 Router Advertisement Option for DNS Configuration

DNS	Title
RFC 6106	IPv6 Router Advertisement Options for DNS Configuration
RFC 7393	Using the Port Control Protocol (PCP) to Update Dynamic DNS
RFC 7672	SMTP Security via Opportunistic DNS-Based Authentication of Named Entities (DANE) Transport Layer Security (TLS)
RFC 8427	Representing DNS Messages in JSON
RFC 8460	SMTP TLS Reporting
RFC 8461	SMTP MTA Strict Transport Security (MTA-STS)
RFC 8484	DNS Queries over HTTPS (DoH)
RFC 8501	Reverse DNS in IPv6 for Internet Service Providers

EIGRP

EIGRP	Title
RFC 7868	Cisco's Enhanced Interior Gateway Routing Protocol (EIGRP)

ESMTP

ESMTP	Title
RFC 3030	SMTP Service Extensions for Transmission of Large and Binary MIME Messages
RFC 4141	SMTP and MIME Extensions for Content Conversion

FDDI

FDDI	Title
RFC 1390	Transmission of IP and ARP over FDDI Networks

Firewall

Firewall	Title
RFC 2588	IP Multicast and Firewalls
RFC 2979	Behavior of and Requirements for Internet Firewalls
RFC 7288	Reflections on Host Firewalls

FTP

FTP	Title
RFC 0412	User FTP Documentation
RFC 0414	File Transfer Protocol (FTP) status and further comments
RFC 0458	Mail retrieval via FTP

FTP	Title
RFC 0475	FTP and Network Mail System
RFC 0751	Survey of FTP mail and MLFL
RFC 0783	TFTP Protocol (revision 2)
RFC 0906	Bootstrap loading using TFTP
RFC 1068	Background File Transfer Program (BFTP)

HTTP

HTTP	Title
RFC 1945	Hypertext Transfer Protocol – HTTP/1.0
RFC 2068	Hypertext Transfer Protocol – HTTP/1.1
RFC 2145	Use and Interpretation of HTTP Version Numbers
RFC 2295	Transparent Content Negotiation in HTTP
RFC 2518	HTTP Extensions for Distributed Authoring – WEBDAV
RFC 2616	Hypertext Transfer Protocol – HTTP/1.1
RFC 2660	The Secure HyperText Transfer Protocol
RFC 2818	HTTP Over TLS
RFC 3143	Known HTTP Proxy/Caching Problems
RFC 4969	IANA Registration for vCard Enumservice
RFC 7804	Salted Challenge Response HTTP Authentication Mechanism
RFC 8336	The ORIGIN HTTP/2 Frame
RFC 8470	Using Early Data in HTTP
RFC 8484	DNS Queries over HTTPS (DoH)

IMAP

IMAP	Title
RFC 6858	Simplified POP and IMAP Downgrading for Internationalized Email

IPv4

IPv4	Title
RFC 0826	An Ethernet Address Resolution Protocol: Or Converting Network Protocol Addresses to 48.bit Ethernet Address for Transmission on Ethernet Hardware
RFC 0894	A Standard for the Transmission of IP Datagrams over Ethernet Networks
RFC 0895	Standard for the transmission of IP datagrams over experimental Ethernet networks
RFC 1234	Tunneling IPX traffic through IP networks
RFC 1577	Classical IP and ARP over ATM

IPv4	Title
RFC 1700	Assigned Numbers
RFC 1812	Requirements for IP Version 4 Routers
RFC 1917	An Appeal to the Internet Community to Return Unused IP Networks (Prefixes) to the IANA
RFC 1933	Transition Mechanisms for IPv6 Hosts and Routers
RFC 2030	Simple Network Time Protocol (SNTP) Version 4 for IPv4, IPv6 and OSI
RFC 3330	Special-Use IPv4 Addresses
RFC 3378	EtherIP: Tunneling Ethernet Frames in IP Datagrams
RFC 3787	Recommendations for Interoperable IP Networks using Intermediate System to Intermediate System (IS-IS)
RFC 3974	SMTP Operational Experience in Mixed IPv4/v6 Environments
RFC 4361	Node-specific Client Identifiers for Dynamic Host Configuration Protocol Version Four (DHCPv4)
RFC 5692	Transmission of IP over Ethernet over IEEE 802.16 Networks
RFC 5994	Application of Ethernet Pseudowires to MPLS Transport Networks
RFC 6883	IPv6 Guidance for Internet Content Providers and Application Service Providers
RFC 7393	Using the Port Control Protocol (PCP) to Update Dynamic DNS
RFC 7608	IPv6 Prefix Length Recommendation for Forwarding
RFC 7775	IS-IS Route Preference for Extended IP and IPv6 Reachability
RFC 8026	Unified IPv4-in-IPv6 Softwire Customer Premises Equipment (CPE): A DHCPv6-Based Prioritization Mechanism
RFC 8115	DHCPv6 Option for IPv4-Embedded Multicast and Unicast IPv6 Prefixes
RFC 8421	Guidelines for Multihomed and IPv4/IPv6 Dual-Stack Interactive Connectivity Establishment (ICE)
RFC 8468	IPv4, IPv6, and IPv4-IPv6 Coexistence: Updates for the IP Performance Metrics (IPPM) Framework
RFC 8539	Softwire Provisioning Using DHCPv4 over DHCPv6

IPv6

IPv6	Title
RFC 1680	IPng Support for ATM Services
RFC 1881	IPv6 Address Allocation Management
RFC 1888	OSI NSAPs and IPv6
RFC 1933	Transition Mechanisms for IPv6 Hosts and Routers
RFC 1972	A Method for the Transmission of IPv6 Packets over Ethernet Networks

IPv6	Title
RFC 2030	Simple Network Time Protocol (SNTP) Version 4 for IPv4, IPv6 and OSI
RFC 2080	RIPng for IPv6
RFC 2081	RIPng Protocol Applicability Statement
RFC 2461	Neighbor Discovery for IP Version 6 (IPv6)
RFC 2464	Transmission of IPv6 Packets over Ethernet Networks
RFC 3646	DNS Configuration options for Dynamic Host Configuration Protocol for IPv6 (DHCPv6)
RFC 3974	SMTP Operational Experience in Mixed IPv4/v6 Environments
RFC 4191	Default Router Preferences and More-Specific Routes
RFC 4193	Unique Local IPv6 Unicast Addresses
RFC 4361	Node-specific Client Identifiers for Dynamic Host Configuration Protocol Version Four (DHCPv4)
RFC 5006	IPv6 Router Advertisement Option for DNS Configuration
RFC 5340	OSPF for IPv6
RFC 5692	Transmission of IP over Ethernet over IEEE 802.16 Networks
RFC 5902	IAB Thoughts on IPv6 Network Address Translation
RFC 5994	Application of Ethernet Pseudowires to MPLS Transport Networks
RFC 6036	Emerging Service Provider Scenarios for IPv6 Deployment
RFC 6085	Address Mapping of IPv6 Multicast Packets on Ethernet
RFC 6106	IPv6 Router Advertisement Options for DNS Configuration
RFC 6883	IPv6 Guidance for Internet Content Providers and Application Service Providers
RFC 7393	Using the Port Control Protocol (PCP) to Update Dynamic DNS
RFC 7503	OSPFv3 Autoconfiguration
RFC 7561	Mapping Quality of Service (QoS) Procedures of Proxy Mobile IPv6 (PMIPv6) and WLAN
RFC 7608	IPv6 Prefix Length Recommendation for Forwarding
RFC 7775	IS-IS Route Preference for Extended IP and IPv6 Reachability
RFC 8026	Unified IPv4-in-IPv6 Softwire Customer Premises Equipment (CPE): A DHCPv6-Based Prioritization Mechanism
RFC 8064	Recommendation on Stable IPv6 Interface Identifiers
RFC 8115	DHCPv6 Option for IPv4-Embedded Multicast and Unicast IPv6 Prefixes
RFC 8362	OSPFv3 Link State Advertisement (LSA) Extensibility
RFC 8415	Dynamic Host Configuration Protocol for IPv6 (DHCPv6)
RFC 8421	Guidelines for Multihomed and IPv4/IPv6 Dual-Stack Interactive Connectivity Establishment (ICE)
RFC 8468	IPv4, IPv6, and IPv4-IPv6 Coexistence: Updates for the IP Performance Metrics (IPPM) Framework
RFC 8501	Reverse DNS in IPv6 for Internet Service Providers

IPv6	Title
RFC 8539	Softwire Provisioning Using DHCPv4 over DHCPv6

IPX

IPX	Title
RFC 1132	Standard for the transmission of 802.2 packets over IPX networks
RFC 1234	Tunneling IPX traffic through IP networks
RFC 1791	TCP And UDP Over IPX Networks With Fixed Path MTU

ISIS

ISIS	Title
RFC 1142	OSI IS-IS Intra-domain Routing Protocol
RFC 1169	Explaining the role of GOSIP
RFC 1195	Use of OSI IS-IS for routing in TCP/IP and dual environments
RFC 1637	DNS NSAP Resource Records
RFC 2966	Domain-wide Prefix Distribution with Two-Level IS-IS
RFC 2973	IS-IS Mesh Groups
RFC 3277	Intermediate System to Intermediate System (IS-IS) Transient Blackhole Avoidance
RFC 3563	Cooperative Agreement Between the ISOC/IETF and ISO/IEC Joint Technical Committee 1/Sub Committee 6 (JTC1/SC6) on IS-IS Routing Protocol Development
RFC 3719	Recommendations for Interoperable Networks using Intermediate System to Intermediate System (IS-IS)
RFC 3786	Extending the Number of Intermediate System to Intermediate System (IS-IS) Link State PDU (LSP) Fragments Beyond the 256 Limit
RFC 3787	Recommendations for Interoperable IP Networks using Intermediate System to Intermediate System (IS-IS)
RFC 3847	Restart Signaling for Intermediate System to Intermediate System (IS-IS)
RFC 4444	Management Information Base for Intermediate System to Intermediate System (IS-IS)
RFC 4971	Intermediate System to Intermediate System (IS-IS) Extensions for Advertising Router Information
RFC 5130	A Policy Control Mechanism in IS-IS Using Administrative Tags
RFC 7775	IS-IS Route Preference for Extended IP and IPv6 Reachability
RFC 8571	BGP - Link State (BGP-LS) Advertisement of IGP Traffic Engineering Performance Metric Extensions

ISP

ISP	Title
RFC 3013	Recommended Internet Service Provider Security Services and Procedures
RFC 3871	Operational Security Requirements for Large Internet Service Provider (ISP) IP Network Infrastructure
RFC 4778	Operational Security Current Practices in Internet Service Provider Environments
RFC 5632	Comcast's ISP Experiences in a Proactive Network Provider Participation for P2P (P4P) Technical Trial
RFC 6036	Emerging Service Provider Scenarios for IPv6 Deployment
RFC 6883	IPv6 Guidance for Internet Content Providers and Application Service Providers
RFC 8501	Reverse DNS in IPv6 for Internet Service Providers

JAVA

JAVA	Title
RFC 2853	Generic Security Service API Version 2 : Java Bindings
RFC 5653	Generic Security Service API Version 2: Java Bindings Update
RFC 8353	Generic Security Service API Version 2: Java Bindings Update

L2TP

L2TP	Title
RFC 3378	EtherIP: Tunneling Ethernet Frames in IP Datagrams
RFC 3817	Layer 2 Tunneling Protocol (L2TP) Active Discovery Relay for PPP over Ethernet (PPPoE)
RFC 4719	Transport of Ethernet Frames over Layer 2 Tunneling Protocol Version 3 (L2TPv3)

Layer2

Layer2	Title
RFC 0826	An Ethernet Address Resolution Protocol: Or Converting Network Protocol Addresses to 48.bit Ethernet Address for Transmission on Ethernet Hardware
RFC 0894	A Standard for the Transmission of IP Datagrams over Ethernet Networks

Layer2	Title
RFC 0895	Standard for the transmission of IP datagrams over experimental Ethernet networks
RFC 1132	Standard for the transmission of 802.2 packets over IPX networks
RFC 1577	Classical IP and ARP over ATM
RFC 1972	A Method for the Transmission of IPv6 Packets over Ethernet Networks
RFC 2358	Definitions of Managed Objects for the Ethernet-like Interface Types
RFC 2464	Transmission of IPv6 Packets over Ethernet Networks
RFC 3378	EtherIP: Tunneling Ethernet Frames in IP Datagrams
RFC 3619	Extreme Networks' Ethernet Automatic Protection Switching (EAPS) Version 1
RFC 3621	Power Ethernet MIB
RFC 3635	Definitions of Managed Objects for the Ethernet-like Interface Types
RFC 3637	Definitions of Managed Objects for the Ethernet WAN Interface Sublayer
RFC 3817	Layer 2 Tunneling Protocol (L2TP) Active Discovery Relay for PPP over Ethernet (PPPoE)
RFC 4448	Encapsulation Methods for Transport of Ethernet over MPLS Networks
RFC 4638	Accommodating a Maximum Transit Unit/Maximum Receive Unit (MTU/MRU) Greater Than 1492 in the Point-to-Point Protocol over Ethernet (PPPoE)
RFC 4719	Transport of Ethernet Frames over Layer 2 Tunneling Protocol Version 3 (L2TPv3)
RFC 4778	Operational Security Current Practices in Internet Service Provider Environments
RFC 5692	Transmission of IP over Ethernet over IEEE 802.16 Networks
RFC 5828	Generalized Multiprotocol Label Switching (GMPLS) Ethernet Label Switching Architecture and Framework
RFC 5994	Application of Ethernet Pseudowires to MPLS Transport Networks
RFC 6004	Generalized MPLS (GMPLS) Support for Metro Ethernet Forum and G.8011 Ethernet Service Switching
RFC 6005	Generalized MPLS (GMPLS) Support for Metro Ethernet Forum and G.8011 User Network Interface (UNI)
RFC 6060	Generalized Multiprotocol Label Switching (GMPLS) Control of Ethernet Provider Backbone Traffic Engineering (PBB-TE)
RFC 6085	Address Mapping of IPv6 Multicast Packets on Ethernet
RFC 8388	Usage and Applicability of BGP MPLS-Based Ethernet VPN

Layer3

Layer3	Title
RFC 4778	Operational Security Current Practices in Internet Service Provider Environments

LSP

LSP	Title
RFC 3847	Restart Signaling for Intermediate System to Intermediate System (IS-IS)
RFC 5130	A Policy Control Mechanism in IS-IS Using Administrative Tags

MIB

MIB	Title
RFC 4444	Management Information Base for Intermediate System to Intermediate System (IS-IS)

MIME

MIME	Title
RFC 1344	Implications of MIME for Internet Mail Gateways
RFC 1894	An Extensible Message Format for Delivery Status Notifications
RFC 4141	SMTP and MIME Extensions for Content Conversion

MobileIP

MobileIP	Title
RFC 3519	Mobile IP Traversal of Network Address Translation (NAT) Devices

MPLS

MPLS	Title
RFC 4448	Encapsulation Methods for Transport of Ethernet over MPLS Networks
RFC 5828	Generalized Multiprotocol Label Switching (GMPLS) Ethernet Label Switching Architecture and Framework

MPLS	Title
RFC 5994	Application of Ethernet Pseudowires to MPLS Transport Networks
RFC 6004	Generalized MPLS (GMPLS) Support for Metro Ethernet Forum and G.8011 Ethernet Service Switching
RFC 6005	Generalized MPLS (GMPLS) Support for Metro Ethernet Forum and G.8011 User Network Interface (UNI)
RFC 6060	Generalized Multiprotocol Label Switching (GMPLS) Control of Ethernet Provider Backbone Traffic Engineering (PBB-TE)
RFC 8503	BGP/MPLS Layer 3 VPN Multicast Management Information Base

MTA

MTA	Title
RFC 0780	Mail Transfer Protocol
RFC 1894	An Extensible Message Format for Delivery Status Notifications
RFC 2505	Anti-Spam Recommendations for SMTP MTAs
RFC 5321	Simple Mail Transfer Protocol
RFC 7672	SMTP Security via Opportunistic DNS-Based Authentication of Named Entities (DANE) Transport Layer Security (TLS)
RFC 8460	SMTP TLS Reporting
RFC 8461	SMTP MTA Strict Transport Security (MTA-STS)

NAT

NAT	Title
RFC 3519	Mobile IP Traversal of Network Address Translation (NAT) Devices
RFC 5902	IAB Thoughts on IPv6 Network Address Translation
RFC 6281	Understanding Apple's Back to My Mac (BTMM) Service
RFC 7393	Using the Port Control Protocol (PCP) to Update Dynamic DNS
RFC 7857	Updates to Network Address Translation (NAT) Behavioral Requirements

NNTP

NNTP	Title
RFC 0977	Network News Transfer Protocol

NSAP

NSAP	Title
RFC 0982	Guidelines for the specification of the structure of the Domain Specific Part (DSP) of the ISO standard NSAP address
RFC 1169	Explaining the role of GOSIP
RFC 1237	Guidelines for OSI NSAP Allocation in the Internet
RFC 1348	DNS NSAP RRs
RFC 1629	Guidelines for OSI NSAP Allocation in the Internet
RFC 1637	DNS NSAP Resource Records
RFC 1706	DNS NSAP Resource Records
RFC 1888	OSI NSAPs and IPv6

NTP

NTP	Title
RFC 0958	Network Time Protocol (NTP)
RFC 2030	Simple Network Time Protocol (SNTP) Version 4 for IPv4, IPv6 and OSI
RFC 4075	Simple Network Time Protocol (SNTP) Configuration Option for DHCPv6

OSI

OSI	Title
RFC 0652	Telnet output carriage-return disposition option
RFC 0654	Telnet output horizontal tab disposition option
RFC 1070	Use of the Internet as a subnetwork for experimentation with the OSI network layer
RFC 1888	OSI NSAPs and IPv6

OSPF

OSPF	Title
RFC 1131	OSPF specification
RFC 1245	OSPF protocol analysis
RFC 1246	Experience with the OSPF Protocol
RFC 1247	"OSPF Version 2"
RFC 1364	BGP OSPF Interaction
RFC 1370	Applicability Statement for OSPF
RFC 1403	BGP OSPF Interaction
RFC 1583	"OSPF Version 2"
RFC 2178	OSPF Version 2

OSPF	Title
RFC 2328	"OSPF Version 2"
RFC 3137	OSPF Stub Router Advertisement
RFC 3509	Alternative Implementations of OSPF Area Border Routers
RFC 5340	OSPF for IPv6
RFC 5649	Advanced Encryption Standard (AES) Key Wrap with Padding Algorithm
RFC 5709	OSPFv2 HMAC-SHA Cryptographic Authentication
RFC 6549	OSPFv2 Multi-Instance Extensions
RFC 6860	Hiding Transit-Only Networks in OSPF
RFC 7503	OSPFv3 Autoconfiguration
RFC 8362	OSPFv3 Link State Advertisement (LSA) Extensibility
RFC 8571	BGP - Link State (BGP-LS) Advertisement of IGP Traffic Engineering Performance Metric Extensions

POP

POP	Title
RFC 1460	Post Office Protocol - Version 3
RFC 1725	Post Office Protocol - Version 3
RFC 1939	Post Office Protocol - Version 3
RFC 5034	The Post Office Protocol (POP3) Simple Authentication and Security Layer (SASL) Authentication Mechanism
RFC 6858	Simplified POP and IMAP Downgrading for Internationalized Email

PPP

PPP	Title
RFC 3817	Layer 2 Tunneling Protocol (L2TP) Active Discovery Relay for PPP over Ethernet (PPPoE)
RFC 4448	Encapsulation Methods for Transport of Ethernet over MPLS Networks
RFC 4638	Accommodating a Maximum Transit Unit/Maximum Receive Unit (MTU/MRU) Greater Than 1492 in the Point-to-Point Protocol over Ethernet (PPPoE)

RIP

RIP	Title
RFC 1058	Routing Information Protocol
RFC 1387	RIP Version 2 Protocol Analysis
RFC 1388	RIP Version 2 Carrying Additional Information

RIP	Title
RFC 1389	RIP Version 2 MIB Extensions
RFC 1723	RIP Version 2 Carrying Additional Information
RFC 2453	RIP Version 2

RIPng

RIPng	Title
RFC 2080	RIPng for IPv6
RFC 2081	RIPng Protocol Applicability Statement

RIPv2

RIPv2	Title
RFC 1387	RIP Version 2 Protocol Analysis
RFC 1388	RIP Version 2 Carrying Additional Information
RFC 1389	RIP Version 2 MIB Extensions

SDH

SDH	Title
RFC 5828	Generalized Multiprotocol Label Switching (GMPLS) Ethernet Label Switching Architecture and Framework

SecureDNS

SecureDNS	Title
RFC 2065	Domain Name System Security Extensions

Security

Security	Title
RFC 2487	SMTP Service Extension for Secure SMTP over TLS
RFC 2853	Generic Security Service API Version 2 : Java Bindings
RFC 3013	Recommended Internet Service Provider Security Services and Procedures
RFC 3871	Operational Security Requirements for Large Internet Service Provider (ISP) IP Network Infrastructure
RFC 4025	A Method for Storing IPsec Keying Material in DNS
RFC 4035	Protocol Modifications for the DNS Security Extensions
RFC 4641	DNSSEC Operational Practices

Security	Title
RFC 4778	Operational Security Current Practices in Internet Service Provider Environments
RFC 4954	SMTP Service Extension for Authentication
RFC 4956	DNS Security (DNSSEC) Opt-In
RFC 4986	Requirements Related to DNS Security (DNSSEC) Trust Anchor Rollover
RFC 5034	The Post Office Protocol (POP3) Simple Authentication and Security Layer (SASL) Authentication Mechanism
RFC 5235	Sieve Email Filtering: Spamtest and Virustest Extensions
RFC 5653	Generic Security Service API Version 2: Java Bindings Update
RFC 6430	Email Feedback Report Type Value: not-spam
RFC 6811	BGP Prefix Origin Validation
RFC 7672	SMTP Security via Opportunistic DNS-Based Authentication of Named Entities (DANE) Transport Layer Security (TLS)
RFC 7804	Salted Challenge Response HTTP Authentication Mechanism
RFC 8314	Cleartext Considered Obsolete: Use of Transport Layer Security (TLS) for Email Submission and Access
RFC 8353	Generic Security Service API Version 2: Java Bindings Update
RFC 8416	Simplified Local Internet Number Resource Management with the RPKI (SLURM)
RFC 8460	SMTP TLS Reporting
RFC 8461	SMTP MTA Strict Transport Security (MTA-STS)
RFC 8470	Using Early Data in HTTP

Sendmail

Sendmail	Title
RFC 0773	Comments on NCP/TCP mail service transition strategy
RFC 0822	STANDARD FOR THE FORMAT OF ARPA INTERNET TEXT MESSAGES

SMTP

SMTP	Title
RFC 0458	Mail retrieval via FTP
RFC 0475	FTP and Network Mail System
RFC 0773	Comments on NCP/TCP mail service transition strategy
RFC 0780	Mail Transfer Protocol
RFC 0821	Simple Mail Transfer Protocol
RFC 0822	STANDARD FOR THE FORMAT OF ARPA INTERNET TEXT MESSAGES

SMTP	Title
RFC 0876	Survey of SMTP implementations
RFC 0974	Mail routing and the domain system
RFC 1047	Duplicate messages and SMTP
RFC 1425	SMTP Service Extensions
RFC 1426	SMTP Service Extension for 8bit-MIMEtransport
RFC 1427	SMTP Service Extension for Message Size Declaration
RFC 1428	Transition of Internet Mail from Just-Send-8 to 8bit-SMTP/MIME
RFC 1460	Post Office Protocol - Version 3
RFC 1648	Postmaster Convention for X.400 Operations
RFC 1651	SMTP Service Extensions
RFC 1652	SMTP Service Extension for 8bit-MIMEtransport
RFC 1653	SMTP Service Extension for Message Size Declaration
RFC 1725	Post Office Protocol - Version 3
RFC 1830	SMTP Service Extensions for Transmission of Large and Binary MIME Messages
RFC 1845	SMTP Service Extension for Checkpoint/Restart
RFC 1846	SMTP 521 Reply Code
RFC 1854	SMTP Service Extension for Command Pipelining
RFC 1869	SMTP Service Extensions
RFC 1870	SMTP Service Extension for Message Size Declaration
RFC 1891	SMTP Service Extension for Delivery Status Notifications
RFC 1939	Post Office Protocol - Version 3
RFC 1985	SMTP Service Extension for Remote Message Queue Starting
RFC 2034	SMTP Service Extension for Returning Enhanced Error Codes
RFC 2197	SMTP Service Extension for Command Pipelining
RFC 2487	SMTP Service Extension for Secure SMTP over TLS
RFC 2505	Anti-Spam Recommendations for SMTP MTAs
RFC 2554	SMTP Service Extension for Authentication
RFC 2635	DON'T SPEW A Set of Guidelines for Mass Unsolicited Mailings and Postings (spam*)
RFC 2645	ON-DEMAND MAIL RELAY (ODMR) SMTP with Dynamic IP Addresses
RFC 2821	Simple Mail Transfer Protocol
RFC 3030	SMTP Service Extensions for Transmission of Large and Binary MIME Messages
RFC 3461	Simple Mail Transfer Protocol (SMTP) Service Extension for Delivery Status Notifications (DSNs)
RFC 3974	SMTP Operational Experience in Mixed IPv4/v6 Environments
RFC 4141	SMTP and MIME Extensions for Content Conversion
RFC 4408	Sender Policy Framework (SPF) for Authorizing Use of Domains in E-Mail, Version 1
RFC 4871	DomainKeys Identified Mail (DKIM) Signatures

SMTP	Title
RFC 4954	SMTP Service Extension for Authentication
RFC 5039	The Session Initiation Protocol (SIP) and Spam
RFC 5068	Email Submission Operations: Access and Accountability Requirements
RFC 5235	Sieve Email Filtering: Spamtest and Virustest Extensions
RFC 5321	Simple Mail Transfer Protocol
RFC 6430	Email Feedback Report Type Value: not-spam
RFC 6647	Email Greylisting: An Applicability Statement for SMTP
RFC 6858	Simplified POP and IMAP Downgrading for Internationalized Email
RFC 7672	SMTP Security via Opportunistic DNS-Based Authentication of Named Entities (DANE) Transport Layer Security (TLS)
RFC 8314	Cleartext Considered Obsolete: Use of Transport Layer Security (TLS) for Email Submission and Access
RFC 8460	SMTP TLS Reporting
RFC 8461	SMTP MTA Strict Transport Security (MTA-STS)

SNMP

SNMP	Title
RFC 2358	Definitions of Managed Objects for the Ethernet-like Interface Types
RFC 3621	Power Ethernet MIB
RFC 3635	Definitions of Managed Objects for the Ethernet-like Interface Types
RFC 3637	Definitions of Managed Objects for the Ethernet WAN Interface Sublayer

SONET

SONET	Title
RFC 5828	Generalized Multiprotocol Label Switching (GMPLS) Ethernet Label Switching Architecture and Framework

SPAM

SPAM	Title
RFC 2635	DON'T SPEW A Set of Guidelines for Mass Unsolicited Mailings and Postings (spam*)
RFC 5039	The Session Initiation Protocol (SIP) and Spam
RFC 5068	Email Submission Operations: Access and Accountability Requirements

SPAM	Title
RFC 5235	Sieve Email Filtering: Spamtest and Virustest Extensions
RFC 6430	Email Feedback Report Type Value: not-spam
RFC 6647	Email Greylisting: An Applicability Statement for SMTP
RFC 8314	Cleartext Considered Obsolete: Use of Transport Layer Security (TLS) for Email Submission and Access
RFC 8461	SMTP MTA Strict Transport Security (MTA-STS)

TCP

TCP	Title
RFC 1791	TCP And UDP Over IPX Networks With Fixed Path MTU

TDM

TDM	Title
RFC 5828	Generalized Multiprotocol Label Switching (GMPLS) Ethernet Label Switching Architecture and Framework

TELNET

TELNET	Title
RFC 0097	First Cut at a Proposed Telnet Protocol
RFC 0137	Telnet Protocol - a proposed document
RFC 0779	Telnet send-location option

TFTP

TFTP	Title
RFC 0783	TFTP Protocol (revision 2)
RFC 0906	Bootstrap loading using TFTP

Tunneling

Tunneling	Title
RFC 8026	Unified IPv4-in-IPv6 Softwire Customer Premises Equipment (CPE): A DHCPv6-Based Prioritization Mechanism

UDP

UDP	Title
RFC 1791	TCP And UDP Over IPX Networks With Fixed Path MTU

VOIP

VOIP	Title
RFC 4759	The ENUM Dip Indicator Parameter for the "tel" URI

VPN

VPN	Title
RFC 8388	Usage and Applicability of BGP MPLS-Based Ethernet VPN

WIFI

WIFI	Title
RFC 5692	Transmission of IP over Ethernet over IEEE 802.16 Networks
RFC 7458	Extensible Authentication Protocol (EAP) Attributes for Wi-Fi Integration with the Evolved Packet Core
RFC 7561	Mapping Quality of Service (QoS) Procedures of Proxy Mobile IPv6 (PMIPv6) and WLAN

Glossary

802.11:   802.11 is a standard specification for wireless Ethernet, a method of packet-based physical communication in a local area network (LAN), which is maintained by the Institute of Electrical and Electronics Engineers (IEEE). Currently there are many different 802.11 standards that include specifications for frequency and speed such as 802.11a, b, g, and n.

802.3:   802.3 is a standard specification for wired Ethernet, a method of packet-based physical communication in a local area network (LAN), which is maintained by the Institute of Electrical and Electronics Engineers (IEEE). In general, 802.3 specifies the physical media and the working characteristics of Ethernet. There are many different standards today that include wired Ethernet, fiber optical Ethernet, and broadband cable.

802.5:   802.5 was developed by IBM as a standard specification for Token Ring, a method of packet-based physical communication in a local area network (LAN), which is maintained by the Institute of Electrical and Electronics Engineers (IEEE). In general, 802.5 specifies the physical media and the working characteristics of Token Ring.

A: see Administrative Authority record (IPv4).

A-PDU: see Application Layer PDU.

AAAA: see Administrative Authority record (IPv6 or NSAP).

ABR: see Area Border Router.

ACK: see Acknowledge transmission.

© Springer Nature Switzerland AG 2020
G. Howser, *Computer Networks and the Internet*,
https://doi.org/10.1007/978-3-030-34496-2

Acknowledge transmission: For protocols with guaranteed delivery, the receiving endpoint must send a message to acknowledge the correct reception of each message. Because low level protocols sometimes do this by sending only the one byte message ACK (ASCII HEX 6), acknowledgments are sometimes called ACKs.

Address Resolution Protocol: Address Resolution Protocol is used, along with the inverse protocol Reverse Address Resolution Protocol or RARP, to maintain a table of IPv4 and MAC addresses. In practice, a Layer 3 device is extremely likely to reply to an IPv4 message in a timely fashion. Therefore maintaining a table of the mapping between IPv4 and MAC addresses leads to large savings in redundant network traffic. As usual, entries in the ARP table eventually expire and ARP needs to be run again for that IPv4 address. In this way ARP learns changes in IPv4–MAC pairs fairly quickly while still reducing network traffic.

adjacency: Two routers form an OSPF adjacency after exchanging "hello" messages if they are connected by working media and are in the same area.

adjacent: Two nodes, x and y, are said to be adjacent in a graph if there exists an edge $e(x,y)$. If a digraph has an arc $a(a,b)$ but not an arc $a(b,a)$, then b is adjacent to a but a is *not* adjacent to b. In layman's terms, if there is a direct path between two nodes, they are adjacent. The notation $adj(a)$ means the set of all vertices that can be reached *directly* from a, such as all the routers in a network with a link from a.

Administrative Authority record (IPv4):   The A record is the DNS record that links the actual name of the device as a FQDN with the IPv4 address of the device NIC. Use CNAME to link an alias to a FQDN.

Administrative Authority record (IPv6 or NSAP):   The AAAA record is the DNS record that links the actual name of the device as a FQDN with the IPv6 address of the device NIC. Use CNAME to link an alias to a FQDN. The Administrative Authority for an NSAP address is also referred to as "AAAA". Unless explicitly noted, for our purposes AAAA refers to the DNS record.

Advanced Research Projects Agency: The original Internet was the ARPANET founded by ARPA. Later ARPA became DARPA, but ARPA's tracks can still be found in some places such as the inverse zone files in DNS [318].

Advanced Research Projects Agency Network: The precursor to the Internet was jump started in the early days of computing history in 1969, with the U.S. Defense Department's Advanced Research Projects Agency Network (ARPANET). ARPA-funded researchers developed many of the protocols used for Internet communication today [320].

AFI: see Authority and Format Identifier (NSAP).

AFXR: see Asynchronous Full Transfer.

alias: In networking one encounters at least two distinct uses of "alias". The first is using an alias to link multiple names with a single physical device using canonical names. The other use is that of email aliases to direct any number of email in–boxes to a single in–box such as guiding "root" and "postmaster" to a specific user. This allows an administrator to transparently change who has the actual responsibility for undeliverable email which is forwarded to "postmaster".

Alpine: Alpine is an updated version of the older package named PINE from the University of Washington. Alpine is an email client designed for use from a Linux terminal and is very simple to use. The simplicity of Alpine should not be taken as a disadvantage as it provides all the functionality a user would ever need. It is especially useful for the administrative user `root`.

American National Standards Institute: The United States standards organization which handled many standards including telecommunications and computing. ANSI was replaced by NIST.

American Standard Code for Information Interchange:
ASCII is one of the first standards for the binary encoding of data for electronic data communications. As ASCII is an eight bit code, it can only support 256 characters and many of those have special meanings such as ACK and NAK. This means that ASCII is not appropriate for languages that use a different alphabet from the 26 letters used by English. There are other, more modern, encoding methods in use today.

ANSI: see American National Standards Institute.

Anycast Address (NSAP): Anycast addresses are IPv6 or NSAP addresses designed to be assigned to a set of routers or Layer 3 switches such that the first available device in the set will receive a transmission for that device. In the case of ATM, anycast addresses may also be assigned to other devices or services for redundancy and/or load balancing. This is beyond the scope of the present text.

API: see Application Program Interface.

APIPA: see Automatic Private IP Addressing.

AppleTalk: Apple's proprietary network protocol. Apple has moved to full support of IP.

AppleTalk network number: A 16 bit number that uniquely defines a network of AppleTalk devices and serves as the address for the entire network.

AppleTalk Node ID: An 8 bit, or 16 bit, number that uniquely defines a single host on a network of AppleTalk devices and serves as the second part of the address for that host.

AppleTalk Zone: A set if AppleTalk devices logically grouped together with the same AppleTalk network number form an AppleTalk Zone, which is analogous to an IP subnetwork.

Application Layer: The Application Layer is Layer 7 of the OSI Model. The Application Layer has two functions:

- To handle Client and service announcements.
- To provide an interface between the OSI stack and the device operating system, APIs, or programs.

The Application Layer also participates with one–to–one, one–to–many, and many–to–many mappings.

Application Layer PDU: In the OSI Model, the Application Layer, running on one of the endpoints of a conversation, acts as if it is communicating directly with the Application Layer running on the other endpoint of the conversation by exchanging Application Layer Protocol Datagram Units (A-PDU) via a virtual connection. The actual A-PDU is passed down to the Presentation Layer 6 and at the other endpoint the exact same A-PDU is passed up from the Presentation Layer 6 to the Application Layer 7. This allows the Application Layer to act independently from the OSI layers below it and hides from it all the details of the actual communications and the application interface (API).

Application Program Interface: A generic term to include any interface with a process, program, or the OS. Many computer languages and operating systems provide formal APIs to relieve the programmer of the requirement to know the details of communicating with many different operating systems and/or devices on the hardware level. APIs are critical in designing applications that can be ported between hardware and software platforms.

Application Specific Integrated Circuit: It is a basic fact that hardware runs faster than software. For some applications such as switching, the time savings gained by using hardware composed of integrated circuits that only perform the required application more than make up for the fact that the hardware cannot perform multiple actions. ASICs typically can only be upgraded by replacing the hardware. This is a distinct disadvantage if the desired function might change. Ethernet switching, for example, is extremely well known and stable so an ASIC to perform Ethernet switching would be well worth the effort to develop.

area: An area in OSPF is a designated group of routers expected to form a common network which will have mostly traffic destined for other networks in that area. Traffic destined for other areas *must* cross the OSPF backbone (area 0.0.0.0) to reach a network in a different area.

Area Border Router: In both OSPF and OSPFv3, an Area Border Router runs two copies of the protocol stack in order to connect a local area with Area Zero to allow for the exchange of packets between areas. As the OSPF network grows larger, the load on the Area Border Routers increases. It is typical for these routers to have more processing power than routers that run only one protocol stack.

ARP: see Address Resolution Protocol.

ARPA: see Advanced Research Projects Agency.

ARPANET: see Advanced Research Projects Agency Network.

AS: see Autonomous System.

ASCII: see American Standard Code for Information Interchange.

ASIC: see Application Specific Integrated Circuit.

ASN: see Autonomous System Number.

asymmetric: Asymmetric is from the Greek for "not same measure" or "not the same on both sides". In networking this is typically applied to a bidirectional connection where the transmit speed in one direction is different from the transmit speed in the other direction. For example, satellite Internet service is high–speed from the Internet and relatively low–speed into the Internet.

asynchronous: Asynchronous is from the Greek for "not at the same time". In digital communications, this term has three main usages:

1. When the sender and receiver are not constrained to send/receive at specific times. This happens most often when there is some inherent mechanism to allow for the sender to sense the message and the sender does not have to send each time a timer expires.
2. The sender and receiver do not have to insure their clocks match in order to properly send the bits of a message. The message itself has a mechanism to allow the receiver to properly determine the start/stop of each bit.
3. Less common is the usage to denote there is no requirement for the receiver

to have prior knowledge in order to interpret the message properly. This usage is most often encountered in security and encryption. For example, the message "My mother's cat died." could be used to denote "Drop everything and leave town." only if the receiver has prior knowledge of the hidden meaning.

Asynchronous Full Transfer: For some implementations of DNS it is possible to transfer the whole zone file via AFXR or only updates via Incremental File Transfer (IFXR). All implementations of DNS must support AFXR in its simplest form.

Asynchronous Transfer Mode: An alternative method to send messages through a non-IP network. ATM uses 53–byte cells rather than frames and packets and NSAP addressing. Unlike most IP networks, ATM networks are switched rather than routed. This means that connection–oriented networking is native to ATM. IP can easily be carried over an ATM network by using LAN emulation to create the ATM version of a vLAN.

ATM: see Asynchronous Transfer Mode.

attenuation: Attenuation is the certainty that a signal's strength and quality will decrease as it is transmitted over *any* media. Attenuation is caused by the characteristics of the specific media including, copper wire, glass fiber, air, or vacuum.

Authority and Format Identifier (NSAP):   The first byte of an NSAP address denotes which authority assigned the address range and what format will be used. Private NSAP addresses start with an AFI of "47".

automagically: A nonsense word used to denote things that happen behind the scenes with no apparent human intervention. Usually the details are not of immediate importance to the topic at hand.

Automatic Private IP Addressing: Automatic Private IP Addressing is used if a network client fails to get an IP address using DHCP or does not have on assigned statically. To get an IPv4 address, the client will select an address at random in the range 169.254.1.0 to 169.254.254.255 (inclusive), with a netmask of 255.255.0.0. The client will then send an ARP packet asking for the MAC address that corresponds to the randomly-generated IPv4 address. If any other machine is using that address, the client will generate another random address and try again.

Autonomous System: An AS is a set of connected Internet Protocol (IP) routing prefixes under the control of a single administrative entity that presents a common, clearly defined routing policy to the Internet. There does not need to be a direct relationship between IP network addresses and an Autonomous System.

Autonomous System Number: A thirty-two bit number assigned by the IANA to an Autonomous System. ASNs 64,512 to 65,534 and 4,200,000,000 to 4,294,967,294 are reserved for private use and should not be connected to the Internet.

Babel: The Babel routing protocol is a distance-vector routing protocol for Internet Protocol packet-switched networks that is designed to be robust and efficient on both wireless mesh networks and wired networks. Babel has provisions for using multiple dynamically computed metrics; by default, it uses hop-count on wired networks and a variant of transmission delay on wireless links, but can be configured to take radio diversity into account or to automatically compute a link's latency and include it in the metric. Babel operates on both IPv4 and IPv6 networks at the same time. It has been reported to be a robust protocol and to have fast convergence properties.

backbone: A backbone is a network connects other networks and important resources. Backbone networks are often high–speed networks but can be the same speed as the networks they connect.

Backup Designated Router: In a local area (not the backbone area zero), OSPF routers can elect one router to be a Designated Router (DR) and one router to be a Backup Designated Router (BDR). For example, on multi–access broadcast networks (such as LANs) routers elect a DR and BDR. DR and BDR serve as the central point for exchanging OSPF routing information. Each non-DR or non-BDR router will exchange routing information only with the DR and BDR, instead of exchanging updates with every router on the network segment. DR will then distribute topology information to every other router inside the same area, which greatly reduces OSPF traffic [16].

Basic Input/Output System: BIOS is the part of the operating system that provides correct access to peripheral devices. This separation frees the CPU from running at the slower speed of peripherals. The BIOS can be updated separately from the OS which is helpful since a badly updated BIOS will not boot the OS. This is called "bricking" the device because it is useless, except as a paperweight, until the BIOS can be corrected.

BDR: see Backup Designated Router.

Because It's Time Network: BITNET was a co-operative U.S. university computer network founded in 1981 by Ira Fuchs at the City University of New York (CUNY) and Greydon Freeman at Yale University. The first network link was between CUNY and Yale. The name BITNET originally meant "Because It's There Network", but it eventually came to mean "Because It's Time Network". A college or university wishing to join BITNET was required to lease a data circuit (phone line) from a site to an existing BITNET node, buy modems for each end of the data circuit, sending one to the connecting point site, and allow other institutions to connect to its site free of charge.

Bellman–Ford Algorithm: Like Dijkstra's Algorithm, the Bellman–Ford Algorithm is used to find the lowest cost path from the local to each node in a weighted graph or digraph. Bellman–Ford can be used to determine the shortest path from one node in a graph to every other node within the same graph data structure, provided that the nodes are reachable from the starting node and the digraph contains no negative weight loops. Negative weights in a network like the Internet would mean the user would get paid every time a packet traversed the connection. This is not likely to happen.

Berkeley Internet Name Domain service: BIND is a flexible and robust name service package that runs on most Linux distributions and Windows. The current version running on Raspbian is BIND9.

BGP: see Border Gateway Protocol.

BIND: see Berkeley Internet Name Domain service.

BIOS: see Basic Input/Output System.

bit: A bit is the smallest possible piece of information and is represented by a binary digit "one" or "zero".

BITNET: see Because It's Time Network.

Bits per second: In networking, the speed of a connection is measured in raw bits per second rather than the desired message bits per second. No attempt is made to analyze the data to determine what is overhead and what is useful data. All bits are counted to arrive at the speed including over–head such as addressing. For example, if the connection is a 100 megabit Ethernet connection it is rated at 100 mega bps even though it is not possible to send that many useful message bits per second. If the "B" is capitalized, the reference is to bytes rather than bits.

black hole: A black hole (sometimes called "the bit bucket" for the keypunch trash can that held the chads from punched cards) in routing can be a path that never gets a packet to its destination; i.e, the packet disappears without a trace. Black holes tend to form during convergence and then disappear when the network becomes stable as it converges. In router configuration, sending packets for a specific destination to a Black Hole forces the router to drop those packets. For example, it is customary for a router on the Internet to send all packets to and from private IP network addresses to either "null" or "black hole".

BNA: see Burroughs Network Architecture.

BOOTP: see Bootstrap Protocol.

Bootstrap Protocol: BOOTP is a UDP/IP-based protocol that allows a booting host to configure itself dynamically, and more significantly, without user supervision. It provides a means to assign a host its IP address, a file from which to download a boot program from some server, that server's address, and (if present) the address of an Internet gateway [73, 98, 106, 112, 114].

Border Gateway Protocol: A protocol commonly used to interconnect two large ISPs or autonomous systems. BGP is used to limit the number of routes advertised by summarizing routes to a very high level. Summarization has the welcome side–effect of hiding the details of the internal networks.

Bps: see Bytes per second.

bps: see Bits per second.

bridge: Any Layer 2 device that has connections to different physical media and moves frames between them when needed. Currently the most common bridge is a WAP[1] which seamlessly connects a wireless network with a wired network.

Broadband: Any media capable of carrying multiple channels for communications as opposed to single band media which can carry only one channel, or conversation, at a time.

Broadcast: Any message sent that is designed to be processed by all devices (NICs) on a shared media is called a broadcast. Broadcasts can appear at Layer 2 or Layer 3.

Broadcast domain: A broadcast domain is defined as all of the devices (NICs) that can exchange broadcasts and unicasts. A broadcast domain can be defined at Layer 2 or Layer 3, but normally is used to denote a Layer 2 broadcast domain which is a LAN. At Layer 3, a broadcast domain is a subnetwork of devices that all share a common network part of their address.

BTOS: see Burroughs Task Operating System.

Burroughs Network Architecture: A set of proprietary networking protocols developed by Burroughs/Unisys to seamlessly interconnect Burroughs mainframes, minicomputers, and network controllers.

---

[1] Wireless Access Point

Burroughs Task Operating System: A microcomputer operating system developed by Convergent Technologies and purchased by Burroughs. Devices could easily be networked and resources shared by pre–pending the node name of the microcomputer, or server, to which the device was attached.

byte: A byte is a group of eight binary digits and is equivalent to two hexadecimal (base 16) digits or a single character. The networking equivalent is an octet.

Bytes per second: Some protocols may report the number of characters transmitted per second rather than the number of raw bits per second. In this case the reported rate is in bytes per second which is abbreviated as Bps. Such numbers should be treated as documentation only as it is not always clear if the rate includes overhead (raw bps) or if it is in terms of useful message bytes.

Byzantine behavior: Byzantine behavior gets its name from the Byzantine Generals problem posed by Leslie Lamport [45] in relationship to clock synchronization. Byzantine behavior occurs when a process or device sends conflicting information to two or more other devices. In this case, the other devices cannot directly reach a consensus as to the correct information in many cases and the system fails. This type of behavior is beyond the scope of this text, but the term might come up in passing when dealing with trouble–shooting a Byzantine failure.

C: C is a programming language loosely based upon Algol and the language used to write most of UNIX and Linux. It was originally designed by Brian Kernighan and Dennis Ritchie [42]. Many languages are in the "C family" including C++, Java, and C#. A knowledge of C is not required for networking, but it could help if you must compile utilities such as daemons.

Camel Case: A method of capitalization popular in object programming. Variable names and functions begin with a lowercase letter and the first letter of each following word is capitalized [29]. For example, minimumPaymentAmount. It is becoming popular among Linux users as well.

Canonical Name: A CNAME record is a type of resource record in the DNS which maps one domain name (an alias) to another (the Canonical Name.) This can prove convenient when running multiple services, like an FTP server and a web server, each running on different ports from a single IP address. One can, for example, point `ftp.example.com` and `www.example.com` to the DNS entry for `example.com`, which in turn has an A record which points to the IP address for a FQDN `host1.example.com`. Then, if the IP address ever changes, one only has to record the change in one place within the network: in the DNS A record for `host1.example.com`. CNAME records must always point to another domain name, never directly to an IP address.

Carrier Sense Media Access/Collision Avoidance: CSMA/CA is a method to control access to a shared media that minimizes the occurrence of collisions while insuring each device eventually acquires access to the media. CSMA/CA uses a special request message to reserve the media for a message while all other devices await their turn. Ethernet typically uses a related protocol CSMA/CD.

Carrier Sense Media Access/Collision Detection: Ethernet is based upon shared media and therefore needs some method to control which device has access to the media at any given time. The method used is CSMA/CD which governs the actions of all NICs on the segment when a collision is detected rather than attempting to avoid collisions altogether. This may sound odd, but there is less overhead in detecting and dealing with collisions than avoiding them in many cases. Both CSMA/CD and the very similar CSMA/CA tend to break down if there are too many devices on a single segment. This leads directly to the popularity of switched Ethernet where a segment has only two devices and there is no chance of collisions.

CAT: see Category (Structured Wiring).

Category (Structured Wiring): CAT is a set of building and wiring standards that govern the exact cable, the connectors, how the wire is strung, how it is terminated, and how it is tested. For data wiring, the most common sets are CAT5, CAT5e, and CAT7. There may be others as well.

Cellular IP: Cellular IP, or Mobile IP, is a set of protocols to bring the Internet to a cell phone or other cellular device which can run IP.

Central Processing Unit: The CPU is the part of a device that does all the numerical or logical control work of the device. CPUs are inside many different pieces of hardware, NICs for example, and come in an amazing variety of forms.

channel: A single, distinct, communications path between two endpoints much like a single lane on a highway or a TV channel. Some media may carry multiple channels by separating them logically, such as TDM, or physically such as a telephone cable.

CIDR: see Classless Inter–Domain Routing.

Cisco Systems: Cisco Systems, Inc. is a multinational corporation headquartered in San Jose, California in the heart of Silicon Valley. Many of the most important advances in IP were either developed with the help of Cisco or were first implemented on Cisco hardware. At one point in the 1990's Cisco was the most valuable corporation in the world. Cisco markets a complete range of routing solutions from home routers (Linksys) to some of the most powerful routers on the market.

Classless Inter–Domain Routing: CIDR is a method of public IP address assignment. It was introduced in 1993 by Internet Engineering Task Force (IETF) with the goal of dealing with the IPv4 address exhaustion problem and to slow down the growth of routing tables on Internet routers [15]. Before CIDR and variable–length subnet masks, all networks were required to use the natural subnet masks determined by the IPv4 class which did not fully address the needs of organizations that required a network that was not flat, i.e. they required a way to divide their network to localize their traffic to a group of devices. With the advent of CIDR, these organizations could give their address space a longer subnet mask than the natural mask and create as many subnetworks as they required.

Client: A client device accesses network shared resources by contacting a service process running on a server. Resources are often in too much demand and too expensive to maintain at each client that requires them. The client/server model reduces the impact of scarce resources for an organization or an individual.

CNAME: see Canonical Name.

Collision domain: A collision domain consists of all the devices (NICs) that could *potentially* interfere, or have a collision, if the devices transmit at exactly the same time. A bridge, or switch, can be used to limit a Layer 2 collision domain. In fact, it is possible to use a switch to limit a collision domain to two devices no matter how large the LAN.

Country Code TLD: A country code TLD is a two character code assigned to a country as a domain. For example, "uk" is the TLD for the United Kingdom and "lu" is the TLD for Luxemburg.

CPU: see Central Processing Unit.

CRC: see Cyclical Redundancy Check.

Cross–platform web server: XAMP servers from Apache Friends consist of the Windows or Linux versions of Apache, MySQL, and PHP. This set of services provides a uniform platform for the development of complex web sites that are easily ported to a different platform and more secure platform. As it is normally installed, XAMP is not secure enough for production work but can be secured with some effort. [32]

CSMA/CA: see Carrier Sense Media Access/Collision Avoidance.

CSMA/CD: see Carrier Sense Media Access/Collision Detection.

Cyclical Redundancy Check: A CRC is a quick method to catch *most* data transmission errors. The message to be sent is used as the input for a well–known algorithm which produces a pattern of binary bits that is dependent upon the input. This result is attached to the message as then both are sent to the other endpoint. At the destination, the endpoint splits the message and result and then runs the message through the same algorithm. If the answer matches the result that was attached to the message, the message was transmitted successfully. If not, there must have been an error. If the error is a single bit changed, it might be possible to correct the message, but usually the receiver simply sends a negative acknowledgment to initiate a re-transmission of the message.

D-PDU: see Data Link Layer PDU.

DARPA: see Defense Advanced Research Projects Agency.

Data Link Layer: The Data Link Layer (OSI Layer 2) is concerned with moving messages across local area networks (LANs or vLANs) by sending D-PDUs (frames). Like the other layers of the OSI Model, the Data Layer operates by logically sending and receiving frames when in fact it accepts bits from the Physical Layer as a frame or passes a frame to the Physical Layer for transmission as bits. Likewise, the Data Link Layer accepts packets from Layer 3 to be used as the payload for a frame that is passed to the Physical Layer and vice versa. While the Data Layer shares interfaces with both the Physical Layer and the Network Layer, all Layer 2 functions are confined to Layer 2. Therefore, Layer 2 is unaware of the details of the layer below it and the layers above it in the OSI stack.

Data Link Layer PDU: Data Link Layer (Layer 2) protocol datagram units are most commonly called frames. The endpoints of a Layer 2 communication exchange messages as the payloads of frames by sending the frame over the Physical Layer one bit at a time. Logically a Layer 2 conversation looks as if the Layer 2 on each endpoint is directly exchanging frames with the other endpoint's Layer 2 with no interaction at Layer 1.

Data Stream 1:   A DS1 consists of 24 DS0s combined via TDM into a single data stream of 1.44 Mbits/second, or a T1. Since no one really cares if a data stream is part digital voice and part data, DS1 is not commonly used and is called a T1 instead.

Data Stream 3:   A DS3 consists of multiple DS1s[2] combined via TDM into a single data stream of 44.736 Mbit/second, or a T3. The only time the term is used

---

[2] Data Stream 1s

instead of the more common T3 is when service guarantees are being discussed.

Data Stream Zero: A DS0 is the bandwidth allocated to a single voice conversation by a Telco. It is equal to 64kbits/second.

DDNS: see Dynamic Domain Name System.

DDOS: see Distributed Denial of Service attack.

Defense Advanced Research Projects Agency: DARPA is an agency of the United States Department of Defense responsible for the development of emerging technologies for use by the military. Originally known as the Advanced Research Projects Agency (ARPA), the agency was created in February 1958 by President Dwight D. Eisenhower in response to the Soviet launching of Sputnik 1 in 1957. DARPA-funded projects have provided significant technologies that influenced many non-military fields, such as computer networking and the basis for the modern Internet, and graphical user interfaces in information technology. [318].

Demilitarized Zone: "In computer security, a DMZ or demilitarized zone (sometimes referred to as a perimeter network or screened subnet) is a physical or logical subnetwork that contains and exposes an organization's external-facing services to an untrusted network, usually a larger network such as the Internet. The purpose of a DMZ is to add an additional layer of security to an organization's local area network (LAN): an external network node can access only what is exposed in the DMZ, while the rest of the organization's network is fire-walled. The DMZ functions as a small, isolated network positioned between the Internet and the private network and, if its design is effective, allows the organization extra time to detect and address breaches before they would further penetrate into the internal networks. The name is derived from the term "demilitarized zone", an area between states in which military operation is not permitted[3] ." [319]

Denial of Service attack: One method to attack a device or service is to flood it with requests for service faster than it can respond thereby tying up the service with a long line of requests and starving out legitimate requests. This is a DOS or denial of services attack. See also DDOS. One irritating side–effect of DOS attacks in the past is that many devices on the Internet will no longer respond to pings and we have lost some valuable tools because of that.

Designated Intermediate System: When ISIS routers begin to form adjacencies, this can lead to an excessively large number of relationships to be maintained. ISIS routers that *could* form adjacencies instead elect one router to be the Designated Intermediate System (DIS) and run a pseudo–node. The routers then form

---

[3] As always, any citation of Wikipedia is inherently suspect. This entry was verified by my own definition and various sources. I happen to like the definition and quoted it as being better than my own attempts.

a single adjacency to the pseudo–node which helps reduce the overhead on the network and the number of the LSPs flooding the LAN. [18]

Designated Router: If the OSPF area is a local area (not the backbone area zero), OSPF routers can elect one router to be a Designated Router (DR) and one router to be a Backup Designated Router (BDR). For example, on multi–access broadcast networks (such as LANs) routers elect a DR and BDR. DR and BDR serve as the central point for exchanging OSPF routing information. Each non-DR or non-BDR router will exchange routing information only with the DR and BDR, instead of exchanging updates with every router on the network segment. The DR will then distribute topology information to every other router inside the same area, which greatly reduces OSPF traffic [16].

DFI: see DSP Format Identifier.

DHCP: see Dynamic Host Configuration Protocol.

DHCP Server daemon: A process that provides IP addresses and other subnet specific information to a NIC in order that it may gain access to resources. Typically each subnetwork is serviced by no more than one DHCP server. It is possible to run many DHCP servers for a single network but this often causes problems. See also Dynamic Host Configuration Protocol.

dig: see Domain Information Groper.

Dijkstra's Algorithm: Dijkstra's Algorithm is used to find the lowest cost path from a chosen "root" to each node in a weighted graph or digraph. Dijkstra's algorithm can be used to determine the shortest path from one node in a graph to every other node within the same graph data structure, provided that the nodes are reachable from the starting node.

directional graph (Digraph):   A directional graph, or digraph, is a non–empty set of vertices (V) and a set of arcs (A). By this definition, a single vertex is also a digraph. We will only make a distinction between a digraph and a graph when it is important to networking. Bear in mind that most networks are digraphs even if all the links are bi–directional because a link *might* fail in one direction only. This type of failure can be difficult to find.

DIS: see Designated Intermediate System.

Disk Operating System: The original Microsoft operating system was disk–oriented with many commands implemented as small programs on the main, or system, disk and was therefore called DOS. There are other disk operating systems for mainframes and devices, but Microsoft trademarked the name. Linux is also a disk–oriented operating system as are most common operating systems today.

Distributed Denial of Service attack: One method to attack a device or service is to flood it with requests for service faster than it can respond thereby tying up the service with a long line of requests and starving out legitimate requests. This is a DOS or denial of services attack. If the attack is launched from a large number of diverse sources it is a distributed denial of services attack and is much harder to defend against.

DIX: Early Ethernet forerunner developed by Digital Equipment Corporation (DEC), Intel, and Xerox.

DMZ: see Demilitarized Zone.

DNS: see Domain Name Service.

DNSSEC: see Secure Domain Name Service.

Domain Information Groper: Domain Information Groper, or `dig`, is a common TCP/IP utility that queries the DNS database and provides information about a specific host given its IP address or vice versa. Dig is similar to the `nslookup` utility, but provides much more information than `nslookup`. Dig is preferred over `nslookup` which is not always supported.

Domain Name Service: The service that translates human–friendly names such as "www.yahoo.com" to the corresponding machine–friendly addresses, much like a contact list relates names to phone numbers. DNS is distributed throughout the Internet to avoid delays caused by outages, over–use of a server, and attacks on servers. Without DNS the modern Internet would be a much less user–friendly place as one would need to memorize large numbers of 32 bit IP addresses instead of names such as "yahoo.com" and network administrators would not have the flexibility to move services from host to host or take advantage of some of the load balancing techniques that have been developed.

Domain Specific Part: The part of the NSAP address that specifies the domain where the address belongs.

dongle: A plug–in adapter. Typically a hardwired, unpowered piece of equipment such as the Raspberry Pi USB/Ethernet dongle. A cable, or a card, is not usually considered a dongle. At least one website claims the name comes from a company called Rainbow Technologies, which manufactured dongles, which claimed that the term was named for its alleged inventor, a certain Don Gall.

DOS: see Disk Operating System.

DOS attack: see Denial of Service attack.

Dotted decimal format: The shorthand method to represent an IPv4 address as four decimal numbers, between $0 - -255$, with decimal points or dots separating each one. For example, the IPv4 loopback address is 127.0.0.1 in dotted decimal.

DR: see Designated Router.

DS0:   see Data Stream Zero.

DS1:   see Data Stream 1.

DS3:   see Data Stream 3.

DSP: see Domain Specific Part.

DSP Format Identifier: The fifth hex digit of an NSAP address (DFI) signifies the format of the domain specific part of the address.

Dynamic Domain Name System: By its very nature, the Internet is very volatile with devices and whole networks appearing an disappearing at random. If the address of a device is not static, it is not possible to configure DNS to correctly translate the device name to a consistent address. The solution is to dynamically enter the name–address relationship into the working files of the name service when the device obtains an address. The set of protocols and processes to do this are called DDNS.

Dynamic Host Configuration Protocol: Devices obtain an IP address and other required configuration information either statically or dynamically. If the device can move from network to network such as laptops or cell phones, it is not convenient to manually configure the device and so the device is configured dynamically by a DHCP service. The service leases an address on the network for a site–specific length of time from an address pool. Other required information such as the address of a DNS or a gateway to leave the network can also be included as part of DHCP.

EBGP: see External BGP session.

Echo Request and Echo Response: One of the most useful aids when trouble–shooting a network is the program **ping** which is provided by almost all operating systems. **ping** sends out an Echo Request message to an IP address. When the device with that address gets the Echo Request, it has the option to reply with an Echo Response to let the sending program know it is active. Unfortunately the prevalence of denial of service attacks using **ping** has lead to most devices on the Internet being configured to *not* reply to `ping`. So we have lost a number of useful utilities because some people cannot act responsibly on the Internet.

EIA: see Electronic Industries Alliance.

EIGRP: see Enhanced Internal Gateway Routing Protocol.

Electronic Industries Alliance: The EIA works with the Telecommunications Industries Association (TIA) on a number of cabling standards such as color–coding of cable pairs, building wiring standards, and pair assignments in jacks and plugs. Most of the EIA standards encountered in networking are maintained working closely with the TIA.

Encapsulation: Often a message for one protocol or OSI layer is sent by placing it in the data payload of a different message. The most common example of this on the Internet is when a packet is the data payload of a frame in order to send it over a LAN. This is called encapsulation.

Enhanced Internal Gateway Routing Protocol: Originally EIGRP was a proprietary routing protocol developed by Cisco and was later made an open standard protocol. Unfortunately Quagga does not support EIGRP at this time, but apparently there is a group working on implementing it.

Enhanced Simple Mail Transfer Protocol: ESMTP is an extension to SMTP to allow for easier integration of tools to help control spam and other email based annoyances. For all intents and purposes ESMTP is simply a better (enhanced) email protocol or extensions to the existing SMTP.

ESMTP: see Enhanced Simple Mail Transfer Protocol.

eth0: The first Ethernet connection on a device running Linux is always `eth0` which is a mmnemonic for Ethernet zero. On the Raspberry Pi, `eth0` is the RJ45 connection.

eth1: The second Ethernet connection on a device running Linux,if present, is always `eth1` which is a mnemonic for Ethernet one. On the Raspberry Pi, `eth1` is the first dongle connection.

eth2: The third Ethernet connection on a device running Linux,if present, is always `eth2` which is a mnemonic for Ethernet two. On the Raspberry Pi, `eth2` is the second dongle connection.

Ethernet: The most common Layer 2 implementation is Ethernet. Ethernet, IEEE Standard 802.3, grew out of early networking efforts by the Xerox Corporation. Thanks to a somewhat unobvious decision by the engineers to use a 48 bit address when a much smaller address would have sufficed, Ethernet MAC addressing is critical to many current networking standards. Thank you, Xerox. [31]

External BGP session: Pairs of routers running BGP set up one of two types of relationships. If both router processes deal only with other routers in the same AS, then the session is an internal session or IBGP. If the relationship is between routing processes in different Autonomous Systems then the session is an external BGP session or EBGP. External sessions require the administrator to configure how, and what, routing information is to be shared.

FAT: see File Allocation Table (16 bit version).

FAT32: see File Allocation Table (32 bit Version).

FCS: see Frame Check Sequence.

FDDI: see Fiber Data Distribution Interface.

Fiber Data Distribution Interface: The original high speed fiber interconnection for servers and meet–points ran as a 100 megabit per second ring. Obviously, 100 megabits is no longer high speed but there are extremely high speed SONET rings to replace FDDI, pronounced "fiddy".

FIFO: see First In, First Out.

File Allocation Table (16 bit version): FAT is the original DOS file allocation protocol for early versions of DOS and Windows. Due to the massive growth of disk sizes and the limitations of FAT, it is no longer in wide use but it is still understood by most file systems. If you are using any current media such as a USB drive or an SD card, the file system will not be FAT; it will be FAT32.

File Allocation Table (32 bit Version): FAT32 is the default file allocation protocol for early 32–bit versions of Windows and DOS. Most operating systems can understand (read/write) FAT32.

File Transfer Protocol: One of the common protocols that run over IP is the File Transfer Protocol. FTP pre–dates HTTP and was once the main protocol for archives of software, images, and program updates. FTP is still in frequent use but now is most often hidden behind a web page.

First In, First Out: A queue is a list that is *strictly* FIFO to guarantee messages are processed in the order received. In a FIFO queue, messages can only be added at the "tail" (enqueue) and removed from the "head" (dequeue). FIFO queues are extremely important in data communications.

Flavor: A flavor of Linux is a specific distribution or "distro". The two terms can be used interchangeably.

FQDN: see Fully Qualified Domain Name.

frame: A frame is a variable length Data Link PDU. The most common frame today are Ethernet frames which have a length between 64 and 1516 bytes. There are other Data Link protocols that use different frame formats and sizes.

Frame Check Sequence: A frame check sequence is a Cyclical Redundancy Check to allow the receiving endpoint to quickly check all frames for errors and then request a retransmission of the frame.

Free Range Routing: FRR is a fork of the Quagga project and operates in a similar manner using the `zebra` routing daemon to insert routes into the Linux kernel tables. At the time of this writing, FRR supports all of the routing protocols covered in this text and others are in production. Some of the awkwardness of Quagga has been addressed while retaining the similarity with remotely configuring a Cisco router.

FRR: see Free Range Routing.

FTP: see File Transfer Protocol.

full mesh: A full mesh graph is one with all possible edges or arcs. In networking, a full mesh network has dedicated media connecting each possible pair of nodes in the network. This is very robust but quickly becomes too expensive in terms of NICs and actual money.

Fully Qualified Domain Name: A FQDN name consists of a hostname, such as *host1*, that refers to exactly one device on the network followed by a "dot (period character)" and the domain name such as "yahoo.com" to give a FQDN of "host1.yahoo.com". Except in special uses beyond the scope of this text, a fully qualified domain name must be unique.

geek humor: The frequent references to "geek humor" are not completely silly. Network administrators often use some of these phrases which can completely confuse a person new to the field (a "newbie"). Another issue is the fact that many problems, algorithms, and odd solutions to problems have names that are "geek humor" so that people will remember them. Although outside the scope of this text, the Byzantine Generals problem [45] [33] was named that way so that it would be remembered. Unfortunately for us in the computer industry, Byzantine behavior is a common source of failures.

Gondor: A mythical city in the book *The Lord of the Rings* by J.R.R. Tolkien [314]. The city used signal fires to ask for aid from the neighboring kingdom of Rohan. This is used as an example of certain types of communication, such as in 1d.

GOSIP: The Federal Networking Council (FNC), the Internet Activities Board(IAB), and the Internet Engineering Task Force (IETF) have a firm commitment to responsible integration of OSI based upon sound network planning. This implies that OSI will be added to the Internet without sacrificing services now available to existing Internet users, and that a multi-protocol environment will exist in the Internet for a prolonged period. Planning is underway within the Internet community to enable integration of OSI, coexistence of OSI with TCP/IP, and interoperability between OSI and TCP/IP. The U.S. Government OSI Profile (GOSIP) is a necessary tool for planning OSI integration. However, concern remains as to how GOSIP should be applied to near-term network planning. [85]

graph: A graph, usually denoted by $G = (V, E)$ is a non–empty set of vertices (**V**) and a set of edges (**E**) which implies that a graph must have at least one vertex. By this definition, a single vertex is a graph. If there is a path from any vertex u in the graph to any vertex v in the graph, the graph is connected. While multiple *connected* graphs can be part of the same graph, we will consider all graphs as connected. It will be pointed out if a graph (or network) is not *connected*[4].

Graphical User Interface: Most modern operating systems offer a user–friendly[5] interface based heavily upon screen output and mouse input. These are still menu systems at heart, but they are called GUIs.

GUI: see Graphical User Interface.

HDMI: see High Definition Multimedia Interface.

Hello message: Many protocols require a device to announce itself to devices that might be neighbors with a message that is called an "hello". These messages are sometimes repeated at regular intervals to verify to its neighbors that a device and process is operating correctly. The format and exact use of the "hello" message depends upon the protocol.

hexadecimal (base 16): Hexadecimal is an easier to remember way to write long strings of binary digits. Each hex digit translates directly to a unique pattern of

---

[4] In this entry, the term connected refers to a graph, or digraph, where every node is reachable from every other node by traversing from node to node via edges or arcs.

[5] Unfortunately user–friendly things are often not at all user–friendly.

four binary digits and vice versa. You can think of hex as a shorthand for binary.

High Definition Multimedia Interface: The Raspberry Pi supports a standard HDMI interface or a mini-HDMI interface depending upon the model.

Hop: A term used to describe the action of a message, typically a packet, moving from one connectivity device to the next. Most commonly used to denote a packet being relayed to the next router in a path through the network.

HTML: see HyperText Markup Language.

HTTP: see Hyper–Text Transfer Protocol.

httpd: Web Server daemon. The most common web server is Apache which can be downloaded for free from https://httpd.apache.org/.

HTTPS: see Secure Hyper–Text Transfer Protocol.

hub: A hub is a Layer 1 device that accepts signals on any port and then retransmits them out all other ports at the correct signal values. A hub is sometimes referred to as a "multi-tailed repeater".

Hyper–Text Transfer Protocol: HTTP and HTTPS are the protocols that control web pages, browsers, and web servers that comprise the World Wide Web (www). Sir Tim Berners-Lee invented the World Wide Web in 1989. He is the Director of the World Wide Web Consortium (W3C), a Web standards organization founded in 1994 which develops interoperable technologies (specifications, guidelines, software, and tools) to lead the Web to its full potential. A graduate of Oxford University, Sir Tim invented the Web while at CERN, the European Particle Physics Laboratory, in 1989. He wrote the first web client and server in 1990. His specifications of URLs, HTTP and HTML were refined as Web technology spread. On 4 April 2017, Sir Tim was awarded the ACM A.M. Turing Prize for inventing the World Wide Web, the first web browser, and the fundamental protocols and algorithms allowing the Web to scale. [5]. If Tim Berners-Lee had not put the World Wide Web into the public domain, there would not be one that would even remotely resemble what we use every day. This is just my opinion, but I suspect I am correct.

HyperText Markup Language: HTML is the source language for web pages. There are many good (free!) sources on the web for more information about HTML such as from the www consortium:
`https://www.w3schools.com/html/`. Web pages can also be built dynamically using PHP (Personal HTML Programming language) using a LAMP server.

I Hear U message: This message is used in Babel as a response to the "hello" message and as a "Keep Alive" message to insure that a neighbor process is still active.

IANA: see Internet Authority for Names and Addresses.

IBGP: see Internal BGP session.

IBM: see International Business Machines.

ICANN: see Internet Corporation for Assigning Names and Numbers.

ICMP: see Internet Control Message Protocol.

ID: see System Identifier (NSAP).

IDI: see Initial Domain Identifier (NSAP).

IDP: see Initial Domain Part (NSAP).

IEEE: see Institute of Electrical and Electronics Engineers.

IETF: see Internet Engineering Task Force.

IFXR: see Incremental Zone Transfer.

IGRP: see Internal Gateway Routing Protocol.

IHU: see I Hear U message.

IMAP: see Internet Message Access Protocol.

Incremental Zone Transfer: Some DNS protocols such as DDNS require frequent zone file transfers. To minimize the overhead from running such protocols, most DNS processes can use incremental zone transfers (IFXR) to send only the updated information. This does not stop the regular transfer of full zone files as needed, but can drastically reduce the overhead in most cases when running DDNS.

Initial Domain Identifier (NSAP):   The IDI is part of the header of an NSAP address and beyond the needs of this book.

Initial Domain Part (NSAP):   The first three bytes of an NSAP address taken together are the IDP. For our purposes only the first byte of the NSAP address is of interest.

Institute of Electrical and Electronics Engineers: The IEEE (pronounced "Eye triple E") is one of the largest standards and research bodies in computing, especially in terms of hardware and protocols. For example, all of the 802. protocols are maintained by the IEEE.

"IEEE and its organizational units engage in coordinated public policy activities at the national, regional, and international levels in order to advance the mission and vision of securing the benefits of technology for the advancement of society." From their website: https://www.ieee.org, and I recommend anyone interested in the Internet or networking to join the IEEE.

Interconnected Networks: The global network known as the Internet is really a web of smaller Layer 3 networks interconnected by routers, IP Forwarders, and Layer 3 switches. In reality the current Internet is composed of a large number of Internet Service Providers (ISPs), global networks, and meet–points with many smaller networks embedded in the large ones. The Internet and its components are usually thought of as clouds because the details of what networks are involved and how they are connected is unknown, unknowable, and not usually important. Messages go into the Internet and magically arrive at their destination, much like phone calls through the phone system.

Intermediate System: The OSI uses the term Intermediate System to mean an autonomous area of routers. Autonomous System and Intermediate System have the same meaning.

Intermediate System to Intermediate System: ISIS is a routing protocol based upon NSAP addressing rather than IP addressing and was developed about the same time as OSPF to solve some of the problems found in the RIP protocol. ISIS is a link–state protocol which converges quickly and can handle a large number of networks with ease. ISIS networks are broken into one Level 2 backbone connecting any number of local Level 1 areas. Routers can be contained in Level 1 or Level 2 and border routers (designated Level 1 and 2) can pass packets between the Level 1 area and the backbone. Because it works well and is relatively easy to configure, ISIS is popular among large ISPs.

Internal BGP session: Pairs of routers running BGP set up one of two types of relationships. If both router processes deal only with other routers in the same AS, then the session is an internal session or IBGP. If the relationship is between routing processes in different Autonomous Systems then the session is an external BGP session or EBGP. External sessions require the administrator to configure how, and what, routing information is to be shared.

Internal Gateway Routing Protocol: IGRP was the first major proprietary routing protocol developed by Cisco for interior routing between Cisco routers.

While IGRP uses a distance–vector method to determine the best route, it was far superior to RIP. Like RIPv1, IGRP did not support variable–length subnet masks and classless addressing. It has been superseded by EIGRP which is now an open standard routing protocol.

International Business Machines: IBM has been one of the major companies in the computer industry since computers became available as mainframes. IBM has been a major innovator in networking and the development of the PC. One of IBM's innovations was IEEE 802.5 Token Ring.

International Standards Organization: The ISO is chartered by the United Nations to promote and administer various international standards. These standards play an important part in networking outside the United States and must be followed by devices which potentially operate outside the USA or interact with devices outside the USA. In short, any device that works on a network must conform to any applicable ISO standards.

Internet: see Interconnected Networks.

Internet Authority for Names and Addresses: The IANA performs three critical functions for the Internet.

- Domain Names: Management of the DNS Root Zone (assignments of ccTLDs and gTLDs) along with other functions such as the .int and .arpa zones.
- Number Resources: Coordination of the global IP and AS number spaces, such as allocations made to Regional Internet Registries.
- Protocol Assignments: The central repository for protocol name and number registries used in many Internet protocols.

(From their website: https://www.iana.org/).

Internet Control Message Protocol: A core protocol in the IP suite that allows a device to notify the sender that something has gone wrong and packets were not delivered. Whether or not to send an ICMP message is almost always optional. Remember that at Layer 3 and below devices do not care if a packet is correctly received. Bad packets may simply be dropped. It is the responsibility of Layer 4 (Transport Layer) to insure delivery. There are other types of ICMP messages such as ping, but these are not of immediate interest to the goals of this text.

Internet Corporation for Assigning Names and Numbers:
ICANN is a not–for–profit public corporation with participants from all over the world dedicated to keeping the Internet secure, stable and interoperable. It promotes competition and develops policy on the Internet's unique identifiers. Through its coordination role of the Internet's naming system, it does have an important impact on the expansion and evolution of the Internet. (From their

web site: https://www.icann.org/)

Internet Engineering Task Force: The IETF maintains all of the RFC standards for the Internet, among other things. This group is very influential in all aspects of the future of the Internet. All of the RFCs used to support the Internet can be found on the IETF website (which also has an excellent search engine for RFCs) with a minimum of effort.

Internet Message Access Protocol: IMAP is one of the two most common protocols used to connect an email client with an email server. Typically email originates on a PC or other device which transfers the email to an email server. Likewise, email is received by the email server and held there until the email client requests it. If the email client and server communicate via IMAP, the email remains on the server until the client requests it to be deleted.

Internet Operating System: The CISCO Router Operating System, or IOS, provides a user–friendly method to configure Cisco equipment. IOS provides a common "look and feel" across all Cisco devices and was, in my opinion, the major reason why Cisco took such an early and commanding lead in the router market. Quagga on the Raspberry Pi was designed to duplicate this "look and feel".

Internet Protocol: IP refers to the suite of Layer 3 protocols which support the Internet and are allowed and routed on the Internet. IP grew out of early networking protocols developed for UNIX and underlies the TCP and UDP upper layer protocols.

Internet Protocol, Next Generation: An early name for IPv6 that is defunct with the possible exception of the name of the RIP daemon for IPv6 which is often seen as RIPng.

Internet Protocol, Version 4:   Currently the most common version of the IP protocols is IPv4. There are many problems with IPv4 addressing, most of which were caused by explosive growth and short–sighted mismanagement of the assignment of address space. IPv6 addresses the current problems with IPv4 and should serve us well into the future.

Internet Protocol, Version 6:   The newest version of internet protocols is designated IPv6 which provides a much larger address space than IPv4 and to promotes route summarization. IPv6 addresses are 128 bits long and divided into a 64bit prefix and a 64 bit host portion. The goal is to provide so many addresses that all possible devices can be supported.

Internet Service Provider: Most networks do not connect directly to the main Internet but instead connect to a regional ISP or a large national ISP. Regional ISPs provide services to home and office networks and relieve them of the need to contract for bandwidth and other telco services.

Internetwork Packet Exchange: IPX is the proprietary Layer 3 addressing scheme developed by Novell for NetWare. As of NetWare 5.0, Novell no longer suggests using IPX but rather running all of their proprietary services over IP. Many routing protocols will still support IPX, but it is only to be backwardly compatible. NetWare is still popular with schools.

Intranet: see Private Internet.

ios: see Internet Operating System.

IP: see Internet Protocol.

IPng: see Internet Protocol, Next Generation.

IPv4:   see Internet Protocol, Version 4.

IPv6:   see Internet Protocol, Version 6.

IPX: see Internetwork Packet Exchange.

IS: see Intermediate System.

IS-IS: see ISIS Inter–Area Routing.

ISIS: see Intermediate System to Intermediate System.

ISIS Inter–Area Routing: IS–IS is used in this text to denote the protocols which are a part of the ISIS routing protocol suite that handle communications between two ISIS autonomous systems or Autonomous Systems. These protocols use NSAP addressing rather than IPv4 or IPv6 addressing to denote routers.

ISO: see International Standards Organization.

ISP: see Internet Service Provider.

Java: A commonly used Object Oriented Programming (OOP) language in the "C" family. Java is a cross–platform language and runs on large computers, small computers, cell phones, watches, and many other devices.

Keep Alive message: Any message sent simply to inform the other endpoint that this endpoint is still viable but has no messages to send at this time. This is done to confirm a process is live and not a looping or dead. "Keep Alives" are used in Babel, OSPF, and other routing processes that do not send updates on a scheduled basis.

kernel: That part of the OS that is loaded when the computer starts and must remain in memory as long as the computer is operating.

L2TP: see Layer 2 Tunneling Protocol.

LAMP: see LAMP Web Server.

LAMP Web Server: LAMP (Linux, Apache, MySQL, and PHP) is a set of services to provide dynamic web pages hosted by a server running a Linux operating system. One of the advantages of LAMP is it can provide a common set of services and programming interfaces for a large number of platforms. At the "human" level, LAMP, WAMP, and XAMP are compatible. With the help of the tutorials on the Apache Friends website [32], XAMP is very easy to install and configure on Windows, Linux, or OS X. However, there may be security issues after the installation that need to be addressed for a production system.

LAN: see Local Area Network.

Last in – First out: A LIFO, or alternately FILO First in – Last out, structure acts like a stack. Messages are placed on the "top" of the LIFO stack and can only be retrieved from the top of the LIFO stack. LIFO structures are not as common in networking as FIFO, or queue, structures.

Layer 2 Tunneling Protocol: Layer 2 Tunneling Protocol (L2TP) is a tunneling protocol used to support virtual private networks (VPNs) or as part of the delivery of services by ISPs. It does not provide any encryption or confidentiality by itself.

Layer1:   see Physical Layer.

Layer2:   see Data Link Layer.

Layer3:   see Network Layer.

Layer4:   see Transport Layer.

Layer5:   see Session Layer.

Layer6:   see Presentation Layer.

Layer7:   see Application Layer.

LED: see Light Emitting Diode.

LIFO: see Last in – First out.

Light Emitting Diode: In networking, LEDs have two main functions. The most common one is to provide flashing lights to indicate activity or important statuses. If a device has blinking LEDs on or near the NIC, it is usually communicating over the network[6]. The other main place LEDs are used in networking is critical to the functioning of short range fiber optic cables. For shorter range fiber, a laser is not necessary and presents an unwelcome hazard, so for multimode fiber the signal source is an infrared LED. However, any fiber connection should be treated as an eye hazard as people have been known to use the wrong cable. Also any fiber port should be treated as a laser light source capable of potentially damaging your eyes. The sources for fiber optic signals operate in the infrared and cannot be seen.

Link State Announcement: LSAs are used by routers running OSPF or ISIS to inform neighbors of a change to a local link. These LSAs are only exchanged when there is a change in the network and therefore produce much less overhead than a typical distance–vector protocol such as RIP. Link–state protocols typically elect a router to receive LSAs and then flood them to the routers in the area or AS. This avoids the possibility of many routers broadcasting the same LSAs and wasting bandwidth.

Link State Database: Link–state routing protocols depend upon each router in the local area or AS having the same view of the network as all the other routers. This is achieved by each router maintaining a database of the current graph of the network. Once the LSD has been formed and updated, the router runs either Dijkstra's Algorithm or the Bellman–Ford Algorithm to quickly determine the best route to each known destination. Once each router has done this, the network has converged and no further updates are required until a network change occurs.

Link State Packet Pseudonode: When a network of routers running ISIS elect a leader to handle LSAs for the network, that router creates a psuedonode to receive all LSAs from routers in the network. This LSP is also the only router that floods the network with updates when it receives an LSA. When the psuedonode has no LSAs to distribute, the network has converged and no further update announcements are needed until a network link changes state. This drastically reduces the overhead of a stable network when compared to a distance–vector protocol such as RIP.

Linux: Linux is an open–source OS that is extremely similar to the look and feel of UNIX. Because all but the kernel of the OS is open–source and public domain,

---

[6] Of course someone had to do it the other way around. If a red LED is lit on a Marconi ATM switch the media is down or disconnected.

anyone can participate in the development of a Linux distribution or even develop their own distribution for special needs. Because Linux must be open–source, many software companies provide special distributions that include proprietary software that either runs on top of Linux or provides enhancements to Linux that are not open–source. For example, IBM offers Linux for their server products and charges for support and enhancements. If you develop a Linux distribution, called a distro, you *must* make a version of it free and open–source as per the licensing agreement for the kernel.

Listserv: List severs are used to automatically allow email users to join or leave a mailing list. Most mailing list servers accept commands via email or web pages and are administered by a person who can add or delete email addresses as well. Most lists on listservs are not monitored, but there is typically a way to complain about inappropriate email. Unfortunately, many people will automatically add your email to their list serve which can be annoying. Typically, if you click on *unsubscribe* and follow the instructions you will actually be removed from the list. LISTSERV is a registered trademark licensed to L-Soft international, Inc. If you are interested, LISTSERV Lite Free Edition is a freeware version of LIST-SERV Lite, limited to a maximum of 10 mailing lists with up to 500 subscribers each. It is available for users who want to run hobby or interest-based email lists and do not derive a profit, directly or indirectly, from using the software. It is available from **http://www.lsoft.com/** for Unix, Linux, and Windows.

lo: The local loopback interface on a device running Linux is named **lo**. This is the adapter for localhost, 127.0.0.1, and :: 1.

Local Area Network: A LAN is actually a Broadcast Domain or the set of devices that can exchange broadcasts over some media. If two devices can receive broadcasts from each other, they are on the same LAN. If they cannot, they are not on the same LAN. With devices such as hubs, bridges, and switches, LANs can encompass many different types of media and extend over large distances. The only requirement is that a broadcast from one device must reach every other device before it times out. This implies that every media has a maximum distance on a LAN.

locality: The Principle of Locality can refer to any type of locality, but typically refers to spacial locality. The principle (spacial) refers to the fact that most access will occur close to the last access. For example, someone sending you a text message is likely to trigger a response from you which could well lead to long text conversation. This principle arises in the design of telephone systems, computer architecture, and computer programming.

LSA: see Link State Announcement.

LSD: see Link State Database.

LSP: see Link State Packet Pseudonode.

MAC: see Media Access Control.

Mail Exchange Resource Record (DNS):   DNS and SMTP are able to work in a coordinated way where information about the organization wide email network is shared via MX records with name service. MX records can control the priority of email services and backup email services to handle overflow situations. The details are beyond the scope of this text, but the priority in an MX record can control whether the email always goes first to a master email service or if it goes to a series of services in a round–robin manner.

Mail Transfer Agent: Typically an MTA is an email server connected 24 hours a day to the Internet, but any service capable of sending and receiving email can act as an MTA. The two MTA services examined in this text are Sendmail and Postfix which both can run on the Raspberry Pi. There are many others, but these seem to be the most popular currently.

Mailman Email List Service: Mailman is a set of processes to integrate support for email lists into most MTAs running on Linux or Windows. Mailman has both an email and web interface for users to subscribe/unsubscribe and manager their preference for multiple email lists.

Media Access Control: The term MAC can refer to one of two things in Layer 2 networking:

- The most common use of MAC is to denote the Layer 2 address of a device which is the unique 48 bit MAC address which is assigned in the hardware of a NIC. The first six hex digits of the MAC address (formally the BlockID) are assigned to the manufacturer and is common to all devices from the manufacturer. The last six hex digits are assigned as desired by the manufacturer as long as they are unique in all the world. This means the MAC address of every NIC manufactured is unique. If this were not so, Layer 2 would not work properly. Unfortunately, not all manufacturers play by the rules and sometimes a network will have duplicate MAC addresses which is unbelievably hard to determine and track down. It does not happen often.
- The other usage of MAC is to describe the lower sub–level of Layer 2 which is the Media Access Control sub–layer which is responsible for attaching the NIC's MAC address to outgoing frames and listening to shared media for frames the NIC must process.

MAC is used in this text as the Layer 2 hardware address of a NIC.

Meet–point: A meet–point is a small, extremely fast network that connects a number of ISPs together to allow packets to move from one ISP to another. Each large ISP maintains at least one of the meet–points.

Micro–fractures: Microscopic cracks that develop in glass fiber from being crushed or bent past the allowed "bend radius." These cracks act as mirrors and cause the light signal to be reflected back towards the source causing signal loss by back–scattering and interference.

MIME: see Multipurpose Internet Mail Extensions.

Minimum Spanning Tree: When a network is considered as a directed graph with weighted arcs it is useful to be able to construct a tree which completely connects all the nodes (devices) and has the lowest complete cost associated with it. This naturally leads to finding the least cost path from the current node to all other nodes in order for a node (router in this case) to quickly determine the shortest path to the destination node or router. This technique is known as Shortest Path First or SPF. Both OSPF and ISIS rely upon SPF.

MobileIP: see Cellular IP.

modem: see Modulator/Demodulator.

Modulator/Demodulator: A modem (modulator/demodulator) is used to convert an incoming digital signal to an analog signal for transmission over an analog medium such as a voice grade phone line. At the other endpoint, the modem demodulates the signal to retrieve the original digital signal.

MPLS: see Multi-Protocol Label Switching.

MST: see Minimum Spanning Tree.

MTA: see Mail Transfer Agent.

Multi-Protocol Label Switching: MPLS is a protocol–independent routing technique designed to speed up and shape traffic flows across enterprise wide area and service provider networks regardless of the underlying network (ATM, IP, Frame Relay, TDM, or whatever). MPLS allows most data packets to be forwarded at Layer 2 – the switching level – rather than having to be passed up to Layer 3 – the routing level. MPLS was created in the late 1990s as a more efficient alternative to traditional IP routing, which requires each router to independently determine a packet's next hop by inspecting the packet's destination IP address before consulting its own routing table. This process consumes time and hardware resources, potentially resulting in degraded performance for real-time applications such as voice and video. In an MPLS network, the very first router to receive a packet determines the packet's entire route upfront, the identity of which is quickly conveyed to subsequent routers using a label in the packet header. While router hardware has improved exponentially since MPLS was first developed – somewhat diminishing its significance as a more efficient traffic management technology– it remains important and popular due to its various other benefits, particularly security, flexibility and traffic engineering.

multicast: Any message sent to a special MAC or IP address on a shared me-dia that is to be processed by a pre–defined group of devices is called a multi-cast. Multicasts can appear at Layer 2 or Layer 3 and are often used for router–to–router communications, streaming media, and other specialized applications where there is one source with a limited number of destinations that have joined a multicast group.

Multipurpose Internet Mail Extensions: The MIME standards define how at-tachments can be transferred as if they were simple ASCII text and still be inter-preted as binary executables or attachments such as video files, photos, or music.

MX: see Mail Exchange Resource Record (DNS).

N-PDU: see Network Layer PDU.

NAK: see Negative Acknowledgment.

Name service: Name service is the service that translates between human–friendly names and IP or NSAP addresses on the Internet. DNS is the full Distributed Name Service which includes protocols to find the appropriate Name Server that is authoritative for a given domain. It is rarely important to distinguish between NS and DNS.

NS can also refer to the NS Resource Record in a Zone file.

Name Service Lookup: NSlookup is a common utility to inquire of a name ser-vice what information it has on a hostname or IP address. While **nslookup** is still supported on all platforms, it is no longer in active development as DIG is much more flexible. However, on many systems **nslookup** has an interactive mode which **dig** lacks.

named: The daemon responsible for providing name service, typically BIND.

nano: **nano** is a very popular text editor found on almost all Linux systems such as Raspbian. **nano** is simple to learn and easy to use even though this text favors **vi**. The only reason this text uses **vi** is that is *my* preference.

NAT: see Network Address Translation.

National Institute of Standards and Technology: NIST is the replacement for the previous standards organization ANSI and deals with all the standards supported by the US Federal Government. NIST is the USA's representative to the ISO.

Negative Acknowledgment: Technically, a NAK is a single byte (hex 15) response to note there was a problem with a transmission, but NAK is used to denote any negative response from a receiving device. Like ACK, a device has the option in some protocols to NAK more than one transmission with a single response. The exact message used to NAK is protocol dependent.

NetBEUI: see NetBIOS Extended User Interface.

NetBIOS: see Network BIOS.

NetBIOS Extended User Interface: Microsoft's extended version of the IBM NetBIOS. Like NetBIOS, NetBEUI requires almost no configuration and uses very few resources to facilitate actions such as file sharing, printer sharing, and others.

Network Address Translation: NAT allows a router to run a protocol to translate between private network requests and Internet requests. The most common example is to translate from a home 192.168.1.0 network to the Internet. NAT boxes can logically connect private networks to the Internet without violating the prohibition against private addresses on the Internet because the NAT box prevents the private addresses from ever touching the INERNET. More sophisticated NAT boxes can also act as web proxy devices to cache all web pages retrieved by the NAT in the event that some device on the private network might request that page before it becomes stale. Without NAT boxes, everyone would need a range of public IP addresses and the Internet would have self–destructed in the mid 90s.

Network BIOS: When DOS and Windows PCs were first networked together on LANs, a method was needed to allow the sharing of resources such as files, folders, or printers. To allow this sharing, Microsoft and IBM extended the existing BIOS by adding Transport and Session layer services to allow the sharing of resources based upon the name of the computer to which they were attached. This has worked so well that it is still the method of choice to share Windows folders, printers, and files even today. The only major problem with NetBIOS is that each device has a monolithic name with no network and host part. This means that NetBIOS cannot be directly routed unless it is encapsulated which is rarely done.

Network Interface Card: Any port that connects a device to a network is a NIC even if it is an infrared port or a switch port on an Ethernet switch. Each NIC has the memory, CPU power, and programming in firmware to run the required protocols for the media it is designed to connect. Early on this was not the case and device drivers had to be loaded for each network protocol the NIC was to understand. Each NIC has a MAC address associated with it (burned into the hardware or on a chip on the NIC) which is unique in all the world. Fiber NICs can have multiple logical connections, but usually a NIC connects to exactly one network.

Network Layer: The main task of the Network Layer is to move packets through a network to their destination network. As Layer 3 of the OSI Model, the Network Layer operates logically by exchanging N-PDUs (packets) with Layer 3 on the other endpoint of the conversation. This is the layer at which routing takes place if the conversation is connectionless as are most conversations on the Internet.

Network Layer PDU: N-PDUs are packets. The size of a packet can vary depending upon the data payload, the detailed format of the packet, and the Layer 3 protocol involved. For the purposes of this text, the only N-PDUs of interest to us are IPv4 and IPv6 packets.

Network News Transfer Protocol: Before the world wide web took the Internet by storm, much of the same kinds of information was shared by posting it to newsgroups that were hosted by NNTP servers. A client was used to browse, read, and even download postings to a local computer. There are still NNTP servers on the Internet, but they now tend to be subscription services and not all ISPs support NNTP.

Network Service Access Point: The preferred Layer 3 addressing on devices running the full OSI Model is NSAP addressing. NSAP addresses have a wide variety of formats and lengths which can be determined by examining the first byte of the NSAP address. To my knowledge, the only protocol on the Raspberry Pi that supports NSAP addressing is ISIS, but there is no reason why the Pi couldn't support other protocols that rely on NSAP addressing.

Network Time Protocol: Clocks gain or lose time constantly and computer clocks are no different from wall clocks, so to keep the clocks on a network in reasonable agreement, one technique that can be used (if high accuracy is not needed) is NTP. A client on the computer contacts an NTP service and requests the time. The client then adjusts the computer's clock to that time. It is customary to use a set of NTP servers in a round–robin or fail–over mode to be assured of eventually getting a response. Most clients will try a different server if the time change would be greater than a threshold value.

NGO: see Non-Governmental Organization.

nibble: A single hex digit of four bits is called a "nibble". A nibble is half a byte. I suspect this is "geek" humor.

NIC: see Network Interface Card.

NIST: see National Institute of Standards and Technology.

NNTP: see Network News Transfer Protocol.

Non-Governmental Organization: NGOs are used in some examples and exercises as typical organizations that might use ".org" instead of the ".com", ".gov", or country code TLD.

NS: see Name service.

NSAP: see Network Service Access Point.

NSAP Selector: The last byte of an NSAP address is the SEL or selector byte and acts much like a port or socket to guide the data to the proper process on the device. The only SEL this text is interested in is 00 which is the routing process (ISIS).

nslookup: see Name Service Lookup.

NTP: see Network Time Protocol.

OC1:   see Optical Carrier 1.

OC12:   see Optical Carrier 12.

OC24:   see Optical Carrier 24.

OC3:   see Optical Carrier 3.

octet: An octet is a group of eight binary digits. For all intents and purposes, an octet is equivalent to a byte except that networking is done in octets while data storage and representation is done in bytes.

Open Shortest Path First (IPv4):   OSPF is an open–standard, link–state routing protocol designed for medium to large networks. OSPF breaks the network into a two–level hierarchy with each local set of networks contained in an OSPF area and all traffic between areas must traverse the backbone area zero (it is *required* to be area zero.). All routers belong to a single area except those connecting a local area with area zero. No interconnections are allowed between areas, or allowed to leave an area, except those going to/from area zero. One of the biggest drawbacks of OSPF is the inherent difficulty in merging two OSPF networks when two large organizations merge.

OSPF is also used when referring to OSPF in general when it is not important whether the underlying protocols are IPv4 or IPv6.

Open Shortest Path First (IPv6):   OSPFv3 is the IPv6 version of OSPF. While the two versions are incompatible there are ways around the minor issues this causes. The simplest is to run both versions on every router.

Open Systems Interchange: The OSI Model is an internationally recognized suite of open standards developed and maintained by ISO. Typically OSI is used to refer to the OSI Model and the seven layers which comprise it.

Operating System: An operating system is a low–level program designed to load programs, control the execution of programs, provide and control access to hardware resources such a memory, and various other tasks. A good OS works behind the scenes and is not noticed by the user. A poor OS causes the device to crash, lose data, burn up (literally), or display a blue screen of death. All of the operating systems discussed in this text are excellent at what they do and have matured nicely.

Optical Carrier 1:   Optical Carrier is an internationally recognized standard rating that indicates the speed of a SONET connection. Like the T-carriers (T1, T2, and T3), SONET speeds are built in blocks of OC1, or 51.84 Mbit/second. The lowest common speed is OC3 or 155.52 Mbit/second. See Table 2.4 for details.

Optical Carrier 12:   Optical Carrier 12, 622.08 Mbits/second, is a common sub-gigabit speed SONET delivered by a Telso to a premise. See Table 2.4 for details.

Optical Carrier 24:   Optical Carrier 24, 1244.16 Mbits/second, is the standard gigabit speed SONET delivered by a Telso to a premise. See Table 2.4 for details.

Optical Carrier 3:   Optical Carrier 3, 155.52 Mbits/second, is the standard lowest speed SONET delivered by a Telso to a premise. See Table 2.4 for details.

OS: see Operating System.

OSI: see Open Systems Interchange.

OSPF: see Open Shortest Path First (IPv4).

OSPFv3:   see Open Shortest Path First (IPv6).

P-PDU: see Presentation Layer PDU.

PC: see Personal Computer.

PDU: see Protocol Datagram Unit.

Personal Computer: PC is a term for a generic computer designed to be used by usually only one person at a time. A PC can be a desktop, laptop, tablet, or a high power workstation.

phishing: A group of attacks that use spam or bogus email to attempt to trick users into revealing sensitive information. Caution: You do not have to open the email to get stung and many email clients open email for previews anyway.

php: see PHP: Hypertext Preprocessor.

PHP: Hypertext Preprocessor: PHP originally stood for Personal Home Page, but it now stands for the recursive "PHP: Hypertext Preprocessor"[7]. It was originally written in 1994 by Rasmus Lerdorf to keep his personal web pages up to date [47]. PHP is the "p" in LAMP and is used to write programs which produce dynamic web pages to be sent to a web browser. There are many other ways to produce dynamic web pages, but PHP is very common.

Physical Layer: The most basic networking is composed of bits traveling across some media. This is the Physical Layer, or Layer 1, of the OSI model. Layer 1 is tasked with putting bits onto, and taking bits off of, a media constrained by the standards that apply to that media. Layer 1 devices are quite common which have no other purpose than to receive weak signals and then retransmit the same bits at the correct levels for that media. If these devices have one input and one output they are called repeaters. A repeater with more than two ports is called a hub.

ping: see Echo Request and Echo Response.

Point–to–Point Protocol: PPPs exist to extend some of the capabilities of a LAN or Layer 3 network across a WAN connection without using routers. PPP is not important to this text.

Point–to–Point Tunneling Protocol: A version of PPP that facilitates the transfer of non–IP and IP through the Internet by hiding the packets inside the data

---

[7] The ugly colon is part of the official name. I do not like it, but it's there.

payload of a packet. The hidden packet can even be encrypted for extra security.

POP3:   see Post Office Protocol.

Port: A two byte suffix for an IPv4 address which links the message to a process which "listens" to that port. Port and socket can be used interchangeably.

Post Office Protocol: Post Office Protocol (Version 3) is one of the two most common protocols used to connect an email client with an email server. Typically email originates on a PC or other device which transfers the email to an email server. Likewise, email is received by the email server and held there until the email client requests it. If the email client and server communicate via POP3, the email is deleted from the server after the client downloads it.

postfix: Postfix is the collective name for one of the most popular email MTA processes and its configuration files. For many implementations, Postfix is much easier to configure and implement than Sendmail. However, both are becoming more powerful and standard configurations are manageable. Many MTAs run Postfix, Sendmail, or a similar protocol.

PPP: see Point–to–Point Protocol.

PPTP: see Point–to–Point Tunneling Protocol.

Preboot eXecution Environment: PXE is an environment that can be used to correct some issues with devices that will not boot. If you need PXE you are in deep trouble that is beyond the scope of this book.

Presentation Layer: The Presentation Layer (Layer 6 of the OSI Model) shares interfaces with Layer 5 (Session Layer) and Layer 7 (Application Layer). The functions of the Presentation Layer are:

- Encoding and Decoding of messages.
- Compression and Decompression of messages.
- Encryption and Decryption of messages.

The Presentation Layer is one of the OSI upper layers and on the Internet is handled by either TCP or UDP.

Presentation Layer PDU: Like other layers of the OSI Model, the conversation between two endpoints at Layer 6 looks like a direct exchange of P-PDU (not to be confused with Physical Layer PDUs which are bits.).

Private Internet: For all intents and purposes, an Intranet is a private network that provides all the required Internet services *without* exposing any of the private resources to the actual Internet. Many Intranets are connected to the Internet using a combination firewall and NAT device.

Protocol: A protocol is an agreed upon set of rules and steps to accomplish a common goal such as to transfer email, start a session, or perform any other task. Protocols may cover hardware, software, and multiple processes. The Internet is built upon a large number of protocols from many different organizations.

Protocol Datagram Unit: Protocols that operate by exchanging messages must use exact formats for these messages which are called Protocol Datagram Units. For example, Ethernet at OSI Layer 2 exchanges frames (Data Layer PDUs) between devices. PDU is a generic term for any message whose format is rigidly defined by a protocol.

Public TTY Client for Windows: `putty` is an application to facilitate both serial connections via a COM port and `ssh` via IP. `putty` is free from http//putty.org for Windows, Linus, and UNIX.

putty: see Public TTY Client for Windows.

PXE: see Preboot eXecution Environment.

QoS: see Quality of Service.

Quagga: Quagga Open Source is an open–source routing software package that includes any number of routing protocols such as RIP, OSPF, or others. Quagga is a fork from the Zebra project, hence the name[8], and supports all the routing protocols in this text. Currently Quagga contains all the packages required to run the `zebra` daemon which directly updates the routing table in the Linux kernel with information obtained from the various routing daemons.

Quality of Service: Many types of message streams, such as voice and video, require packets to arrive within rigid delay parameters. Unlike ATM, IP does *not* have a built–in mechanism to provide QoS. QoS must be provided by some other protocols on top of IP. QoS is not addressed in this text at this time.

Queue: A queue is a data structure with a strictly "first in, first out" or FIFO ordering. Elements can only be added to the end of the queue called the tail and only be retrieved (dequeued) from the other end called the head.
There is another similar structure which constantly re–orders the elements in the structure that is called a priority queue but is a completely different structure. A priority queue returns the lowest (or highest) priority item in the structure and the priority of elements can be changed while they are in the priority queue. Priority

---

[8] A quagga is an extinct close relative of the zebra. Apparently quaggas were hunted to extinction around 1878, but the whole thing is a bit murky.

queues are used in shortest path algorithms.

RARP: see Reverse Address Resolution Protocol.

Raspberry Pi Hobby Computer: The current Raspberry Pi microcomputers are amazing little machines. With the addition of an HMDI monitor (or conversion dongles), a keyboard, and mouse the Pi is a very powerful computer capable of running different Linux distributions. In fact, there are kits available to make the Pi into a laptop, a credit–card sized touch screen tablet, router, wireless access point (WAP), or desktop PC. With a little effort and some extra equipment, the Pi can be a wearable computer that controls flashing lights on your clothing. The Pi is cheap enough to play with and powerful enough to be useful. All in all a remarkable machine.

Raspbian: The Raspberry Pi operating system is a version of Debian Linux compiled specifically for the Pi microcomputer. This has a number of advantages as Debian is a lean Linux distribution but does support the popular Linux desktops such as Gnome. As a direct result of basing Raspbian on Debian, the number and range of packages available for direct installation is astonishing. As an indirect consequence, if you wish to install software and can't find a Raspbian tutorial you can usually follow a Debian tutorial without problems. As a last resort, sources typically compile on the Pi if they compile with Debian. This feature is fantastic if you get stuck compiling from sources.

RD: see Routing Domain Identifier.

Registered Jack 45:   A Telco standard eight position jack and plug commonly used for Ethernet connections from 10 megabits and higher speeds. RJ45s are very common connectors and have proven to be very reliable provided they tested good when installed.

Request For Comments: Standards for the Internet are composed of Request For Comments, or RFCs, maintained by the IETF on their website. As stated before, RFCs are not easy to read and cover topics in such minute detail that they are not much use in learning how a protocol works as part of the big picture. Typically, RFCs are of more use when one is writing adapter drivers or low–level services for a protocol. RFCs are highly technical documents which are usually put forward as "drafts" and then finalized after the networking community has had a chance to comment on the proposal. While it sometimes seems chaotic, there is a very rigid procedure to create a new RFC. This procedure is covered by a number of different RFCs[9].

---

[9] In my opinion, this is both circular and humorously appropriate.

Resource Record: DNS zone files, which contain all the of the information needed to contact devices and services in a domain, are made up of resource records

Resource Reservation Protocol: Cisco IOS Software supports two fundamental Quality of Service (QoS) architectures: Differentiated Services (DiffServ) and Integrated Services (IntServ). In the DiffServ model a packet's "class" can be marked directly in the packet, which contrasts with the IntServ model where a signaling protocol is required to tell the routers which flows of packets requires special QoS treatment. DiffServ achieves better QoS scalability, while IntServ provides a tighter QoS mechanism for real-time traffic. These approaches can be complimentary and are not mutually exclusive. [19][10]

Reverse Address Resolution Protocol: RARP is used, along with the its inverse protocol ARP, to maintain a table of IPv4 and MAC addresses. In practice, a Layer 3 device is extremely likely to reply to an IPv4 message in a timely fashion. Therefore maintaining a table of the mapping between IPv4 and MAC addresses leads to large savings in redundant network traffic.

RFC: see Request For Comments.

RIP: see Route Interchange Protocol.

RIPng: see Route Interchange Protocol for IPv6.

RIPv1:   see Route Interchange Protocol, Version 1.

RIPv2:   see Route Interchange Protocol, Version 2.

RJ45:   see Registered Jack 45.

Rohan: A mythical kingdom in the book *The Lord of the Rings* by J.R.R. Tolkien [314]. The kingdom used signal fires to ask for aid from the neighboring kingdom of Gondor. This is used as an example of certain types of communication, such as in 1d.

root: In networking, root can have multiple meanings. `root` is the administrative user for Linux and Raspbian. In graph theory, root is the top of a tree. Bear in mind, the root of a graph is the vertex we have chosen to be root. There is nothing special about root except that some algorithms require a starting vertex of a graph which we call "root".

---

[10] url: https://www.cisco.com/c/en/us/products/ios-nx-os-software/resource-reservation-protocol-rsvp/index.html

round robin: A common technique to insure that all items get the same attention such as data conversations. The items are taking in a repeating pattern such as "ABCDABCDABCD...".

route flapping: Certain types of network link failures can cause a route to a destination to go up and down fairly often. Each time this happens routers must update their route tables which can lead to network slowdowns and even failure. All routing protocols have at least some issues dealing with route flapping, but some are not able to handle this well. For example, RIP is notoriously bad at dealing with route flapping while Babel is designed to handle it well.

Route Interchange Protocol: RIP is a generic catch all for the Route Interchange Protocols RIPv1, RIPv2, and RIPng. RIP is a distance–vector routing protocol that is well suited for small networks because of its ease of configuration. However, as networks grow larger and more complex, RIP announcements begin to cause severe overhead which only gets worse as the network continues to grow. If the network includes both IPv4 and IPv6 addresses, both RIPv2 and RIPng must run to route the two sets of addresses. There is no single RIP for both sets of Layer 3 addresses.

Route Interchange Protocol for IPv6:   A routing protocol designed to extend RIP to handle IPv6 and its entirely new addressing structure. In order to run RIP for both IPv4 and IPv6 , a router must run two entirely separate protocols as the two versions of IP are in no way compatible. There are efforts being made to somehow bridge, or at least translate, between IPv4 and IPv6 but at this writing the two are still completely separate Layer 3 addressing schemes. This issue exists with all protocols that route IP.

Route Interchange Protocol, Version 1:   The earliest version of RIP was introduced *before* the advent of CIDR and its variable–length subnets. RIPv1 *enforces* classful subnetworking on all packets. For example, an address of 1.1.1.5 absolutely must have a subnet mask of 255.0.0.0. Enforcing classful subnetworks did not allow organizations to logically construct networks that met their needs. Unless forced to, do not use RIPv1 and treat it as an historical curiosity. When using RIP, always use RIPv2 as it has the same functionality as RIPv1 but is *much* more flexible and efficient.

Route Interchange Protocol, Version 2:   When it became obvious that classful subnetworking as enforced by RIPv1 was inadequate for most organizations, a new version of RIP was introduced as RIPv2. It is backwards compatible with version 1, but it allows for variable–length subnet masks. RIPv2 is completely compatible with CIDR.

Routing Domain Identifier: When used in ISIS routing, the RD part of the NSAP address must match for two routers to form a relationship and exchange routes. This is critical in large ISP networks but is not going to be encountered in smaller networks that often.

RR: see Resource Record.

RSVP: see Resource Reservation Protocol.

S-PDU: see Session Layer PDU.

SD Association: The SD Association promotes the correct use of SD and microSD cards. The SD format utility provided on their website is the preferred method to format an SD card. Windows, Linus, and Unix *might* format an SD card correctly but there could be problems.

SDA: see SD Association.

SDH: see Synchronous Digital Hierarchy.

Secure Domain Name Service: DNSSEC[11] is a suite of security protocols designed to protect DNS from attack over a network. This includes such things as public key/private key encryption as well as other techniques to thwart hacking.

Secure Hyper–Text Transfer Protocol: HTTPS is the secure version of HTTP. There is no need to distinguish between HTTP and HTTPS for the purposes of this text unless you are interested in securing your web traffic.

Secure Shell (ssh):   SSH allows terminal access to a remote machine as if one were at the console. Telnet has the same basic functionality, but is lacking in any security while SSH uses strong encryption. Most operating systems give the user the option of **ssh** from the command line or from a desktop application. For example, Windows users can use Putty [6] for a desktop application that allows for customized colors and fonts or **ssh** from a DOS prompt. Putty was used to generate the Pi screen captures for this text because of the ability to do black text on a white background.

security: Security in this text is understood to mean best practices to make it difficult for an outsider to disrupt the system, not a means to completely lock out all attackers completely. A good network administrator must be aware that keeping everything secure is a constant, and often losing, battle.

SEL: see NSAP Selector.

---

[11] DNSSEC.bis and DNSSEC are the same protocols.

sendmail: Sendmail is the collective name for one of the most popular email MTA processes and its configuration files. Sendmail itself can be so complex that entire books are written, and frequently rewritten, to cover how to configure and implement it as the heart of an email system. Most SMTP servers will run either Sendmail or Postfix. Other MTA exist but are less common.

Sequenced Packet Exchange: SPX is the proprietary upper layer protocols owned by Novell for NetWare.

Serial Line Internet Protocol: SLIP was introduced to allow a device to establish an asynchronous connection to another device using IPv4 over a serial link. SLIP was very useful but was made obsolete by PPP.

Server: A server is any device that exists to share resources with other devices on a network. Servers are usually devices with more memory, CPU power, high speed connections, and non–volitele memory that other devices. Servers share resource by processes known as services.

service: A service is a process that runs on a Server to provide Clients with access to shared network resources. Many people use the terms "service" and "server" interchangeably.

Session Layer: The Session Layer is Layer 5 of the OSI Model and is tasked with:

- Initializing a session, which includes allocating resources to this session.
- Maintaining a session, which includes re-establishing a dropped session.
- Terminating a session, which includes releasing local resources for reuse.

In some cases an endpoint may disallow a session due to security issues or lack of resources.

Session Layer PDU: Two endpoints communicate at the Session Layer (Layer 5) by logically exchanging S-PDUs. In reality, the S-PDU is passed down through the Transport Layer on down to the Physical Layer, transferred as bits, and on the other endpoint passed up through the Transport Layer to the other Session Layer.

shim (OS shim):   In networking and OSs a shim is a small piece of code which is inserted into either the OS or some other running code in the same way that many viruses are inserted. Shims have to be very small and transparent to the "host" code. This technique is still in legitimate use, but has many pitfalls. Shims with errors can cause side–effects that are difficult to diagnose and correct.

Shortest Path First: The network can be considered a graph with each edge assigned a weight or cost. Each node (typically a router) uses information learned from the network to construct a tree with itself as the root using Dijkstra's algorithm or will user Bellman–Ford to construct the shortest paths. This allows the device to choose the least cost or shortest path to a destination network based upon the total cost and not just the next hop. Using a graph of the network to facilitate taking the best path is known as Shortest Path First (SPF).

Simple Mail Transfer Protocol: SMTP is the traditional server–to–server email transfer protocol provided by an MTA that is available on the network 24 x 7. SMTP is not typically run on a client PC due to the need to be available to receive email at any time; however, email that cannot be delivered is retried at intervals until the SMTP service determines it cannot be sent. It would be possible to run SMTP on a client, but it does cause extra overhead so clients typically run POP3 or IMAP to transfer email to the server. Due to the rise in spam and phishing email, SMTP and ESMTP are constantly evolving and must have current patches applied on a regular basis. This is another reason to avoid SMTP on a client device.

SLIP: see Serial Line Internet Protocol.

Small Office/Home Office: SOHO networks are typically rather small, private networks connected to the Internet via a router that performs NAT and acts as an IP Forwarder. These networks tend to be a single IP network within the 192.168.0.0 private network space. If the network contains multiple routers or multiple subnetworks, it is not usually considered a SOHO network.

SMTP: see Simple Mail Transfer Protocol.

Snail Mail: Internet slang for delivery of mail–type messages by physically sending hard–copy via the USPS, UPS, DHL, FedEx, or a similar service.

Sneaker Net: Internet slang for delivering data by literally walking over to the recipient with the data. Sneaker Net or Snail Mail are usually used as the slowest reasonable speed to transmit data.

SOA: see Start Of Authority.

Socket: A two byte suffix for an IPv4 address which links the message to a process which "listens" to that socket or port. Port and socket can be used interchangeably.

SOHO: see Small Office/Home Office.

SONET: see Synchronous Optical Network.

spam: Any unwanted mass email, much like junk mail. Spam is usually annoying but should always be viewed as a potential security risk. Delete spam before you open it.

SPF: see Shortest Path First.

SPX: see Sequenced Packet Exchange.

SQL: see Standard Query Language.

ssh: see Secure Shell (ssh).

Standard Query Language: SQL (pronounced either "see–quil" or SQL) is the result of the attempt to develop a common set of queries and update messages for databases. Before SQL a programmer needed to learn a new language and programming extensions for each database. This also meant that database support was difficult to integrate with programming languages. With SQL the database engine is independent of the messages that use it and the actual engine could easily be implemented using the client/service model. In many cases, the database engine can be changed to a different vendor without forcing existing code to be compiled again.

Start Of Authority: The SOA record is required at the start of every DNS zone file. It contains contact information about the domain administrator which must be current, timing settings to control how often certain actions happen, and the critical serial number (SERIAL) for the zone file. The serial number controls when the zone file is transferred between the primary and secondary name servers. The SERIAL must be updated when the zone files are edited on the primary name server in order for the secondary name servers to be notified to initiate a zone file transfer. The SOA area also contains DNS information for the primary and secondary name servers, especially their IP addresses.

sudo: see sudo.

sudo: As UNIX and Linux matured, it became obvious that administering the machine as **root** had some hidden pitfalls. The solution is the **sudo** command which is an abbreviation of the commands: su (become **root** with administrative rights), do (do a single command), then give back administrative rights. Some Linux administrators, myself included, are not great fans of **sudo**, but prefer it to accidentally making changes that are difficult to correct because we have forgotten we were **root**. Better safe than sorry.

synchronous: Synchronous is from the Greek for "same time". In digital communications, this term has three main usages:
1. When the sender and receiver are constrained to send/receive at specific times. When a timer expires, the sender *must* send even if there is no new data to send. When this happens the sender sends a message of all binary zeros, or nulls. This

occurs in SONET very frequently.

2. The sender and receiver *must* insure their clocks match in order to properly send the bits of a message otherwise the receiver might be unable to correctly find the beginning of the message.

3. Less common is the usage to denote the receiver *must* have prior knowledge in order to interpret the message properly. This usage is most often encountered in security and encryption. For example, the message "My mother's cat died." could be used to denote "Drop everything and leave town." only if the receiver has prior knowledge of the hidden meaning.

Synchronous Digital Hierarchy: SDH is the suite of protocols that is the European equivalent of SONET and is *not* directly interoperable with SONET. Like SONET, SDH runs over fiber optic rings, segments, trees, and other topologies and has enormous bandwidth potential.

Synchronous Optical Network: SONET is a TDM based high–speed, high–bandwidth transport capable of sending bits for long distances. For most intents and purposes, SONET is a Physical Layer protocol, but from a SONET viewpoint it can be divided into Layers 1, 2, and 3. SONET works in multiples of 51.84 Mbit/s (OC-1) but usually does not use speeds lower than 155 mbits/second (OC-3). The upper limit of SONET is extremely high and increases as new equipment is produced. SONET has very low error rates and rarely fails due to the care spent in installing the fiber and equipment.

System Identifier (NSAP):   The System Identifier (ID) is a six byte field of the NSAP address that uniquely identifies a device on the network. Typically this is the MAC address of one of the device's NICs.

T–Carrier 1:   A T1 consists of 24 digital voice/data 64 kbit/second streams combined via TDM into a single data stream of 1.44 Mbits/second, or a T1. Typically there is little reason to distinguish between a digital voice or data message on a T–Carrier. Machines need not distinguish at all as bits are bits.

T–Carrier 2:   A T2 consists of multiple digital voice/data 64 kbit/second streams combined via TDM into a single data stream of 6.32 Mbits/second. T2 is not common at the time of this writing..

T–Carrier 3:   A T3 consists of multiple T1s combined via TDM into a single data stream of 44.736 Mbit/second. Telcos typically build their network in terms of T3 services and only break their services down to lower bandwidth inside the customer's premises.

T-PDU: see Transport Layer PDU.

T1: see T–Carrier 1.

T2: see T–Carrier 2.

T3: see T–Carrier 3.

TCP: see Transaction Control Protocol.

TCP/IP: see Transaction Control Protocol over IP.

TDM: see Time Division Multiplexing.

Telco: Telco is used as shorthand for telephone or telecommunications company. Generally used to refer to a company that leases bandwidth.

Telecommunications Industry Association: The TIA is responsible for cabling standards such as color–coding of cable pairs, building wiring standards, and pair assignments in jacks and plugs. Without the TIA standards, network wiring would be utter chaos. The most common standard from the TIA is the order of the color–coded pairs of wire in an ethernet cable terminated in an RJ45 which has two incompatible versions: TIA/EIA-568A and TIA/EIA-568B. Most TIA standards are produced with its sister organization the Electronic Industries Alliance (EIA).

Telnet: Remote "Dumb" Terminal Protocol to provide a screen and keyboard only interface to a host device. Originally Telnet emulated a teletype with a screen (glass teletype).

TFTP: see Trivial File Transfer Protocol.

The Onion Router: TOR, or Onion routing and browsing, gets its name from the idea that if a message bounces around with changes at each bounce, it is very hard to peel the layers of the message to find out where it originated. Originally designed for sending messages and email anonymously, TOR now comprises a darkish, semi–hidden part of the Internet. A Raspberry Pi can easily be converted into an onion router to make it much harder to trace activity back to a specific IP address and device. Be advised that when messages leave an onion router for the real Internet it is obvious that someone is trying to hide something.[12]

TIA: see Telecommunications Industry Association.

Time Division Multiplexing: TDM is a very old, extremely powerful technique to carry a number of lower speed connections over a single medium. First de-

---

[12] Read the Purloined Letter by Edgar Allen Poe.

veloped to carry multiple phone connections over a single pair of copper wires, TDM is the basis for all telco services including high–speed SONET. The bandwidth is divided into time slices which are allocated to the slower connections in a round–robin fashion. Connections transmit when it is their time slice and are quiet when it is not. Because the media has at least enough bandwidth to carry all the slower connections, there is no delay for the slower connections. Common TDM services in the USA include T1, T3, and SONET. Outside the USA one might encounter TDM in E1, E3, and SDH.

Time To Live: TTL is the limit on the time information is still relative. When used with a table entry or DNS, TTL is a true time usually specified in seconds. Once the entry has been active for longer than the TTL it is discarded or an update is requested. TTL has a slightly different meaning when applied to an IP packet. Each time the packet is forwarded by a router the TTL is decremented by 1. A packet with a TTL of zero is automatically dropped by a router which helps to prevent routing loops as eventually a packet must reach its destination or be dropped. Information that has exceeded its TTL is sometime said to have "turned into a pumpkin" or "pumpkin–ed out" which are geek humor derived from the fairy tale Cinderella.

TLD: see Top Level Domain.

Token Bus: A token bus network is much like a token-ring network except that the ends of the network do not meet to form the ring. Instead, the network is still terminated at both ends. A token is still required before a node can use the network. Like in a token-ring, it needs to include the address of the destination along with the data it needs to send. Although in the token bus, it implements a virtual ring on the coaxial cable. Though both topologies use tokens, the similarities end there, because token bus uses a different topology and the token-passing protocol is different. In a token-ring network, the token or data is passed to the next physical node along the ring, but in a token bus network, it does not matter where the nodes are physically located since token-passing is done via a numeric sequence of node addresses. The token or data is passed to the next sequential node address no matter if the physical location of that node is at the very end of the bus network. This is the virtual ring; the physical layout of the network will not change it. Token bus networks are defined by the IEEE 802.4 protocol. [313]

Token Ring: A token ring network is a local area network (LAN) topology where nodes/stations are arranged in a ring topology. Data passes sequentially between nodes on the network until it returns to the source station. To prevent congestion and collision, a token ring topology uses a token to ensure that only one node/station on the line is used at a time, thereby easily denoting media users of its activity. A token ring LAN is physically wired as a star topology but configured as a ring topology. The token ring LAN system was standardized by the Institute of Electrical and Electronics Engineers as IEEE 802.5. [313]

Top Level Domain: TLDs are the highest level category used to distinguish between domain names. For instance, "example.org" and "example.com" are two
different domains. For the first one the TLD is "com" and for the second the TLD
is "org" which means they are completely separate domains. A TLD is sometimes
referred to as the suffix of the Fully Qualified Domain Name or FQDN.

TOR: see The Onion Router.

Transaction Control Protocol: TCP can be thought of as the guaranteed delivery equivalent of the upper layers (4 - 7) of the OSI Model. Unfortunately the
functions of TCP are not as well structured as those of the OSI Model and there
are "hooks" from function–to–function that shouldn't be there, but it works.

Transaction Control Protocol over IP: TCP/IP is the full stack of protocols
used to transfer messages over the Internet and is analogous to the OSI stack.
TCP is comprised of a jumble of interconnected protocols that perform roughly
the same functions as the upper layers of the OSI seven layer model. IP in this
case refers to all three lower layers taken as a single, unified layer even though
this is not a helpful way to look at the lower layers.

transparency: A protocol is transparent when it does not place any constraints
on transmitted data. It means that headers as well as data must be transported
unmodified end to end.

Transport Layer: The Transport Layer is Layer 4 of the OSI Model and is tasked
with providing guaranteed delivery or "best effort" delivery. On the Internet,
guaranteed delivery is provided by the TCP version of the upper layers while
"best effort" is provided by the UDP version of the upper layers. Layer 4 typically runs only on the endpoint of the conversation and does not take part in the
routing of packets.

Transport Layer PDU: On the endpoints of a conversation, Transport Layers
(Layer 4) communicate logically by exchanging T-PDUs. Of course, this is actually done by passing the T-PDU down through the OSI stack to the physical
medium, passing the bits along however they get through the network, and then
passing the bits back up the other endpoint's OSI stack to the other Transport
Layer.

tree: In graph theory a tree is a connected, simple, acyclic graph. For networking
purposes a tree is a subset of the network that includes all the nodes and a path
to each node *with no cycles or loops*. Trees are critical in building networks to
insure reachability and to insure the network has no active loops. This is typically
done by temporarily turning off connections, idling connections, or designating

connections as preferred.

Trivial File Transfer Protocol: TFTP is related to FTP and is closely associated with BOOTP. TFTP is designed to quickly transfer files to and from a TFTP server. Typically TFTP is used to backup and restore configuration files or to download an operating system to a device. Unlike FTP, TFTP is run as command instead of a dialogue which makes it ideal for single file transfers initiated by a device rather than a human.

TTL: see Time To Live.

tunneling: A very common technique for sending a protocol over a network by hiding the protocol's messages as the payload of another protocol. For additional security and privacy, the protocol's messages may be encrypted before being made the payload of the enclosing protocol. VPNs are done via tunneling.

typo–squating: Typo–squatting is when hijackers register misspelled versions of your domain name to send the traffic to malicious sites. Registering all possible versions of your domain name including singular and plural versions, all common domain extensions and hyphenated and non hyphenated word compounds can allow a hacker to acquire traffic intended for your site [52]. In the early days of the Internet, an attempt to contact a non–existing FQDN would receive a generic "not found" message. Now the response is an ad asking you to register the domain.

UDP: see User Datagram Protocol.

unicast: A message sent from one source to exactly one destination. Unicasts typically appear at Layer 2 or Layer 3.

Universal Resource Locator: A part of the World–Wide–Web, the URL denotes the location and type of a resource. Web pages have a URL starting with either "http" or "https" while FTP requests have URLs starting with "ftp". URLs can be absolute and contain all the information to find a resource or they can be relative to the current location.

Universal Serial Bus: USB is a standard external bus that can be used to connect multiple types of peripherals, including mice, NICs, and keyboards, to a computer or other device. Most current computers support the USB 2.0 and 3.0 standards. USB ports are typically backwards compatible. The current Raspberry Pi typically has four USB ports while older models only have two.

UNIX Operating System: UNIX is a trademarked proprietary family of multitasking, multiuser computer operating systems that derive from the original

AT&T UNIX development starting in the 1970s at the Bell Labs research center by Ken Thompson, Dennis Ritchie, and others. Linux is basically an open–source version of UNIX.

Upper OSI layers: Layers 4 through 7 of the OSI Model are often taken together as the "Upper Layers" for two main reasons. First, the upper layers only run on the endpoints of a conversation and second, these layers correspond roughly to TCP, UDP, or ICMP on the Internet.

URL: see Universal Resource Locator.

USB: see Universal Serial Bus.

User Datagram Protocol: UDP does not guarantee delivery of messages from one endpoint to the other. It is most definitely a "best effort" protocol that em-phasizes speed of delivery and is useful for applications where a re–transmission of a corrupted packet of data is either not needed or impossible. UDP is often used for streaming services and announcements such as RIP updates that will be repeated soon anyway.

Variable Envelope Return Paths: The fundamental problem in managing a large mailing list is matching bounced email messages to subscription addresses. Vari-able envelope return paths completely eliminate this problem. They automati-cally and reliably identify the subscription address relevant to each bounce mes-sage. They provide the address in a form that is trivial for automated bounce handlers to parse. They require support from the local mailer, but they do not require support from any other hosts.

Variable Length Subnet Mask: Originally IP addresses were required to use the natural, or default, subnet mask which lead to problems for some organizations and a terrible waste of IP addresses. In response to this problem, CIDR was de-veloped which allows for any valid subnet mask that is at least as long as the natural mask. This allows organizations to use their assigned address space to create hierarchical networks that better fit their needs and that waste fewer ad-dresses. VLSM and CIDR allowed the Internet to function until NAT devices were able to pseudo–connect private networks to the Internet. It would be hard to find a current NIC that did not support VLSM and CIDR.

VERP: see Variable Envelope Return Paths.

vertex (also called node or point): A directional graph, or digraph, is a non–empty set of vertices (V) and a set of arcs (A). By this definition, a single vertex is also a digraph. We will only make a distinction between a digraph and a graph when it is important to networking. Bear in mind that most networks are digraphs even if all the links are bi–directional because a link *might* fail in one direction only. This type of failure can be hard to find.

vi: see vi text editor.

vi text editor: One of the most commonly used Linux text editors is **vi** or the enhanced version **vim**. Most of the work in this text is done using **vi** rather than **nano** simply because the author prefers vi. You should always use the editor you are most comfortable with using but be aware there are other easy to use text editors out there. For the work in this text, I suggest you use an editor from the command line rather than the desktop because some text editors use a different encoding for the end of a line. Either **vi** or **nano** will not cause you problems when editing configuration files on the Raspberry Pi.

Virtual Local Area Network: A virtual LAN is a virtual broadcast domain and behaves exactly like any other LAN. The only difference is that a VLAN exists only because of the configuration of a switch or other device and the links in the VLAN cannot all be seen as they can in a LAN. Ethernet switches can often be configured to act as two or more virtual switches *which are not connected*. The result is that not all ports on the same switch share broadcasts and therefore are on different LANs. In short, a VLAN is a LAN.

Virtual Private Network: A VPN is used to securely join a remote network by encrypting the traffic between the remote network and the device. There are many different techniques to do this and they do not always inter-operate. Packets for the remote connection are encrypted as the payload for an unencrypted packet which is then routed in the normal way. It can get complicated and is beyond the scope of this text; however, the Raspberry Pi definitely supports a number of VPN clients.

Virtual Terminal Shell: Routing on the Raspberry Pi is provided by Quagga which was developed (as Zebra) to emulate the experience of configuring a Cisco device. Obviously it should be possible to use a text editor to change the config-uration files, but that is not in keeping with the Cisco experience. Instead **vtysh** is used to simulate remotely configuring a Cisco router.

vlan: see Virtual Local Area Network.

VLSM: see Variable Length Subnet Mask.

Voice Over Internet Protocol: Telcos and other companies have developed the ability for phones to be IP addressable and carry voice traffic over the Internet as normal packets. To my knowledge, the Raspberry Pi does not directly support VOIP simply because no one seems to have tried it yet. With external equipment,

I suspect VOIP on the Pi would be possible.

VOIP: see Voice Over Internet Protocol.

VPN: see Virtual Private Network.

vtysh: see Virtual Terminal Shell.

WAMP: see Windows web server.

WAN: see Wide Area Network.

WAP: see Wireless Access Point.

Wide Area Network: If a link in a network uses physical equipment that could provide a long distance connection or a link leased from a telco, the network is a WAN. There are a range of networks larger than a LAN and a variety of buzz words to distinguish them such as MAN,VAN,BAN, and so on. For our purposes a network is a LAN or WAN. As long as the connections work, the distinctions are not material.

WiFi: see Wireless Network.

Windows web server: WAMP servers consist of the Windows versions of Apache, MySQL, and PHP. This set of services provide a uniform platform for the development of complex web sites that are easily ported to a Linux platform.

Wireless Access Point: A WAP is a device that bridges a WiFi and wired network into one network. The Raspberry Pi makes a great WAP. WAPs can provide additional services such as DHCP and even routing as needed. Technically a WAP only provides bridging, but often the difference between a true WAP and a router with one NIC being wired and one being WiFi is not that important.

Wireless Local Area Network: A WLAN is a wireless LAN and works exactly the same way. The term `wlan` refers to the wireless NIC of a device running Linux such as the Raspberry Pi. Care must be taken when enabling the `wlan` interface to avoid creating an unauthorized network and impacting someone elses security. Rogue WiFi networks may present an illegal attack on a network.

Wireless Network: WiFi refers to any wireless connection such as 802.11a, 802.11b, and the others. A WiFi network is like any other network except the devices are mobile.

WLAN: see Wireless Local Area Network.

wlan0: The first WiFi connection on a device running Linux is always **wlan0**. This interface is only present if Linux discovers a wireless device even if the device is not configured properly. The newer Raspberry Pi has a built in wireless NIC.

World Wide Web: The World Wide Web is a subset of the Internet consisting of all the web servers and browsers connected to the Internet. As a web page can contain links to pages on another server anywhere on the Internet, it truly is a web.

www: see World Wide Web.

XAMP: see Cross–platform web server.

Zebra: When capitalized in this text, Zebra is meant to refer to the Zebra project to develop an open–source router with a Cisco Systems–like interface. The Quagga project is a direct descendent of the Zebra project and both use the zebra daemon.

zebra routing daemon: When not capitalized in this text, **zebra** refers to the routing daemon that receives routes from the various routing protocols and uses that information to update the routing tables in the Linux kernel. It must be running or the routing daemons such as **ospfd** will not work.

zone file: Zone files contain all the information required for the **named** or **bind** daemon to provide DNS. Zone files are edited on a primary name server and transferred to all secondary name servers as needed to keep all services up to date.

References

1. Ada: Howto: Install lamp stack (on raspbian stretch June 2018). on-line (2018). URL `https://www.raspberrypi.org/forums/`. Accessed: February 25, 2019 376
2. Aitchison, R.: Pro DNS and Bind, 1rst edn. Apress (2006) 337, 340, 343, 344, 354, 367, 369, 371
3. Aitchison, R.: Pro DNS and BIND. Zytrax.com (2019). URL `http://www.zytrax.com/books/dns/`. On-Line. Accessed: February 13, 2019 343, 344, 355, 369
4. Balena: Balena etcher download. on line. (2019). URL `https://www.balena.io/etcher/`. Accessed: August 2, 2019. Linked from Ada (Raspberry Pi). 120
5. Berners-Lee, T.: Tim berners-lee biography. On line. (2019). URL `https://www.w3.org/People/Berners-Lee/`. Access: JUly 28, 2019 472
6. Bitvise: putty.org. On line. (2019). URL `https://www.putty.org/`. Accessed: October 12, 2019 494
7. Breuning, S.: The raspberry pi webserver homepage. Online (2014). URL `http://raspberrywebserver.com/`. Accessed: March 13, 2019 376
8. CanaKit: Canakit raspberry pi kits. On-line (2018). URL `https://www.canakit.com/raspberry-pi/`. Accessed: November 30, 2018 125, 126
9. Carroll, L.: Alice in Wonderland, special edition edn. Random House (1946) 5
10. Carroll, L.: Through the Looking Glass: and What Alice Found There, special edition edn. Random House (1946) 331
11. Charrett, A.: Bind on windows. On line (2013). URL `https://alex.charrett.com/bind-on-windows`. Accessed: January 31, 2019 344
12. Cicso: General Routing Information. On line (2009). URL `http://cisconet.com/`. Accessed: January 4, 2019 234
13. Cisco: Advanced Cisco Router Configuration: Student Guide (Cisco IOS Release 11.2). Cisco Systems (1996) 314
14. Cisco: OSPF Design Guide. On Line (2005). URL `https://www.cisco.com/`. Accessed: January 4, 2019 272, 279
15. Cisco: CIDR (classless inter-domain routing). On line. (2016). URL `https://study-ccna.com/`. Accessed: July 29, 2019 462
16. Cisco: Designated & Backup Designated Router. On line. (2016). URL `https://study-ccna.com/`. Accessed: July 29, 2019 457, 465
17. Cisco: IP Routing: BGP Configuration Guide. on-line (2018). URL `https://www.cisco.com/`. Accessed: July 9, 2019 313

© Springer Nature Switzerland AG 2020
G. Howser, *Computer Networks and the Internet*,
https://doi.org/10.1007/978-3-030-34496-2

18. Cisco: IS-IS DIS and Pseudonode. On line. (2019). URL `https://networklessons.com/`. Accessed: August 3, 2019 465

19. Cisco: Resource Reservation Protocol (RSVP). On line. (2019). URL `https://www.cisco.com/`. Accessed: August 3, 2019 492

20. Cisco, Martey, A.: Integrated is-is routing protocol concepts. On-line (2002). URL `http://www.ciscopress.com/articles/` 303

21. Clarke, A.C.: Hazards of Prophecy: The Failure of Imagination. Harper & Row (1973) vii

22. CNN: Worm strikes down Windows 2000 systems. On line. (2005). URL `http://www.cnn.com/2005/TECH/internet/08/16/` 369

23. CompTIA: Comptia network+. On line (2018). URL `https://certification.comptia.org/`. Accessed: August 18, 2018 vii

24. Contributors: Apache webserver homepage. On line. (2019). URL `https://httpd.apache.org/`. Accessed: March 13, 2019 376

25. Cormen, T., Leiserson, C., Rivest, R.L., Stein, C.: Introduction to Algorithms (Third ed.). MIT Press (2009) 50, 62, 205, 208, 210, 213, 246, 323

26. Costales, B., Allman, E.: Sendmail, 2 edn. O'Reily (1997) 388

27. Dean, T.: Network+ Guide to Networks, 4th edn. Thompson Course Technology (2006) vii, 42

28. Debian: https://www.debian.org/. On line (2018). Accessed July 26, 2018 viii, 139

29. Deitel, P., Deitel, H.: C++ How to Program, 8th edn. Prentice Hall (2012) 460

30. Dijkstra, E.W.: A note on two problems in connexion with graphs. Numerische mathematik **1**(1), 269–271 (1959) 209, 278

31. Freiberger, P., Swaine, M.: The pirates of silicon valley. Movie (1999) 469

32. Friends, A.: Xampp apache + mariadb + php + perl. On line. (2019). URL `https://www.apachefriends.org/index.html`. Accessed: March 13, 2019 462, 478

33. Ghosh, S.: Distributed systems: an algorithmic approach. Chapman and Hall/CRC (2014) 470

34. GNU Communityr: Site administrator documentation. On line. (2015). URL `http://www.gnu.org/software/mailman/site.html`. Accessed: August 21, 2019 409

35. Harris, D.: Pegasus email. on line. URL `http://www.pmail.com/`. Accessed: August 17, 2019 404

36. Huggins, D.: Exam/cram 70-291: Implementing, Managing, and Maintaining a Windows Server 2003 Network Infrastructure. Que Certification (2006) 361

37. Hutzler, R.C., et al: Email Submission Operations: Access and Accountability Requirements. RFC 5068 (2007). DOI 10.17487/RFC5068. URL `https://rfc-editor.org/rfc/rfc5068.txt`. Accessed: April 4, 2019 412

38. ICANN: What does ICAnn do? On line. (2019). URL `https://www.icann.org/`. Accessed: January 24, 2019. 334

39. Incognito: Dhcp options. On line (2019). URL `https://www.incognito.com/tutorials/`. Accessed: January 18, 2019 220

40. Internet Society: Introduction to ipv6. on line (2019). URL `https://www.internetsociety.org/tutorials/`. Accessed: Aubust 8, 2019 78, 80, 84

41. Joshi, V.: Finding the shortest path, with a little help from dijkstra. (on line) (2017). URL `https://medium.com/basecs/`. Accessed: July 23, 2019 210

42. Kernighan, B.W., Ritchie, D.M., Tondo, C.L., Gimpel, S.E.: The C programming language, 2 edn. prentice-Hall Englewood Cliffs, NJ (1988) 460

43. Kunihiro Ishiguro, e.a.: Quagga. On-line (2017). URL `https://www.quagga.net/docs.html`. Note: Quagga Version 1.2.0 186, 231, 279, 307

44. Kunihiro Ishiguro, et al.: Zebra project homepage. on-line (2018). Accessed: December 6, 2018 186

45. Lamport, L., Shostak, R., Pease, M.: The byzantine generals problem. ACM Transactions on Programming Languages and Systems (TOPLAS) **4**(3), 382–401 (1982) 460, 470

46. Lasker, L., Parkes, W.F.: Wargames. DVD (1983). Directed by John Badham 40

47. Meloni, J.C., Telles, M.A.: PHP 6 fast & easy web development. Cengage Learning (2008) 488

48. MicroSoft Support: Nslookup. On line. (2017). URL https://docs.microsoft.com/en-us/. Search entry: nslookup. Accessed: August 20, 2019 365

49. Moreu, R.: Hackers. DVD (1995). Directed by Iain Softley 40

50. Mozilla: Thunderbird. On line. URL https://www.thunderbird.net/en-US/. Accessed: August 17, 2019 404

51. Murdocca, M., Heuring, V.: Computer Architecture and Organization. New York, NY, USA: John Wiley & Sons (2007) 271

52. Name Cheap: Domain phishing and other security attacks. Online (2019). URL https://www.namecheap.com/security/. Accessed: January 24, 2019 334, 502

53. NIST: Fips pub 146-1 u.s. government open systems interconnection profile (draft 2). Tech. rep., National Institute of Standards and Technology (1989) 302

54. NS1: Ipv6 dns: Understanding ipv6 and a quick implementation guide. On line. (2019). URL https://ns1.com/resources/. Accessed: August 13, 2019 345

55. Paquet, C., Teare, D.: Configuring IS-IS Protocol. On-line (2003). URL http://www.ciscopress.com/. Accessed: June 19, 2019 303

56. Project, A.F.: Ada fruit website. On-line (2018). URL https://www.adafruit.com/. Accessed: November 21, 2018 124, 126

57. Project, A.F.: Learn ada fruit. on-line (2018). URL https://learn.adafruit.com/. Accessed: November 25, 2018 130

58. raspberrypi.org: Raspberry pi products. On Line (2018). URL https://www.raspberrypi.org/products/. Accessed June 23, 2018 125

59. RFC 0097, Melvin, J.T., Watson, R.: First Cut at a Proposed Telnet Protocol. RFC 97 (1971). DOI 10.17487/RFC0097. URL https://rfc-editor.org/rfc/rfc97.txt. Accessed: June 12, 2019 423

60. RFC 0137, O'Sullivan, T.C.: Telnet Protocol - a proposed document. RFC 137 (1971). DOI 10.17487/RFC0137. URL https://rfc-editor.org/rfc/rfc137.txt. Accessed: June 12, 2019 423

61. RFC 0527, COVILL, D.: ARPAWOCKY. RFC 527 (1973). DOI 10.17487/RFC0527. URL https://rfc-editor.org/rfc/rfc527.txt. Accessed: June 12, 2019 68

62. RFC 0773, Cerf, V.: Comments on NCP/TCP mail service transition strategy. RFC 773 (1980). DOI 10.17487/RFC0773. URL https://rfc-editor.org/rfc/rfc773.txt. Accessed: March 20, 2019 411

63. RFC 0779, Unknown: Telnet send-location option. RFC 779 (1981). DOI 10.17487/RFC0779. URL https://rfc-editor.org/rfc/rfc779.txt. Accessed: February 28, 2019 423

64. RFC 0799, Mills, D.L.: Internet Name Domains. On-line (1981). DOI 10.17487/RFC0799. URL https://rfc-editor.org/rfc/rfc799.txt. Accessed: January 17, 2019 372

65. RFC 0821, Postel, J.: Simple Mail Transfer Protocol. RFC 821 (1982). DOI 10.17487/RFC0821. URL https://rfc-editor.org/rfc/rfc821.txt. Accessed: March 20, 2019 411

66. RFC 0822, Cargille, A.: STANDARD FOR THE FORMAT OF ARPA INTERNET TEXT MESSAGES. RFC 822 (1982). DOI 10.17487/RFC0822. URL https://rfc-editor.org/rfc/rfc822.txt. Accessed: March 4, 2019 389, 411

67. RFC 0826, unknown: An Ethernet Address Resolution Protocol: Or Converting Network Protocol Addresses to 48.bit Ethernet Address for Transmission on Ethernet Hardware. RFC 826 (1982). DOI 10.17487/RFC0826. URL https://rfc-editor.org/rfc/rfc826.txt. Accessed: February 26, 2019 52, 84

68. RFC 0876, Smallberg, D.: Survey of SMTP implementations. RFC 876 (1983). DOI 10. 17487/RFC0876. URL `https://rfc-editor.org/rfc/rfc876.txt`. Accessed: March 03, 2019 411

69. RFC 0882, Mockapetris, P.: DOMAIN NAMES - CONCEPTS and FACILITIES. On-line (1983). DOI 10.17487/RFC0882. URL `https://rfc-editor.org/rfc/rfc882.txt`. Accessed: January 22, 2019 372

70. RFC 0883, Mockapetris, P.: DOMAIN NAMES - IMPLEMENTATION and SPECIFICATIONS. On-line (1983). DOI 10.17487/RFC0883. URL `https://rfc-editor.org/rfc/rfc883.txt`. Accessed: January 22, 2019 372

71. RFC 0894, unknown: A Standard for the Transmission of IP Datagrams over Ethernet Networks. RFC 894 (1984). DOI 10.17487/RFC0894. URL `https://rfc-editor.org/rfc/rfc894.txt`. Accessed: February 26, 2019 52, 84

72. RFC 0895, unknown: Standard for the transmission of IP datagrams over experimental Ethernet networks. RFC 895 (1984). DOI 10.17487/RFC0895. URL `https://rfc-editor.org/rfc/rfc895.txt`. Accessed: February 01, 2019 52, 84

73. RFC 0951, Croft, B., Gilmore, J.: BOOTSTRAP PROTOCOL (BOOTP). On-line (1985). DOI 10.17487/RFC0951. URL `https://tools.ietf.org/html/rfc951`. Accessed: January 17, 2019 218, 228, 459

74. RFC 0973, Mockapetris, P.: DOMAIN NAMES - IMPLEMENTATION and SPECIFICATIONS. On-line (1986). DOI 10.17487/RFC0973. URL `https://tools.ietf.org/html/rfc8973`. Accessed: January 22, 2019 372

75. RFC 0974, Partridge, C.: Mail routing and the domain system. RFC 974 (1986). DOI 10.17487/RFC0974. URL `https://tools.ietf.org/html/rfc0974`. Accessed: January 15, 2019 351, 372, 411

76. RFC 1034, Mockapetris, P.: DOMAIN NAMES - CONCEPTS and FACILITIES. On-line (1987). DOI 10.17487/RFC1034. URL `https://tools.ietf.org/html/rfc1034`. Accessed: January 22, 2019 372

77. RFC 1035, Mockapetris, P.: Domain names - implementation and specification. RFC 1035 (1987). DOI 10.17487/RFC1035. URL `https://rfc-editor.org/rfc/rfc1035.txt`. Accessed: January 22, 2019 351, 372

78. RFC 1047, Partridge, C.: Duplicate messages and SMTP. RFC 1047 (1988). DOI 10.17487/ RFC1047. URL `https://rfc-editor.org/rfc/rfc1047.txt`. Accessed: March 03, 2019 411

79. RFC 1058, Unknown: Routing Information Protocol. RFC 1058 (1988). DOI 10.17487/ RFC1058. URL `https://rfc-editor.org/rfc/rfc1058.txt`. Accessed: February 14, 2019 269

80. RFC 1070, Rose, D.M.T.: Use of the Internet as a subnetwork for experimentation with the OSI network layer. RFC 1070 (1989). DOI 10.17487/RFC1070. URL `https://rfc-editor.org/rfc/rfc1070.txt`. Accessed: June 8, 2019 32

81. RFC 1101, Mockapetris, P.: DNS Encoding of Network Names and Other Types. On-line (1989). DOI 10.17487/RFC1101. URL `https://tools.ietf.org/html/rfc1101`. Accessed: January 22, 2019 372

82. RFC 1131, unknown: OSPF specification. RFC 1131 (1989). DOI 10.17487/RFC1131. URL `https://rfc-editor.org/rfc/rfc1131.txt`. Accessed: March 30, 2019 297

83. RFC 1132, Leo J. McLaughlin III: Standard for the transmission of 802.2 packets over IPX networks. RFC 1132 (1989). DOI 10.17487/RFC1132. URL `https://rfc-editor.org/rfc/rfc1132.txt`. Accessed: March 30, 2019 52, 84

84. RFC 1142, Unknown: OSI IS-IS Intra-domain Routing Protocol. RFC 1142 (1990). DOI 10. 17487/RFC1142. URL `https://rfc-editor.org/rfc/rfc1142.txt`. Accessed: February 19, 2019 317

85. RFC 1169, Mills, K.L., Cerf, D.V.G.: Explaining the role of GOSIP. RFC 1169 (1990). DOI 10.17487/RFC1169. URL `https://rfc-editor.org/rfc/rfc1169.txt`. Accessed: June 25, 2019 302, 317, 471

86. RFC 1178, Libes, D.: Choosing a Name for Your Computer. Request For Comments (1990). DOI 10.17487/RFC1178. URL `https://tools.ietf.org/html/rfc1178`. Accessed: December 10, 2018 129, 372

87. RFC 1195, Callon, R.: Use of OSI IS-IS for routing in TCP/IP and dual environments. Request For Comments (1990). DOI 10.17487/RFC1195. URL `https://tools.ietf.org/html/rfc1195`. Accessed: December 12, 2018 317

88. RFC 1234, Provan, D.: Tunneling IPX traffic through IP networks. RFC 1234 (1991). DOI 10. 17487/RFC1234. URL `https://rfc-editor.org/rfc/rfc1234.txt`. Accessed: March 30, 2019 84

89. RFC 1245, Maloy, J.: OSPF protocol analysis. Request For Comments (1991). DOI 10.17487/ RFC1245. URL `https://tools.ietf.org/html/rfc1245`. Accessed: December 29, 2018 271, 297

90. RFC 1246, Maloy, J.: Experience with the OSPF Protocol. On line (1991). DOI 10.17487/ RFC1246. URL `https://tools.ietf.org/html/rfc1246`. Accessed: December 29, 2018 297

91. RFC 1247, Moy, J.: "OSPF Version 2". Request For Comments (1991). DOI 10.17487/ RFC1247. URL `https://tools.ietf.org/html/rfc1247`. Accessed: December 29, 2018 271, 297

92. RFC 1266, Rekhter, Y.: Experience with the BGP Protocol. RFC 1266 (1991). DOI 10. 17487/RFC1266. URL `https://rfc-editor.org/rfc/rfc1266.txt`. Accessed: March 21, 2019 318

93. RFC 1348, Manning, B.: DNS NSAP RRs. On-line (1992). DOI 10.17487/RFC1348. URL `https://tools.ietf.org/html/rfc1348`. Accessed: January 23, 2019 372

94. RFC 1364, Varadhan, K.: BGP OSPF Interaction. RFC 1364 (1992). DOI 10.17487/ RFC1364. URL `https://rfc-editor.org/rfc/rfc1364.txt`. Accessed: March 21, 2019 297, 318

95. RFC 1370, Chapin, I.A.B.C.L.: Applicability Statement for OSPF. On line (1992). DOI 10.17487/RFC1370. URL `https://tools.ietf.org/html/rfc1370`. Accessed: December 29, 2018 297

96. RFC 1387, Malkin, G.S.: RIP Version 2 Protocol Analysis. RFC 1387 (1993). DOI 10.17487/ RFC1387. URL `https://rfc-editor.org/rfc/rfc1387.txt`. Accessed: March 14, 2019 269

97. RFC 1388, Malkin, G.S.: RIP Version 2 Carrying Additional Information. RFC 1388 (1993). DOI 10.17487/RFC1388. URL `https://rfc-editor.org/rfc/rfc1388.txt`. Accessed: March 30, 2019 269

98. RFC 1395, Reynolds, J.K.: BOOTP Vendor Information Extensions. On-line (1993). DOI 10.17487/RFC1395. URL `https://tools.ietf.org/html/rfc1395`. Accessed: January 17, 2019 228, 459

99. RFC 1397, Haskin, D.L.: Default Route Advertisement In BGP2 and BGP3 Version of The Border Gateway Protocol. RFC 1397 (1993). DOI 10.17487/RFC1397. URL `https://rfc-editor.org/rfc/rfc1397.txt`. Accessed: March 21, 2019 198, 318

100. RFC 1403, unknown: BGP OSPF Interaction. RFC 1403 (1993). DOI 10.17487/RFC1403. URL `https://rfc-editor.org/rfc/rfc1403.txt`. Accessed: March 21, 2019 297, 318

101. RFC 1425, Crocker, D., Freed, N., Klensin, D.J.C., Rose, D.M.T., Stefferud, E.A.: SMTP Service Extensions. RFC 1425 (1993). DOI 10.17487/RFC1425. URL `https://rfc-editor.org/rfc/rfc1425.txt`. Accessed: March 03, 2019 411

102. RFC 1426, Crocker, D., Freed, N., Klensin, D.J.C., Rose, D.M.T., Stefferud, E.A.: SMTP Service Extension for 8bit-MIMEtransport. RFC 1426 (1993). DOI 10.17487/RFC1426. URL https://rfc-editor.org/rfc/rfc1426.txt. Accessed: March 03, 2019 411

103. RFC 1427, Freed, N., Klensin, D.J.C., Moore, K.: SMTP Service Extension for Message Size Declaration. RFC 1427 (1993). DOI 10.17487/RFC1427. URL https://rfc-editor.org/rfc/rfc1427.txt. Accessed: March 3, 2019 411

104. RFC 1428, Vaudreuil, G.: Transition of Internet Mail from Just-Send-8 to 8bit-SMTP/MIME. RFC 1428 (1993). DOI 10.17487/RFC1428. URL https://rfc-editor.org/rfc/rfc1428.txt. Accessed: March 3, 2019 411

105. RFC 1460, Rose, D.M.T.: Post Office Protocol - Version 3. RFC 1460 (1993). DOI 10.17487/RFC1460. URL https://rfc-editor.org/rfc/rfc1460.txt. Accessed: March 16, 2019 411

106. RFC 1497, Reynolds, J.K.: BOOTP Vendor Information Extensions. On-line (1993). DOI 10.17487/RFC1497. URL https://tools.ietf.org/html/rfc1497. Accessed: January 17, 2019 228, 459

107. RFC 1504, Oppenheimer, A.B.: Appletalk Update-Based Routing Protocol: Enhanced Appletalk Routing. RFC 1504 (1993). DOI 10.17487/RFC1504. URL https://rfc-editor.org/rfc/rfc1504.txt. Accessed: June 14, 2019 57

108. RFC 1517, Hinden, B.: Applicability Statement for the Implementation of Classless Inter-Domain Routing (CIDR). RFC 1517 (1993). DOI 10.17487/RFC1517. URL https://rfc-editor.org/rfc/rfc1517.txt. Accessed: July 29, 2019 85

109. RFC 1518, Rekhter, Y., Li, T.: An Architecture for IP Address Allocation with CIDR. RFC 1518 (1993). DOI 10.17487/RFC1518. URL https://rfc-editor.org/rfc/rfc1518.txt. Accessed: July 29, 2019 85

110. RFC 1519, Fuller, V., Li, T.: Classless Inter-Domain Routing (CIDR): an Address Assignment and Aggregation Strategy. RFC 1519 (1993). DOI 10.17487/RFC1519. URL https://rfc-editor.org/rfc/rfc1519.txt. Accessed: July 29, 2019 85

111. RFC 1520, Rekhter, Y., Topolcic, C.M.: Exchanging Routing Information Across Provider Boundaries in the CIDR Environment. RFC 1520 (1993). DOI 10.17487/RFC1520. URL https://rfc-editor.org/rfc/rfc1520.txt. Accessed: July 29, 2019 85, 318

112. RFC 1532, Wimer, W.: Clarifications and Extensions for the Bootstrap Protocol. On-line (1993). DOI 10.17487/RFC1532. URL https://tools.ietf.org/html/rfc1532. Accessed: January 17, 2019 228, 459

113. RFC 1541, Droms, R.: Dynamic Host Configuration Protocol. On-line (1997). DOI 10.17487/RFC1541. URL https://tools.ietf.org/html/rfc2131. Obsoletes: 1541. Updated by: 3396, 4361, 5494, 6842. 217, 228

114. RFC 1542, Wimer, W.: Clarifications and Extensions for the Bootstrap Protocol. On-line (1993). DOI 10.17487/RFC1542. URL https://tools.ietf.org/html/rfc1542. Accessed: January 17, 2019 228, 459

115. RFC 1577, Laubach, M.E.: Classical IP and ARP over ATM. RFC 1577 (1994). DOI 10.17487/RFC1577. URL https://rfc-editor.org/rfc/rfc1577.txt. Accessed: March 21, 2019 52, 84

116. RFC 1583, Moy, J.: "OSPF Version 2". Request For Comments (1991). DOI 10.17487/RFC1583. URL https://tools.ietf.org/html/rfc1583. Accessed: December 12, 2018 233, 271, 297

117. RFC 1591, Postel, D.J.: Domain Name System Structure and Delegation. RFC 1591 (1994). DOI 10.17487/RFC1591. URL https://rfc-editor.org/rfc/rfc1591.txt. Accessed: February 13, 2019 372

118. RFC 1637, Colella, R.P., Manning, B.: DNS NSAP Resource Records. RFC 1637 (1994). DOI 10.17487/RFC1637. URL https://rfc-editor.org/rfc/rfc1637.txt. Accessed: February 15, 2019 317, 372

119. RFC 1648, Cargille, A.: Postmaster Convention for X.400 Operations. RFC 1648 (1994). DOI 10.17487/RFC1648. URL https://rfc-editor.org/rfc/rfc1648.txt. Accessed: April 30, 2019 394, 411

120. RFC 1651, Freed, N., Klensin, D.J.C., Rose, D.M.T., Crocker, D., Stefferu, E.A.: SMTP Service Extensions. RFC 1651 (1994). DOI 10.17487/RFC1651. URL https://rfc-editor.org/rfc/rfc1651.txt. Accessed: March 03, 2019 411

121. RFC 1652, Freed, N., Klensin, D.J.C., Rose, D.M.T., Crocker, D., Stefferud, E.A.: SMTP Service Extension for 8bit-MIMEtransport. RFC 1652 (1994). DOI 10.17487/RFC1652. URL https://rfc-editor.org/rfc/rfc1652.txt. Accessed: March 03, 2019 411

122. RFC 1653, Freed, N., Klensin, D.J.C., Moore, K.: SMTP Service Extension for Message Size Declaration. RFC 1653 (1994). DOI 10.17487/RFC1653. URL https://rfc-editor.org/rfc/rfc1653.txt. Accessed: March 03, 2019 411

123. RFC 1680, Brazdziunas, C.: IPng Support for ATM Services. RFC 1680 (1994). DOI 10.17487/RFC1680. URL https://rfc-editor.org/rfc/rfc1680.txt. Accessed: March 21, 2019 86

124. RFC 1700, Reynolds, J.K., Postel, D.J.: Assigned Numbers. RFC 1700 (1994). DOI 10.17487/RFC1700. URL https://rfc-editor.org/rfc/rfc1700.txt. Accessed: August 17, 2018 84

125. RFC 1700, Reynolds, J.K., Postel, D.J.: Assigned Numbers. RFC 1700 (1994). DOI 10.17487/RFC1700. URL https://rfc-editor.org/rfc/rfc1700.txt. Accessed: August 17, 2018 100

126. RFC 1706, Colella, R.P.: DNS NSAP Resource Records. On-line (1994). DOI 10.17487/RFC1706. URL https://tools.ietf.org/html/rfc1706. Accessed: January 23, 2019 372

127. RFC 1723, Malkin, G.: RIP Version 2 Carrying Additional Information. Request For Comments (1994). DOI 10.17487/RFC1723. URL https://tools.ietf.org/html/rfc1723. Accessed: December 12, 2018 269

128. RFC 1725, Myers, J.G., Rose, D.M.T.: Post Office Protocol - Version 3. RFC 1725 (1994). DOI 10.17487/RFC1725. URL https://rfc-editor.org/rfc/rfc1725.txt. Accessed: March 16, 2019 411

129. RFC 1794, Brisco, T.P.: DNS Support for Load Balancing. RFC 1794 (1995). DOI 10.17487/RFC1794. URL https://rfc-editor.org/rfc/rfc1794.txt. Accessed: February 15, 2019 372

130. RFC 1812, Baker, F.: Requirements for IP Version 4 Routers. RFC 1812 (1995). DOI 10.17487/RFC1812. URL https://rfc-editor.org/rfc/rfc1812.txt. Accessed: July 29, 2019 84, 85

131. RFC 1817, Rekhter, Y.: CIDR and Classful Routing. RFC 1817 (1995). DOI 10.17487/RFC1817. URL https://rfc-editor.org/rfc/rfc1817.txt. Accessed: July 29, 2019 85

132. RFC 1830, Vaudreuil, G.: SMTP Service Extensions for Transmission of Large and Binary MIME Messages. RFC 1830 (1995). DOI 10.17487/RFC1830. URL https://rfc-editor.org/rfc/rfc1830.txt. Accessed: March 3, 2019 411

133. RFC 1845, Crocker, D., Freed, N.: SMTP Service Extension for Checkpoint/Restart. RFC 1845 (1995). DOI 10.17487/RFC1845. URL https://rfc-editor.org/rfc/rfc1845.txt. Accessed: March 03, 2019 411

134. RFC 1846, Dupont, F., Durand, A.: SMTP 521 Reply Code. RFC 1846 (1995). DOI 10.17487/RFC1846. URL https://rfc-editor.org/rfc/rfc1846.txt. Accessed: March 03, 2019 412

135. RFC 1854, Freed, N.: SMTP Service Extension for Command Pipelining. RFC 1854 (1995). DOI 10.17487/RFC1854. URL https://rfc-editor.org/rfc/rfc1854.txt. Accessed: March 03, 2019 412

136. RFC 1869, Crocker, D., Klensin, D.J.C., Stefferud, E.A., Rose, D.M.T., Freed, N.: SMTP Service Extensions. RFC 1869 (1995). DOI 10.17487/RFC1869. URL https://rfc-editor.org/rfc/rfc1869.txt. Accessed: March 03, 2019 412

137. RFC 1870, Klensin, D.J.C., Freed, N., Moore, K.: SMTP Service Extension for Message Size Declaration. RFC 1870 (1995). DOI 10.17487/RFC1870. URL https://rfc-editor.org/rfc/rfc1870.txt. Accessed: March 03, 2019 412

138. RFC 1881, unknown: IPv6 Address Allocation Management. RFC 1881 (1995). DOI 10.17487/RFC1881. URL https://rfc-editor.org/rfc/rfc1881.txt. Accessed: March 30, 2019 86

139. RFC 1888, Bound, J., Carpenter, B.E., Harrington, D., Houldsworth, J., Lloyd, A.: OSI NSAPs and IPv6. RFC 1888 (1996). DOI 10.17487/RFC1888. URL https://rfc-editor.org/rfc/rfc1888.txt. Accessed: March 4, 2019 32, 86

140. RFC 1891, Moore, K.: SMTP Service Extension for Delivery Status Notifications. RFC 1891 (1996). DOI 10.17487/RFC1891. URL https://rfc-editor.org/rfc/rfc1891.txt. Accessed: March 03, 2019 412

141. RFC 1894, Vaudreuil, G., Moore, K.: An Extensible Message Format for Delivery Status Notifications. RFC 1894 (1996). DOI 10.17487/RFC1894. URL https://rfc-editor.org/rfc/rfc1894.txt. Accessed: March 30, 2019 411

142. RFC 1912, Barr, D.: Common DNS Operational and Configuration Errors. On line. (1996). DOI 10.17487/RFC1912. URL https://tools.ietf.org/html/rfc1912. Accessd on: February 11, 2019 354, 372

143. RFC 1917, II, P.J.N.: An Appeal to the Internet Community to Return Unused IP Networks (Prefixes) to the IANA. RFC 1917 (1996). DOI 10.17487/RFC1917. URL https://rfc-editor.org/rfc/rfc1917.txt. Accessed: July 29, 2019. Interesting attempt. 84, 85

144. RFC 1918, Moskowitz, R., Karrenberg, D., Rekhter, Y., Lear, E., de Groot, G.J.: Address Allocation for Private Internets. RFC 1918 (1996). DOI 10.17487/RFC1918. URL https://rfc-editor.org/rfc/rfc1918.txt. Accessed: February 15, 2019 372

145. RFC 1933, Nordmark, E., Gilligan, R.E.: Transition Mechanisms for IPv6 Hosts and Routers. RFC 1933 (1996). DOI 10.17487/RFC1933. URL https://rfc-editor.org/rfc/rfc1933.txt. Accessed: March 30, 2019 84, 86

146. RFC 1939, Rose, D.M.T., Myers, J.G.: Post Office Protocol - Version 3. RFC 1939 (1996). DOI 10.17487/RFC1939. URL https://rfc-editor.org/rfc/rfc1939.txt. Accessed: March 16, 2019 412

147. RFC 1945, Nielsen, H.F., Berners-Lee, T.: Hypertext Transfer Protocol – HTTP/1.0. RFC 1945 (1996). DOI 10.17487/RFC1945. URL https://rfc-editor.org/rfc/rfc1945.txt. Accessed: February 25, 2019 383

148. RFC 1972, Crawford, D.M.: A Method for the Transmission of IPv6 Packets over Ethernet Networks. RFC 1972 (1996). DOI 10.17487/RFC1972. URL https://rfc-editor.org/rfc/rfc1972.txt. Accessed: February 26, 2019 52, 86

149. RFC 1982, Bush, R., Elz, R.: Serial Number Arithmetic. RFC 1982 (1996). DOI 10.17487/RFC1982. URL https://rfc-editor.org/rfc/rfc1982.txt. Accessed: February 15, 2019 354, 372

150. RFC 1985, Winter, J.D.: SMTP Service Extension for Remote Message Queue Starting. RFC 1985 (1996). DOI 10.17487/RFC1985. URL https://rfc-editor.org/rfc/rfc1985.txt. Accessed: March 03, 2019 412

151. RFC 1995, Ohta, D.M.: Incremental Zone Transfer in DNS. RFC 1995 (1996). DOI 10.17487/RFC1995. URL https://rfc-editor.org/rfc/rfc1995.txt. Accessed: January 23, 2019 359, 372

152. RFC 1996, Vixie, P.A.: A Mechanism for Prompt Notification of Zone Changes (DNS NOTIFY). RFC 1996 (1996). DOI 10.17487/RFC1996. URL `https://rfc-editor.org/rfc/rfc1996.txt`. Accessed: February 15, 2019 372

153. RFC 2030, Mills, P.D.L.: Simple Network Time Protocol (SNTP) Version 4 for IPv4, IPv6 and OSI. RFC 2030 (1996). DOI 10.17487/RFC2030. URL `https://rfc-editor.org/rfc/rfc2030.txt`. Accessed: March 30, 2019 84, 86

154. RFC 2034, Freed, N.: SMTP Service Extension for Returning Enhanced Error Codes. RFC 2034 (1996). DOI 10.17487/RFC2034. URL `https://rfc-editor.org/rfc/rfc2034.txt`. Accessed: March 03, 2019 412

155. RFC 2036, Huston, G.: Observations on the use of Components of the Class A Address Space within the Internet. RFC 2036 (1996). DOI 10.17487/RFC2036. URL `https://rfc-editor.org/rfc/rfc2036.txt`. Accessed: July 29, 2019 85

156. RFC 2065, Eastlake, D., 3rd: Domain Name System Security Extensions. On-line (1997). DOI 10.17487/RFC2065. URL `https://tools.ietf.org/html/rfc2065`. Accessed: January 23, 2019 372

157. RFC 2068, Fielding, R.T., et al.: Hypertext Transfer Protocol – HTTP/1.1. RFC 2068 (1997). DOI 10.17487/RFC2068. URL `https://rfc-editor.org/rfc/rfc2068.txt`. Accessed: February 24, 2019 383

158. RFC 2080, Malkin, G.S., Minnear, R.E.: RIPng for IPv6. RFC 2080 (1997). DOI 10.17487/RFC2080. URL `https://rfc-editor.org/rfc/rfc2080.txt`. Accessed: March 30, 2019 86, 269

159. RFC 2081, Malkin, G.S.: RIPng Protocol Applicability Statement. RFC 2081 (1997). DOI 10.17487/RFC2081. URL `https://rfc-editor.org/rfc/rfc2081.txt`. Accessed: March 30, 2019 86, 269

160. RFC 2131, Droms, R.: Dynamic Host Configuration Protocol. On-line (1997). DOI 10.17487/RFC2131. URL `https://tools.ietf.org/html/rfc2131`. Accessed: January 21, 2019 228

161. RFC 2132, Alexander, S.: DHCP Options and BOOTP Vendor Extensions. On-line (1997). DOI 10.17487/RFC2132. URL `https://tools.ietf.org/html/rfc2132`. Accessed: January 21, 2019 228

162. RFC 2136, Vixie, P.A., et al.: Dynamic Updates in the Domain Name System (DNS UPDATE). On-line (1997). DOI 10.17487/RFC2136. URL `https://tools.ietf.org/html/rfc2136`. Accessed: January 23, 2019 372

163. RFC 2145, Nielsen, H.F., et al.: Use and Interpretation of HTTP Version Numbers. RFC 2145 (1997). DOI 10.17487/RFC2145. URL `https://rfc-editor.org/rfc/rfc2145.txt`. Accessed: February 24, 2019 383

164. RFC 2178, Moy, J.: OSPF Version 2. Request For Comments. On-line. (1997). DOI 10.17487/RFC2178. URL `https://tools.ietf.org/html/rfc2178`. Accessed: January 25, 2019 297

165. RFC 2181, Elz, R., Bush, R.: Clarifications to the DNS Specification. Request For Comments (1998). DOI 10.17487/RFC2181. URL `https://tools.ietf.org/html/rfc2181`. Accessed: January 22, 2019 372

166. RFC 2197, Freed, N.: SMTP Service Extension for Command Pipelining. RFC 2197 (1997). DOI 10.17487/RFC2197. URL `https://rfc-editor.org/rfc/rfc2197.txt`. Accessed: March 3, 2019 412

167. RFC 2295, Mutz, D.A., Holtman, K.: Transparent Content Negotiation in HTTP. RFC 2295 (1998). DOI 10.17487/RFC2295. URL `https://rfc-editor.org/rfc/rfc2295.txt`. Accessed: February 25, 2019 383

168. RFC 2317, Eidnes, H., et. al.: Classless IN-ADDR.ARPA delegation. RFC 2317 (1998). DOI 10.17487/RFC2317. URL `https://rfc-editor.org/rfc/rfc2317.txt`. Accessed: February 15, 2019 372

169. RFC 2328, Moy, J.: "OSPF Version 2". Request For Comments (1991). DOI 10.17487/RFC2328. URL https://tools.ietf.org/html/rfc2328. Accessed: January 25, 2019 297

170. RFC 2358, Johnson, J.T., Flick, J.W.: Definitions of Managed Objects for the Ethernet-like Interface Types. RFC 2358 (1998). DOI 10.17487/RFC2358. URL https://rfc-editor.org/rfc/rfc2358.txt. Accessed: February 27, 2019 52

171. RFC 2453, Malkin, G.: RIP Version 2. RFC 2453 (1998). DOI 10.17487/RFC2453. URL https://tools.ietf.org/html/rfc2453. Accessed: December 12, 2018 247, 269

172. RFC 2461, Narten, D.T., Simpson, W.A., et al: Neighbor Discovery for IP Version 6 (IPv6). RFC 2461 (1998). DOI 10.17487/RFC2461. URL https://rfc-editor.org/rfc/rfc2461.txt. Accessed: March 30, 2019 86, 198

173. RFC 2464, Crawford, D.M.: Transmission of IPv6 Packets over Ethernet Networks. RFC 2464 (1998). DOI 10.17487/RFC2464. URL https://rfc-editor.org/rfc/rfc2464.txt. Accessed: February 27, 2019 52, 86

174. RFC 2487, Hoffman, P.E.: SMTP Service Extension for Secure SMTP over TLS. RFC 2487 (1999). DOI 10.17487/RFC2487. URL https://rfc-editor.org/rfc/rfc2487.txt. Accessed: March 3, 2019 412

175. RFC 2505, Lindberg, G.: Anti-Spam Recommendations for SMTP MTAs. RFC 2505 (1999). DOI 10.17487/RFC2505. URL https://rfc-editor.org/rfc/rfc2505.txt. Accessed: March 3, 2019 411, 412

176. RFC 2518, Carter, S., et al.: HTTP Extensions for Distributed Authoring – WEBDAV. RFC 2518 (1999). DOI 10.17487/RFC2518. URL https://rfc-editor.org/rfc/rfc2518.txt. Accessed: February 25, 2019 383

177. RFC 2535, Arends, R., 3rd, D.E.E.: Domain Name System Security Extensions. On-line (2005). DOI 10.17487/RFC2535. URL https://tools.ietf.org/html/rfc2535. Accessed: January 23, 2019 351

178. RFC 2554, Myers, J.G.: SMTP Service Extension for Authentication. RFC 2554 (1999). DOI 10.17487/RFC2554. URL https://rfc-editor.org/rfc/rfc2554.txt. Accessed: March 3, 2019 412

179. RFC 2606, Donald E. Eastlake 3rd, Panitz, A.R.: Reserved Top Level DNS Names. RFC 2606 (1999). DOI 10.17487/RFC2606. URL https://rfc-editor.org/rfc/rfc2606.txt. Accessed: February 15, 2019 372

180. RFC 2616, Nielsen, H.F., et al.: Hypertext Transfer Protocol – HTTP/1.1. RFC 2616 (1999). DOI 10.17487/RFC2616. URL https://rfc-editor.org/rfc/rfc2616.txt. Accessed: February 24, 2019 383

181. RFC 2635, Hambridge, S., Lunde, A.: DON'T SPEW A Set of Guidelines for Mass Unsolicited Mailings and Postings (spam*). RFC 2635 (1999). DOI 10.17487/RFC2635. URL https://rfc-editor.org/rfc/rfc2635.txt. Accessed: April 4, 2019 412

182. RFC 2645, Gellens, R.: ON-DEMAND MAIL RELAY (ODMR) SMTP with Dynamic IP Addresses. RFC 2645 (1999). DOI 10.17487/RFC2645. URL https://rfc-editor.org/rfc/rfc2645.txt. Accessed: March 3, 2019 412

183. RFC 2660, Rescorla, E., Schiffman, A.M.: The Secure HyperText Transfer Protocol. RFC 2660 (1999). DOI 10.17487/RFC2660. URL https://rfc-editor.org/rfc/rfc2660.txt. Accessed: February 25, 2019 383

184. RFC 2671, Vixie, P.A.: Extension Mechanisms for DNS (EDNS0). RFC 2671 (1999). DOI 10.17487/RFC2671. URL https://rfc-editor.org/rfc/rfc2671.txt. Accessed: February 15, 2019 372

185. RFC 2782, Gulbrandsen, A., Esibov, D.L.: A DNS RR for specifying the location of services (DNS SRV). RFC 2782 (2000). DOI 10.17487/RFC2782. URL https://rfc-editor.org/rfc/rfc2782.txt. Accessed: February 15, 2019 372

186. RFC 2821, Klensin, D.J.C.: Simple Mail Transfer Protocol. RFC 2821 (2001). DOI 10.
17487/RFC2821. URL `https://rfc-editor.org/rfc/rfc2821.txt`. Accessed:
March 3, 2019 412

187. RFC 2842, Scudder, J., Chandra, R.: Capabilities Advertisement with
BGP-4. RFC 2842 (2000). DOI 10.17487/RFC2842. URL
`https://rfc-editor.org/rfc/rfc2842.txt`. Accessed: March 21, 2019
318

188. RFC 2858, Rekhter, Y., Bates, T.J.: Multiprotocol Extensions for BGP-4. RFC 2858 (2000).
DOI 10.17487/RFC2858. URL `https://rfc-editor.org/rfc/rfc2858.txt`.
Accessed: March 21, 2019 318

189. RFC 2872, Bernet, Y., Pabbati, R.B.: Application and Sub Application Identity Policy
Element for Use with RSVP. RFC 2872 (2000). DOI 10.17487/RFC2872. URL
`https://rfc-editor.org/rfc/rfc2872.txt`. Accessed: January 15, 2019 351

190. RFC 2874, Huitema, C., Crawford, D.M.: DNS Extensions to Support IPv6 Address
Aggregation and Renumbering. RFC 2874 (2000). DOI 10.17487/RFC2874. URL
`https://rfc-editor.org/rfc/rfc2874.txt`. Accessed: February 15, 2019 351,
372

191. RFC 2966, Li, T., et al.: Domain-wide Prefix Distribution with Two-
Level IS-IS. RFC 2966 (2000). DOI 10.17487/RFC2966. URL
`https://rfc-editor.org/rfc/rfc2966.txt`. Accessed: February 19, 2019 317

192. RFC 2973, Parker, J., et al.: IS-IS Mesh Groups. RFC 2973 (2000). DOI 10.17487/RFC2973.
URL `https://rfc-editor.org/rfc/rfc2973.txt`. Accessed: February 21, 2019
317

193. RFC 3013, Killalea, T.: Recommended Internet Service Provider Security Services and Procedures. RFC 3013 (2000). DOI 10.17487/RFC3013. URL
`https://rfc-editor.org/rfc/rfc3013.txt`. Accessed: May 25, 2019 319

194. RFC 3030, Vaudreuil, G.: SMTP Service Extensions for Transmission of Large and
Binary MIME Messages. RFC 3030 (2000). DOI 10.17487/RFC3030. URL
`https://rfc-editor.org/rfc/rfc3030.txt`. Accessed: March 3, 2019 412,
413

195. RFC 3046, Patrick, M.W.: DHCP Relay Agent Information Option. On-line (2001). DOI
10.17487/RFC3046. URL `https://tools.ietf.org/html/rfc3046`. Accessed:
January 21, 2019 228

196. RFC 3137, Retana, A., McPherson, D.R., et al: OSPF Stub Router Advertisement. RFC 3137 (2001). DOI 10.17487/RFC3137. URL
`https://rfc-editor.org/rfc/rfc3137.txt`. Accessed: March 30, 2019
297

197. RFC 3143, Dilley, J., Cooper, I.: Known HTTP Proxy/Caching Problems. RFC 3143 (2001).
DOI 10.17487/RFC3143. URL `https://rfc-editor.org/rfc/rfc3143.txt`.
Accessed: March 11, 2019 383

198. RFC 3277, anny R. McPherson: Intermediate System to Intermediate System (IS-IS)
Transient Blackhole Avoidance. RFC 3277 (2002). DOI 10.17487/RFC3277. URL
`https://rfc-editor.org/rfc/rfc3277.txt`. Accessed: February 21, 2019 317

199. RFC 3315, Perkins, C.E., et al.: Dynamic Host Configuration Protocol
for IPv6 (DHCPv6). On-line (2013). DOI 10.17487/RFC3315. URL
`https://tools.ietf.org/html/rfc3315`. Accessed: January 21, 2019 228

200. RFC 3330, IANA: Special-Use IPv4 Addresses. RFC 3330 (2002). DOI 10.17487/RFC3330.
URL `https://rfc-editor.org/rfc/rfc3330.txt`. Accessed: March 30, 2019
63, 84

201. RFC 3363, Bush, R., et al.: Representing Internet Protocol version 6 (IPv6) Addresses in
the Domain Name System (DNS). RFC 3363 (2002). DOI 10.17487/RFC3363. URL
`https://rfc-editor.org/rfc/rfc3363.txt`. Accessed: February 15, 2019 372

202. RFC 3378, Housley, R., Hollenbeck, S.: EtherIP: Tunneling Ethernet Frames
in IP Datagrams. RFC 3378 (2002). DOI 10.17487/RFC3378. URL

`https://rfc-editor.org/rfc/rfc3378.txt`. Accessed: February 24, 2019 52, 85

203. RFC 3396, Lemon, T., Cheshire, D.S.D.: Encoding Long Options in the Dynamic Host Configuration Protocol (DHCPv4). On-line (2002). DOI 10.17487/RFC3396. URL `https://tools.ietf.org/html/rfc3396`. Accessed: January 21, 2019 228

204. RFC 3401, Mealling, M.H.: Dynamic Delegation Discovery System (DDDS) Part One: The Comprehensive DDDS. RFC 3401 (2002). DOI 10.17487/RFC3401. URL `https://rfc-editor.org/rfc/rfc3401.txt`. Accessed: February 15, 2019 372

205. RFC 3402, Mealling, M.H.: Dynamic Delegation Discovery System (DDDS) Part Two: The Algorithm. RFC 3402 (2002). DOI 10.17487/RFC3402. URL `https://rfc-editor.org/rfc/rfc3402.txt`. Accessed: February 15, 2019 372

206. RFC 3403, Mealling, M.H.: Dynamic Delegation Discovery System (DDDS) Part Three: The Domain Name System (DNS) Database. RFC 3403 (2002). DOI 10.17487/RFC3403. URL `https://rfc-editor.org/rfc/rfc3403.txt`. Accessed: February 15, 2019 372

207. RFC 3404, Mealling, M.H.: Dynamic Delegation Discovery System (DDDS) Part Four: The Uniform Resource Identifiers (URI). RFC 3404 (2002). DOI 10.17487/RFC3404. URL `https://rfc-editor.org/rfc/rfc3404.txt`. Accessed: February 15, 2019 373

208. RFC 3405, Mealling, M.H.: Dynamic Delegation Discovery System (DDDS) Part Five: URI.ARPA Assignment Procedures. RFC 3405 (2002). DOI 10.17487/RFC3405. URL `https://rfc-editor.org/rfc/rfc3405.txt`. Accessed: February 15, 2019 373

209. RFC 3425, Lawrence, D.C.: Obsoleting IQUERY. RFC 3425 (2002). DOI 10.17487/RFC3425. URL `https://rfc-editor.org/rfc/rfc3425.txt`. Accessed: February 15, 2019 373

210. RFC 3461, Moore, K.: Simple Mail Transfer Protocol (SMTP) Service Extension for Delivery Status Notifications (DSNs). RFC 3461 (2003). DOI 10.17487/RFC3461. URL `https://rfc-editor.org/rfc/rfc3461.txt`. Accessed: March 3, 2019 412

211. RFC 3484, Draves, R.P.: Default Address Selection for Internet Protocol version 6 (IPv6). RFC 3484 (2003). DOI 10.17487/RFC3484. URL `https://rfc-editor.org/rfc/rfc3484.txt`. Accessed: February 15, 2019 373

212. RFC 3509, Yeung, D.M., Lindem, A., et al: Alternative Implementations of OSPF Area Border Routers. RFC 3509 (2003). DOI 10.17487/RFC3509. URL `https://rfc-editor.org/rfc/rfc3509.txt`. Accessed: March 30, 2019 297

213. RFC 3519, Levkowetz, H., Vaarala, S.: Mobile IP Traversal of Network Address Translation (NAT) Devices. RFC 3519 (2003). DOI 10.17487/RFC3519. URL `https://rfc-editor.org/rfc/rfc3519.txt`. Accessed: February 26, 2019 423

214. RFC 3596, Ksinant, V., et al.: DNS Extensions to Support IP Version 6. RFC 3596 (2003). DOI 10.17487/RFC3596. URL `https://rfc-editor.org/rfc/rfc3596.txt`. Accessed: February 15, 2019 351, 373

215. RFC 3597, Gustafsson, A.: Handling of Unknown DNS Resource Record (RR) Types. RFC 3597 (2003). DOI 10.17487/RFC3597. URL `https://rfc-editor.org/rfc/rfc3597.txt`. Accessed: February 15, 2019 373

216. RFC 3619, Yip, M., Shah, S.: Extreme Networks' Ethernet Automatic Protection Switching (EAPS) Version 1. RFC 3619 (2003). DOI 10.17487/RFC3619. URL `https://rfc-editor.org/rfc/rfc3619.txt`. Accessed: February 27, 2019 52

217. RFC 3621, Romascanu, D., Berger, A.: Power Ethernet MIB. RFC 3621 (2003). DOI 10.17487/RFC3621. URL `https://rfc-editor.org/rfc/rfc3621.txt`. Accessed: February 22, 2019 52

218. RFC 3635, Flick, J.W.: Definitions of Managed Objects for the Ethernet-like Interface Types. RFC 3635 (2003). DOI 10.17487/RFC3635. URL `https://rfc-editor.org/rfc/rfc3635.txt`. Accessed: February 27, 2019 52

219. RFC 3637, Heard, M.: Definitions of Managed Objects for the Ethernet WAN Interface Sublayer. RFC 3637 (2003). DOI 10.17487/RFC3637. URL `https://rfc-editor.org/rfc/rfc3637.txt`. Accessed: February 27, 2019 52

220. RFC 3646, Droms, R.: DNS Configuration options for Dynamic Host Configuration Protocol for IPv6 (DHCPv6). RFC 3646 (2003). DOI 10.17487/RFC3646. URL `https://rfc-editor.org/rfc/rfc3646.txt`. Accessed: March 21, 2019 86, 228, 373

221. RFC 3719, Parker, J.: Recommendations for Interoperable Networks using Intermediate System to Intermediate System (IS-IS). RFC 3719 (2004). DOI 10.17487/RFC3719. URL `https://rfc-editor.org/rfc/rfc3719.txt`. Accessed: January 20, 2019 317

222. RFC 3787, Parker, J.: Recommendations for Interoperable IP Networks using Intermediate System to Intermediate System (IS-IS). RFC 3787 (2004). DOI 10.17487/RFC3787. URL `https://rfc-editor.org/rfc/rfc3787.txt`. Accessed: March 9, 2019 85, 317

223. RFC 3817, Townsley, M., da Silva, R.: Layer 2 Tunneling Protocol (L2TP) Active Discovery Relay for PPP over Ethernet (PPPoE). RFC 3817 (2004). DOI 10.17487/RFC3817. URL `https://rfc-editor.org/rfc/rfc3817.txt`. Accessed: February 27, 2019 52

224. RFC 3833, Atkins, D.: Threat Analysis of the Domain Name System (DNS). RFC 3833 (2004). DOI 10.17487/RFC3833. URL `https://rfc-editor.org/rfc/rfc3833.txt`. Accessed: February 13, 2019 373

225. RFC 3847, Shand, M., Ginsberg, L.: Restart Signaling for Intermediate System to Intermediate System (IS-IS). RFC 3847 (2004). DOI 10.17487/RFC3847. URL `https://rfc-editor.org/rfc/rfc3847.txt`. Accessed: March 9, 2019 317

226. RFC 3871, Allman, E.P., et al.: DomainKeys Identified Mail (DKIM) Signatures. RFC 4871 (2007). DOI 10.17487/RFC4871. URL `https://rfc-editor.org/rfc/rfc4871.txt`. Accessed: February 16, 2019 351, 373, 412

227. RFC 3871, Jones, G.M.: Operational Security Requirements for Large Internet Service Provider (ISP) IP Network Infrastructure. RFC 3871 (2004). DOI 10.17487/RFC3871. URL `https://rfc-editor.org/rfc/rfc3871.txt`. Accessed: May 25, 2019 319

228. RFC 3974, Nakamura, M., ichiro Itoh, J.: SMTP Operational Experience in Mixed IPv4/v6 Environments. RFC 3974 (2005). DOI 10.17487/RFC3974. URL `https://rfc-editor.org/rfc/rfc3974.txt`. Accessed: March 3, 2019 85, 86, 412

229. RFC 4033, Arends, R.: DNS Security Introduction and Requirements. On-line (2005). DOI 10.17487/RFC4033. URL `https://tools.ietf.org/html/rfc4035`. Accessed: January 22, 2019 373

230. RFC 4034, Arends, R., et al.: Resource Records for the DNS Security Extensions. On-line (2005). DOI 10.17487/RFC4034. URL `https://tools.ietf.org/html/rfc4034`. Accessed: January 23, 2019 351

231. RFC 4141, Crocker, D., Toyoda, K.: SMTP and MIME Extensions for Content Conversion. RFC 4141 (2005). DOI 10.17487/RFC4141. URL `https://rfc-editor.org/rfc/rfc4141.txt`. Accessed: March 3, 2019 412, 413

232. RFC 4191, Thaler, D., Draves, R.P.: Default Router Preferences and More-Specific Routes. RFC 4191 (2005). DOI 10.17487/RFC4191. URL `https://rfc-editor.org/rfc/rfc4191.txt`. Accessed: March 30, 2019 86, 198

233. RFC 4193, Haberman, B., Hinden, B.: Unique Local IPv6 Unicast Addresses. RFC 4193 (2005). DOI 10.17487/RFC4193. URL `https://rfc-editor.org/rfc/rfc4193.txt`. Accessed: March 30, 2019 86, 163

234. RFC 4343, Arends, R.: Domain Name System (DNS) Case Insensitivity Clarification. On-line (2005). DOI 10.17487/RFC4343. URL `https://tools.ietf.org/html/rfc4343`. Accessed: January 22, 2019 373

235. RFC 4361, Lemon, T.: Node-specific Client Identifiers for Dynamic Host Configuration Protocol Version Four (DHCPv4). On-line (2006). DOI 10.17487/RFC4361. URL `https://tools.ietf.org/html/rfc4361`. Accessed: January 21, 2019 85, 86, 228

236. RFC 4367, Rosenberg, J.: What's in a Name: False Assumptions about DNS Names.   RFC 4367 (2006).   DOI 10.17487/RFC4367.   URL `https://rfc-editor.org/rfc/rfc4367.txt`. Accessed: February 13, 2019 373

237. RFC 4408, Schlitt, W.: Sender Policy Framework (SPF) for Authorizing Use of Domains in E-Mail, Version 1.   RFC 4408 (2006).   DOI 10.17487/RFC4408.   URL `https://rfc-editor.org/rfc/rfc4408.txt`. Accessed: February 13, 2019 412

238. RFC 4448, Martini, L., et al.: Encapsulation Methods for Transport of Ethernet over MPLS Networks.   RFC 4448 (2006).   DOI 10.17487/RFC4448.   URL `https://rfc-editor.org/rfc/rfc4448.txt`. Accessed: February 27, 2019 52

239. RFC 4472, Durand, A.: Operational Considerations and Issues with IPv6 DNS.   RFC 4472 (2006).   DOI 10.17487/RFC4472.   URL `https://rfc-editor.org/rfc/rfc4472.txt`. Accessed: February 13, 2019 373

240. RFC 4501, Josefsson, S.: Domain Name System Uniform Resource Identifiers.   RFC 4501 (2006).   DOI 10.17487/RFC4501.   URL `https://rfc-editor.org/rfc/rfc4501.txt`.   Accessed: January 15, 2019 373

241. RFC 4592, Lewis, E.P.: The Role of Wildcards in the Domain Name System.   RFC 4592 (2006).   DOI 10.17487/RFC4592.   URL `https://rfc-editor.org/rfc/rfc4592.txt`. Accessed: February 13, 2019 373

242. RFC 4632, Fuller, V., Li, T.: Classless Inter-domain Routing (CIDR): The Internet Address Assignment and Aggregation Plan. RFC 4632 (2006). DOI 10.17487/RFC4632. URL `https://rfc-editor.org/rfc/rfc4632.txt`. Accessed: July 29, 2019 85

243. RFC 4638, Duckett, M., et al.: Accommodating a Maximum Transit Unit/Maximum Receive Unit (MTU/MRU) Greater Than 1492 in the Point-to-Point Protocol over Ethernet (PPPoE).   RFC 4638 (2006).   DOI 10.17487/RFC4638.   URL `https://rfc-editor.org/rfc/rfc4638.txt`. Accessed: February 27, 2019 52

244. RFC 4697, Barber, P., Larson, M.: Observed DNS Resolution Misbehavior. RFC 4697 (2006). DOI 10.17487/RFC4697.   URL `https://rfc-editor.org/rfc/rfc4697.txt`. Accessed: February 13, 2019 373

245. RFC 4702, Rekhter, Y., et al.: The Dynamic Host Configuration Protocol (DHCP) Client Fully Qualified Domain Name (FQDN) Option. RFC 4702 (2006). DOI 10.17487/RFC4702. URL `https://rfc-editor.org/rfc/rfc4702.txt`. Accessed: February 21, 2019 228, 373

246. RFC 4703, Volz, B., Stapp, M.: Resolution of Fully Qualified Domain Name (FQDN) Conflicts among Dynamic Host Configuration Protocol (DHCP) Clients. RFC 4703 (2006). DOI 10.17487/RFC4703.   URL `https://rfc-editor.org/rfc/rfc4703.txt`. Accessed: March 8, 2019 228, 373

247. RFC 4719, Townsley, M., et al.: Transport of Ethernet Frames over Layer 2 Tunneling Protocol Version 3 (L2TPv3).   RFC 4719 (2006).   DOI 10.17487/RFC4719.   URL `https://rfc-editor.org/rfc/rfc4719.txt`. Accessed: February 27, 2019 52

248. RFC 4778, Kaeo, M.: Operational Security Current Practices in Internet Service Provider Environments.   RFC 4778 (2007).   DOI 10.17487/RFC4778.   URL `https://rfc-editor.org/rfc/rfc4778.txt`. Accessed: May 25, 2019 53, 84, 319

249. RFC 4893, Chen, E., Vohra, Q.: BGP Support for Four-octet AS Number Space.   RFC 4893 (2007).   DOI 10.17487/RFC4893.   URL `https://rfc-editor.org/rfc/rfc4893.txt`.   Accessed: March 21, 2019 318

250. RFC 4954, Siemborski, R., Melnikov, A.: SMTP Service Extension for Authentication.   RFC 4954 (2007).   DOI 10.17487/RFC4954.   URL `https://rfc-editor.org/rfc/rfc4954.txt`.   Accessed: March 03, 2019 412

251. RFC 4955, Blacka, D.: DNS Security (DNSSEC) Experiments. RFC 4955 (2007). DOI 10.17487/RFC4955. URL `https://rfc-editor.org/rfc/rfc4955.txt`. Accessed: February 15, 2019 373

252. RFC 4969, Mayrhofer, A.: IANA Registration for vCard Enumservice. RFC 4969 (2007). DOI 10.17487/RFC4969. URL https://rfc-editor.org/rfc/rfc4969.txt. Accessed: February 15, 2019 383

253. RFC 4971, Vasseur, J., et al.: Intermediate System to Intermediate System (IS-IS) Extensions for Advertising Router Information. RFC 4971 (2007). DOI 10.17487/RFC4971. URL https://rfc-editor.org/rfc/rfc4971.txt. Accessed: March 9, 2019 317

254. RFC 5006, Beloeil, L., Madanapalli, S., Park, S.D., Jeong, J.: IPv6 Router Advertisement Option for DNS Configuration. RFC 5006 (2007). DOI 10.17487/RFC5006. URL https://rfc-editor.org/rfc/rfc5006.txt. Accessed: March 21, 2019 86, 373

255. RFC 5039, Rosenberg, J., Jennings, C.: The Session Initiation Protocol (SIP) and Spam. RFC 5039 (2008). DOI 10.17487/RFC5039. URL https://rfc-editor.org/rfc/rfc5039.txt. Accessed: April 4, 2019 412

256. RFC 5130, Shand, M., Previdi, S.: A Policy Control Mechanism in IS-IS Using Administrative Tags. RFC 5130 (2008). DOI 10.17487/RFC5130. URL https://rfc-editor.org/rfc/rfc5130.txt. Accessed: March 9, 2019 317

257. RFC 5235, Daboo, C.: Sieve Email Filtering: Spamtest and Virustest Extensions. RFC 5235 (2008). DOI 10.17487/RFC5235. URL https://rfc-editor.org/rfc/rfc5235.txt. Accessed: April 4, 2019 412

258. RFC 5321, Klensin, D.J.C.: Simple Mail Transfer Protocol. RFC 5321 (2008). DOI 10.17487/RFC5321. URL https://rfc-editor.org/rfc/rfc5321.txt. Accessed: April 27, 2019 411, 412

259. RFC 5340, Coltun, R.: OSPF for IPv6. Request For Comments, On-line (2008). DOI 10.17487/RFC5340. URL https://tools.ietf.org/html/rfc5340. Accessed: January 25, 2019 86, 297

260. RFC 5396, Huston, G., Michaelson, G.G.: Textual Representation of Autonomous System (AS) Numbers. RFC 5396 (2008). DOI 10.17487/RFC5396. URL https://rfc-editor.org/rfc/rfc5396.txt. Accessed: July 9, 2019 318

261. RFC 5692, Riegel, M., et al.: Transmission of IP over Ethernet over IEEE 802.16 Networks. RFC 5692 (2009). DOI 10.17487/RFC5692. URL https://rfc-editor.org/rfc/rfc5692.txt. Accessed: February 27, 2019 53, 85, 86

262. RFC 5828, Berger, L., et al.: Generalized Multiprotocol Label Switching (GMPLS) Ethernet Label Switching Architecture and Framework. RFC 5828 (2010). DOI 10.17487/RFC5828. URL https://rfc-editor.org/rfc/rfc5828.txt. Accessed: February 27, 2019 53

263. RFC 5902, Zhang, L., et al.: IAB Thoughts on IPv6 Network Address Translation. RFC 5902 (2010). DOI 10.17487/RFC5902. URL https://rfc-editor.org/rfc/rfc5902.txt. Accessed: February 28, 2019 86, 423

264. RFC 5993, Cherukuri, R., et al.: Application of Ethernet Pseudowires to MPLS Transport Networks. RFC 5994 (2010). DOI 10.17487/RFC5994. URL https://rfc-editor.org/rfc/rfc5994.txt. Accessed: February 27, 2019 53, 85, 86

265. RFC 6004, Fedyk, D., Berger, L.: Generalized MPLS (GMPLS) Support for Metro Ethernet Forum and G.8011 Ethernet Service Switching. RFC 6004 (2010). DOI 10.17487/RFC6004. URL https://rfc-editor.org/rfc/rfc6004.txt. Accessed: February 27, 2019 53

266. RFC 6005, Fedyk, D., Berger, L.: Generalized MPLS (GMPLS) Support for Metro Ethernet Forum and G.8011 User Network Interface (UNI). RFC 6005 (2010). DOI 10.17487/RFC6005. URL https://rfc-editor.org/rfc/rfc6005.txt. Accessed: February 27, 2019 53

267. RFC 6060, Bitar, D.N.N., et al.: Generalized Multiprotocol Label Switching (GMPLS) Control of Ethernet Provider Backbone Traffic Engineering (PBB-TE). RFC 6060 (2011). DOI

10.17487/RFC6060. URL https://rfc-editor.org/rfc/rfc6060.txt. Accessed: February 27, 2019 53

268. RFC 6085, Dec, W., et al.: Address Mapping of IPv6 Multicast Packets on Ethernet. RFC 6085 (2011). DOI 10.17487/RFC6085. URL https://rfc-editor.org/rfc/rfc6085.txt. Accessed: February 27, 2019 53, 86

269. RFC 6106, Madanapalli, S., Jeong, J.: IPv6 Router Advertisement Options for DNS Configuration. RFC 6106 (2010). DOI 10.17487/RFC6106. URL https://rfc-editor.org/rfc/rfc6106.txt. Accessed: March 21, 2019 86, 373

270. RFC 6126, Chroboczek, J.: The Babel Routing Protocol. RFC 6126 (2011). DOI 10.17487/RFC6126. URL https://rfc-editor.org/rfc/rfc6126.txt. Accessed: February 28, 2019 321, 322, 325, 327

271. RFC 6430, Leiba, B., Li, K.: Email Feedback Report Type Value: not-spam. RFC 6430 (2011). DOI 10.17487/RFC6430. URL https://rfc-editor.org/rfc/rfc6430.txt. Accessed: April 4, 2019 412

272. RFC 6549, Lindem, A.: OSPFv2 Multi-Instance Extensions. Request For Comments, On-line (2012). DOI 10.17487/RFC6549. URL https://tools.ietf.org/html/rfc6549. Accessed: January 25, 2019 297

273. RFC 6647, Kucherawy, M., Crocker, D.: Email Greylisting: An Applicability Statement for SMTP. RFC 6647 (2012). DOI 10.17487/RFC6647. URL https://rfc-editor.org/rfc/rfc6647.txt. Accessed: March 4, 2019 412

274. RFC 6760, Cheshire, S., Krochmal, M.: Requirements for a Protocol to Replace the AppleTalk Name Binding Protocol (NBP). RFC 6760 (2013). DOI 10.17487/RFC6760. URL https://rfc-editor.org/rfc/rfc6760.txt. Accessed: June 20, 2019 57

275. RFC 6811, Mohapatra, P., Scudder, J.: BGP Prefix Origin Validation. RFC 6811 (2013). DOI 10.17487/RFC6811. URL https://rfc-editor.org/rfc/rfc6811.txt. Accessed: March 21, 2019 318

276. RFC 6842, Swamy, N.: Client Identifier Option in DHCP Server Replies. On-line (2013). DOI 10.17487/RFC6842. URL https://tools.ietf.org/html/rfc6842. Accessed: January 21, 2019 228

277. RFC 6858, Gulbrandsen, A.: Simplified POP and IMAP Downgrading for Internationalized Email. RFC 6858 (2013). DOI 10.17487/RFC6858. URL https://rfc-editor.org/rfc/rfc6858.txt. Accessed: March 16, 2019 412

278. RFC 7298, Ovsienko, D.: Babel Hashed Message Authentication Code (HMAC) Cryptographic Authentication. RFC 7298 (2014). DOI 10.17487/RFC7298. URL https://rfc-editor.org/rfc/rfc7298.txt. Accessed: February 28, 2019 327

279. RFC 7393, Deng, X., Boucadair, M., Zhao, Q., Huang, J., Zhou, C.: Using the Port Control Protocol (PCP) to Update Dynamic DNS. RFC 7393 (2014). DOI 10.17487/RFC7393. URL https://rfc-editor.org/rfc/rfc7393.txt. Accessed: March 8, 2019 85, 86, 373, 423

280. RFC 7503, Lindem, A., Arkko, J.: OSPFv3 Autoconfiguration. RFC 7503 (2015). DOI 10.17487/RFC7503. URL https://rfc-editor.org/rfc/rfc7503.txt. Accessed: March 30, 2019 86, 297

281. RFC 7557, Chroboczek, J.: Extension Mechanism for the Babel Routing Protocol. RFC 7557 (2015). DOI 10.17487/RFC7557. URL https://rfc-editor.org/rfc/rfc7557.txt. Accessed: February 28, 2019 327

282. RFC 7608, Boucadair, M., Petrescu, A., Baker, F.: IPv6 Prefix Length Recommendation for Forwarding. RFC 7608 (2015). DOI 10.17487/RFC7608. URL https://rfc-editor.org/rfc/rfc7608.txt. Accessed: July 29, 2019 85, 86

283. RFC 7672, Dukhovni, V., Hardaker, W.: SMTP Security via Opportunistic DNS-Based Authentication of Named Entities (DANE) Transport Layer Security (TLS). RFC 7672 (2015). DOI 10.17487/RFC7672. URL https://rfc-editor.org/rfc/rfc7672.txt. Accessed: March 4, 2019 373, 411, 413

284. RFC 7775, Ginsberg, L., Litkowski, S.: IS-IS Route Preference for Extended IP and IPv6 Reachability. RFC 7775 (2016). DOI 10.17487/RFC7775. URL `https://rfc-editor.org/rfc/rfc7775.txt`. Accessed: March 30, 2019 85, 86, 317

285. RFC 7804, Melnikov, A.: Salted Challenge Response HTTP Authentication Mechanism. RFC 7804 (2016). DOI 10.17487/RFC7804. URL `https://rfc-editor.org/rfc/rfc7804.txt`. Accessed: March 12, 2019 383

286. RFC 7857, Penno, R., Perreault, S., Boucadair, M., Sivakumar, S., Naito, K.: Updates to Network Address Translation (NAT) Behavioral Requirements. RFC 7857 (2016). DOI 10.17487/RFC7857. URL `https://rfc-editor.org/rfc/rfc7857.txt`. Accessed: February 28, 2019 423

287. RFC 7868, Savage, D., Ng, J., Moore, S., Slice, D., Paluch, P., White, R.: Cisco's Enhanced Interior Gateway Routing Protocol (EIGRP). On-line (2016). DOI 10.17487/RFC7868. URL `https://tools.ietf.org/html/rfc7868`. Access: December 13, 2018 233

288. RFC 7969, Lemon, T., Mrugalski, T.: Customizing DHCP Configuration on the Basis of Network Topology. RFC 7969 (2016). DOI 10.17487/RFC7969. URL `https://rfc-editor.org/rfc/rfc7969.txt`. Accessed: March 18, 2019 229

289. RFC 8064, Gont, F., Cooper, A., Thaler, D., LIU, W.S.: Recommendation on Stable IPv6 Interface Identifiers. RFC 8064 (2017). DOI 10.17487/RFC8064. URL `https://rfc-editor.org/rfc/rfc8064.txt`. Accessed: February 17, 2019 86

290. RFC 8115, Boucadair, M., Qin, J., Tsou, T., Deng, X.: DHCPv6 Option for IPv4-Embedded Multicast and Unicast IPv6 Prefixes. RFC 8115 (2017). DOI 10.17487/RFC8115. URL `https://rfc-editor.org/rfc/rfc8115.txt`. Accessed: March 18, 2019 85, 86, 229

291. RFC 8314, Moore, K., Newman, C.: Cleartext Considered Obsolete: Use of Transport Layer Security (TLS) for Email Submission and Access. RFC 8314 (2018). DOI 10.17487/RFC8314. URL `https://rfc-editor.org/rfc/rfc8314.txt`. Accessed: April 4, 2019 413

292. RFC 8336, Nottingham, M., Nygren, E.: The ORIGIN HTTP/2 Frame. RFC 8336 (2018). DOI 10.17487/RFC8336. URL `https://rfc-editor.org/rfc/rfc8336.txt`. Accessed: March 12, 2019 383

293. RFC 8353, Upadhyay, M.D., Malkani, S.: Generic Security Service API Version 2: Java Bindings Update. RFC 8353 (2018). DOI 10.17487/RFC8353. URL `https://rfc-editor.org/rfc/rfc8353.txt`. Accessed: March 18, 2019 229

294. RFC 8362, Lindem, A., Roy, A.: OSPFv3 Link State Advertisement (LSA) Extensibility. RFC 8362 (2018). DOI 10.17487/RFC8362. URL `https://rfc-editor.org/rfc/rfc8362.txt`. Accessed: March 30, 2019 86, 297

295. RFC 8388, Rabadan, J., Palislamovic, S., Henderickx, W., Sajassi, A., Uttaro, J.: Usage and Applicability of BGP MPLS-Based Ethernet VPN. RFC 8388 (2018). DOI 10.17487/RFC8388. URL `https://rfc-editor.org/rfc/rfc8388.txt`. Accessed: February 27, 2019 53, 318

296. RFC 8415, Mrugalski, T., Siodelski, M.: Dynamic Host Configuration Protocol for IPv6 (DHCPv6). RFC 8415 (2018). DOI 10.17487/RFC8415. URL `https://rfc-editor.org/rfc/rfc8415.txt`. Accessed: March 18, 2019 87, 229

297. RFC 8460, Margolis, D., Brotman, A., Ramakrishnan, B., Jones, J., Risher, M.: SMTP TLS Reporting. RFC 8460 (2018). DOI 10.17487/RFC8460. URL `https://rfc-editor.org/rfc/rfc8460.txt`. Accessed: March 4, 2019 373, 411, 413

298. RFC 8461, Margolis, D., Risher, M., Ramakrishnan, B., Brotman, A., Jones, J.: SMTP MTA Strict Transport Security (MTA-STS). RFC 8461 (2018). DOI 10.17487/RFC8461. URL `https://rfc-editor.org/rfc/rfc8461.txt`. Accessed: March 4, 2019 373, 411, 413

299. RFC 8468, Morton, A., Fabini, J., Elkins, N., mackermann@bcbsm.com, Hegde, V.: IPv4, IPv6, and IPv4-IPv6 Coexistence: Updates for the IP Performance Metrics (IPPM) Framework. RFC 8468 (2018). DOI 10.17487/RFC8468. URL https://rfc-editor.org/rfc/rfc8468.txt. Accessed: March 30, 2019 85, 87

300. RFC 8470, Thomson, M., Nottingham, M., Tarreau, W.: Using Early Data in HTTP. RFC 8470 (2018). DOI 10.17487/RFC8470. URL https://rfc-editor.org/rfc/rfc8470.txt. Accessed: March 12, 2019 383

301. RFC 8484, Hoffman, P.E., McManus, P.: DNS Queries over HTTPS (DoH). RFC 8484 (2018). DOI 10.17487/RFC8484. URL https://rfc-editor.org/rfc/rfc8484.txt. Accessed: March 12, 2019 373, 383

302. RFC 8501, Howard, L.: Reverse DNS in IPv6 for Internet Service Providers. RFC 8501 (2018). DOI 10.17487/RFC8501. URL https://rfc-editor.org/rfc/rfc8501.txt. Accessed: May 25, 2019 87, 319, 373

303. RFC 8503, Tsunoda, H.: BGP/MPLS Layer 3 VPN Multicast Management Information Base. RFC 8503 (2018). DOI 10.17487/RFC8503. URL https://rfc-editor.org/rfc/rfc8503.txt. Accessed: March 21, 2019 318

304. RFC 8539, Farrer, I., Sun, Q., Cui, Y., Sun, L.: Softwire Provisioning Using DHCPv4 over DHCPv6. RFC 8539 (2019). DOI 10.17487/RFC8539. URL https://rfc-editor.org/rfc/rfc8539.txt. Accessed: March 18, 2019 85, 87, 229

305. RFC 8571, Ginsberg, L., Previdi, S., Wu, Q., Tantsura, J., Filsfils, C.: BGP - Link State (BGP-LS) Advertisement of IGP Traffic Engineering Performance Metric Extensions. RFC 8571 (2019). DOI 10.17487/RFC8571. URL https://rfc-editor.org/rfc/rfc8571.txt. Accessed: March 21, 2019 297, 317, 318

306. Roberts, F., Tesman, B.: Applied Combinatorics, second edn. Chapman and Hall/CRC (2009) 205

307. Robinson, P.A., Lasker, L.: Sneakers. DVD (1992). Direceted by: Phil Alden Robinson 40

308. Sahni, S.: Data Structures, Algorithms and Applications in C++. Computer Science, Singapore: McGraw-Hill (1998) 205

309. Sahni, S.: Data structures, algorithms, and applications in Java. Universities Press (2005) 205

310. SDA: SD Memory Card Formatter 5.0.1 for SD/SDHC/SDXC. On-Line (2018). URL https://www.sdcard.org/downloads/formatter_4/. Accessed: November 19, 2018 120, 127

311. Software, J.: Simpld dns plus. On line. URL https://simpledns.com/private-ipv6. Accessed: August 10, 2019 163

312. Tanenbaum, A.S., Wetherall, D.J.: Computer Networks, 5 edn. Prentice Hall (2011) 11

313. Techopedia: www.techopedia.com. On Line (2019). URL https://www.techopedia.com/. Accessed: July 19, 2019 500, 501

314. Tolkien, J.: The Return of the King. Lord of the rings. Houghton Mifflin Co. (1965) 177, 471, 492

315. Unbuntu Community: Mailman. On line. (2017). URL https://help.ubuntu.com/community/Mailman. Accessed: August 10, 2019 406

316. Various: Win32diskimager. On-line (2018). URL https://sourceforge.net/projects/win32diskimager/. Access: November 19, 2018 120

317. Various: You tube. On-line (2018). Accessed: 14 November 2018 125

318. Wikipedia: https://en.wikipedia.org/wiki/darpa. On line. (2019). URL https://en.wikipedia.org/wiki/DARPA. Access: August 7, 2019 452, 464

319. Wikipedia, Contributors: DMZ — Wikipedia, the free encyclopedia (2019). URL https://en.wikipedia.org/wiki/DMZ_(computing). [Online; accessed: February 20, 2019] 464

320. Zimmermann, K.A., Emspak, J.: Internet history timeline: Arpanet to the world wide web. On line. (2017). URL https://www.livescience.com/. Accessed: July 28, 2019 452

Index

© Springer Nature Switzerland AG 2020
G. Howser, *Computer Networks and the Internet*,
https://doi.org/10.1007/978-3-030-34496-2

527

Printed in the United States
By Bookmasters